Natalie Kyriacou OAM is an award-winning environmentalist, writer, professional public speaker and company director with a passion for harnessing curiosity to solve nature crises. Natalie was awarded the Medal of the Order of Australia and the *Forbes* 30 Under 30 honour for her services to wildlife and environmental conservation in 2018 and was recognised as one of *The Australian's* Top 100 Innovators in 2022. She was the United Nations Environment Programme's Young Champions of the Earth finalist for her innovation in wildlife and environmental conservation and is LinkedIn's Top Green Voice.

She is a board director at the Foundation for National Parks & Wildlife, a board committee member at CARE Australia, the founder and chair of My Green World, a UNESCO Green Citizens Pathfinder, a member of the XPrize Brain Trust for Biodiversity and Conservation, and an Australian delegate and Climate Justice Lead at the W20 (the official engagement group of the G20). She is a National Ambassador for the Australian Conservation Foundation.

Praise for Nature's Last Dance

'In this racy, raucous and riveting new book, Natalie Kyriacou makes a compelling case for falling in love – with kākāpōs, rhinos and nature in general. Genetically speaking, after all, we are related to everything from banana slugs to bamboo trees, so this is a clarion call to reimagine how we think about and how we treat our extended family. As Kyriacou makes clear, either humans change our values, systems and behaviours, or we will join dodos on the list of extinct species. Magically, this book delivers a dire warning in ways that will make you laugh – the literary equivalent of medicine that actually tastes good.'
– Dr David R Boyd, former UN special rapporteur on the human right to a healthy environment

'Natalie Kyriacou nails it! *Nature's Last Dance* is an astonishingly good book. Saving nature to the fore, it never gives up hope. I couldn't put it down.'
– Dr Bob Brown, environmentalist and former leader of the Australian Greens party

'A lyrical call to awaken our love for the wild before the music stops.'
– Dame Christiana Figueres, former executive secretary of the UNFCCC, architect of the Paris Agreement and Costa Rican diplomat

'I loved this book. It's that rare kind of work that is fascinating and full of insight while being genuinely important – all told in a way that's completely gripping. A vital read.'
– Peter FitzSimons AM, author and journalist

'A beautiful, terrible, humbling, hopeful, captivating and illuminating call to action. It's also very funny. Please rush out and buy this wonderful book in bulk – then send one to every influential person you know.'
– Professor Emerita Lesley Hughes, academic and climate scientist

'With biting wit and boundless curiosity, Natalie Kyriacou brings both levity and urgency to one of the greatest crises of our time. *Nature's Last Dance* is not just a catalogue of extinction – it is also a bold celebration of the creatures still with us and a compelling reminder of how much we stand to lose. Kyriacou writes like a scientist with a soul and a storyteller with purpose.'
– Cristina 'Mitty' Mittermeier, conservationist, photographer and author

'Part Shakespearean-like tragedy, part bawdy romance, part political manifesto, this engrossing, quirky and at times risqué book evokes curiosity. It makes you laugh and then makes you cry as it traces our dysfunctional relationship with nature, along with inspired efforts to mend it. In charting a creative course, it offers a must-read glimmer of hope for the future of humanity.'
– John E Scanlon AO, former secretary-general of CITES, director of the IUCN Environmental Law Centre and principal advisor of UNEP

'I didn't expect a book to make me laugh out loud, feel deeply and reimagine my relationship with nature, but *Nature's Last Dance* did just that. It's hilarious, wise, raw and inspiring. Natalie Kyriacou brings a bold, fresh voice that speaks to the heart and the head. This book is a powerful reminder of what we stand to lose – and why it's worth fighting for.'
– Judy Slatyer, president of WWF-Australia

'At a time when our instincts might tell us to look away from the devastating destruction of our planet, Natalie Kyriacou provides a stunning case for doing the opposite. Balancing loss and hope, destruction and resilience, *Nature's Last Dance* is an arresting love letter to the diversity of life on Earth. Kyriacou has masterfully painted a world that feels impossible … and yet, it's ours. Poetic, curious and with a contagious sense of wonder, Kyriacou is a unique voice for an environment worth fighting for.'
– Clare Stephens, writer, podcaster and author

'*Nature's Last Dance* is a bold and vibrant work, brimming with rigorous research and wise insights. This is storytelling at its finest, lyrically written with a sense of wonder and deep vein of humour. Natalie Kyriacou OAM has achieved an astonishing feat: creating a page-turner that reinvigorates the fight against climate change. But she doesn't stop there. Kyriacou dives into the patriarchal world of Charles Darwin, astutely questioning male dominance hierarchies, and celebrating matriarchal societies like the bonobo apes. At its heart, this book is a love letter to the natural world; a rallying cry to save our precious planet. Put simply, this is the most important environmental book of our times.'
– Tracey Spicer AM, Walkley Award-winning author, journalist and broadcaster

'Natalie is a powerful voice for nature: fierce, funny and unforgettable.'
– Keith Tuffley AM, explorer, CEO of Race to Belém and board director of the Villars Institute and the E.O. Wilson Biodiversity Foundation

'This book is a fabulous cocktail shaken with biting wit and served with a garnish of urgent hope. Kyriacou makes a compelling, often hilarious, case for falling in love with the organisms whose planet we share, from kākāpōs to banana slugs. It reads like a love letter to a world on the brink, blending the soul of a poet and the precision of a scientist. You will laugh out loud, you will feel a profound pang for what we stand to lose, but you will also leave surging with a desire to make the planet a better place.

Nature's Last Dance doesn't let you look away from the crisis; it makes you lean in and reminds you fiercely why every strange, wonderful creature is worth the fight!'
– Professor Adam Hart, scientist, author and broadcaster

'The most breathtaking reminder of the wild and miraculous waking dream that is our natural world, and a glorious tribute to the wondrous souls who fight to keep that dream alive.'
– Trent Dalton, author of *Boy Swallows Universe*

THE BOOK SOCIAL, AN IMPRINT OF LEGEND TIMES GROUP LTD
3rd Floor
86-90 Paul Street
London EC2A 4NE
www.thebooksocial.co.uk

First published in Australia by Affirm Press in 2025
This edition first published in UK by Legend Times in 2026

© Natalie Kyriacou, 2025

Cover design: Nada Backovic

The right of the author to be identified as the author and translator of this work has been asserted in accordance with the Copyright, Designs and Patents Act 1988. British Library Cataloguing in Publication Data available.

Printed by Akcent Media, 5 The Quay, St Ives, Cambs, PE27 5AR

ISBN: 9781918291902

All rights reserved. No part of this publication may be reproduced, stored in or introduced into a retrieval system, or transmitted, in any form or by any means (electronic, mechanical, photocopying, recording or otherwise), without the prior written permission of the publisher. This book is sold subject to the condition that it shall not be resold, lent, hired out or otherwise circulated without the express prior consent of the publisher.

Nature's Last Dance: Tales of Wonder in an Age of Extinction

Natalie Kyriacou OAM

Acknowledgement of Country

I wish to acknowledge the Traditional Owners of the lands on which this book was written: the Gadigal people of the Eora Nation. I pay my deepest respects to their Elders – past and present – and extend this respect to all Aboriginal and Torres Strait Islander peoples across this vast continent.

For 65,000 years, First Nations peoples have cared for Country – its lands, skies and waters – with deep knowledge and traditions that have been passed down through generations.

This always was, and always will be, Aboriginal land.

Dedication

For my parents, who raised me with so much love it gave me a confidence that is likely disproportionate to my ability. For Judy and Peter, my second set of parents. For my sister, Nicole, who is always by my side, advocating for me.

And for future generations. We owe you a planet worth inheriting.

Preface

Many, many years ago, humans emerged from primordial ooze and promptly declared themselves the rulers of nature and the pinnacle of evolution. It is a decision that, in retrospect, has not aged well, with many considering us one of nature's more regrettable experiments.

Nevertheless, here we are. And now we must wrestle with our relationship with the natural world. Today, we find ourselves grappling with an awkward truth: our relationship with the natural world is, at best, strained. As it turns out, we may not be the masters of nature we once imagined ourselves to be.

For me, this humbling lesson was learned early in life. When I was five years old, I decided to befriend frogs. Not just any frogs – all frogs. I saw myself as the mother of all frogs, the benevolent ruler of the amphibian kingdom. Unfortunately, the frogs didn't share my vision. They scrambled away with a mix of anxiety and disdain, leaving me with muddy hands and the distinct impression that my leadership potential was not as compelling as I'd previously thought.

This marked the beginning of my lifelong fascination with the natural world – not just its beauty but its chaos, its cruelty, its fragility, its resilience. And it is this fascination that forms the heart of *Nature's Last Dance*.

You see, nature is not the backdrop to our lives; it *is* our lives. Wars have been fought over it, empires built upon it. Leaders have risen and fallen by its hand. It has shaped economies, forged legal systems and fed nations. Yet it rarely gets the credit it deserves. Nature is the force that shapes our lives and powers the world.

However, this isn't really a book about nature – or at least, not in the way you might think. This book is an ode to nature, but it's also a slightly irreverent, somewhat absurd glimpse into humanity, culture, economics, philosophy, politics, science and history – all seen through the lens of the natural world. It's about birds and bats and sharks and trees, about ecosystems and environments. But it's also about how militaries are being tested and bested by nature, how tiny insects have disrupted modern legal systems and global trade, and how animal mating behaviours are challenging everything we thought we knew about human evolution. It charts the rise and fall of civilisations; it reveals how humanity is altering the course of nature and evolution; and it examines how economies, political arenas and legal systems are being up-ended by nature.

I hope this book makes you think and makes you feel. But most importantly, I hope it sparks your curiosity and helps you find new ways to see the world, and maybe even protect it. That is, after all, the overall intent. I cover a lot across these chapters, and some people will surely disagree with some of the things I suggest. But I think there is something joyful and thought-provoking for everyone. I hope you find it.

A Note to Readers

Admittedly, when I started writing this book, I was operating under the comforting assumption that nobody would ever read it. Thus, I was quite certain you would never learn of the terrible things I secretly thought of you. Alas, I am now faced with the troubling realisation that some people might not take kindly to being likened to venereal disease within the pages of this book.

If I've offended you, it was likely unintentional. Except for the times it was most certainly intentional.

Nevertheless, here it is, printed, bound and packaged on paper that may one day end up on sale for 50 cents at a garage sale.

But a quick word before we dive in: throughout this book, I occasionally use sarcasm, irony and dry humour to make a point or to offer a bit of levity while exploring serious topics.

I want to acknowledge that readers approach books with diverse learning styles and perspectives, and I deeply respect that. My intention is never to confuse or alienate, but I know that humour can sometimes get lost in translation, especially for those who prefer things to be direct and literal. So, if you find yourself wondering, 'Wait, was that a joke?' – the answer is probably yes. Or at least, it was an attempt at one.

Regardless, I hope the meaning behind the words and the intent of the book still resonates with you.

This book draws comprehensively on the insights and findings of many leading experts, brilliant minds and scientific studies. However, the conclusions, interpretations and any potential annoyances you feel are entirely my doing and do not necessarily reflect the positions of any affiliated organisations or interviewees. In other words, if this book irritates you, please aim your indignation – politely, if possible – solely at me.

Above all, I'm grateful you're here, however you choose to read this book.

Contents

Introduction: Welcome to the Anthropocene: The Vanishing Act 17

PART 1: Origins: Extinction, Seduction and Survival
Chapter 1: Courtship and Seduction: A Twinkle in the Eye of a Kākāpō 25
Chapter 2: A Legacy of Extinction: The Sound of Silence 43
Chapter 3: A Spectacle of Survival: A Humbled Army 58
Chapter 4: Unravelling Ecosystems: Hippos from Hell 71

PART 2: A High-Stakes Game of Extinction and Survival
Chapter 5: Desperation: The Loneliest Rhino in the World 87
Chapter 6: A Call to Arms: The Greatest Rescue in History 100

PART 3: The Tangled Fates of Us and Them
Chapter 7: The Miracle Makers: A Drunk Moose and an Economist 119
Chapter 8: Wolves of Wall Street: Cowboys and Whale Poo 139
Chapter 9: Invisible Creatures: The Curse of Ugly 156
Chapter 10: The Rights of Nature: The River That Is a Person 176

PART 4: Battlegrounds: Politics, Power and Nature
Chapter 11: Dumping Grounds: Let Them Sink 201
Chapter 12: Conspicuous Consumption: The Emperor's Jewels 220
Chapter 13: Tales of Deceit and Cunning: The Propaganda Playbook 251
Chapter 14: The Herpes Agenda: When Politics and Nature Collide 279

PART 5: Solutions: The Tempest, the Fury and the Birdwatcher	299
Chapter 15: Giants Among Us: The Fight of Our Lives	301
Chapter 16: The Delight and Scandal of a Birdwatcher: To Fall in Love	317
Epilogue: The Way Forward	328
Acknowledgements	346
Sources	348

Introduction

Welcome to the Anthropocene: The Vanishing Act

> How lonely it will be here, when it's just us.
> Charlotte McConaghy[1]

The last great auk, it is said, was strangled unceremoniously in its sleep in 1844. Plump and penguin-like, the great auk had survived for thousands of years until humans discovered the utility of its soft down feathers, eggs and meat. Great auks mate for life, and it was on Eldey island in Iceland where the final pair on Earth met their fate at the hands of three fishermen who fell upon them.[2]

'I took him by the neck and he flapped his wings. He made no cry. I strangled him,' said the man who killed the last of a species.[3]

Extinction is rarely loud. It is a whisper, so quiet that you might just miss it.

The great auk wasn't the first to fall to extinction at the hands of humanity, and it certainly wasn't the last. The dodo, almost two centuries prior, became the symbol of human-induced extinction. Another flightless wonder, the dodo heralded the beginning of the modern extinction era. The dodo is, for all intents and purposes, an icon of extinction: it is and was the (futile) canary in the coalmine that sounded the alarm to what would become a tragic chapter in Earth's 4.54 billion-year history, written in blood and wrought by humanity's hand.

But the dodo's demise did little to stem the rising tide of extinction that would plunge the world into an era where the force of humanity would rival the cataclysm that wiped out dinosaurs.

So many more extinctions were to come, most of them going quietly, slipping softly into the annals of obscure scientific journals. The smooth handfish. The Yangtze river dolphin. The quagga. The Pyrenean ibex. The Chiriqui harlequin frog. The Rocky Mountain locust.

This is the canvas upon which we etch our legacy today. Welcome to the Anthropocene: the reign of the human.

Land and sea both bear the scars of humanity's rule. Today, one million plant and animal species face extinction. Wild mammals make up less than 6 per cent of the total mammal biomass on Earth, dwarfed by humans and livestock. That is, by weight, humans and their food dominate and devour the globe. The children born today face the very real possibility of a world emptied of much of its wildlife.[4] This is a catastrophic legacy. The tales of human-driven wildlife extinction are of unfathomable horror: lands blackened with bodies of bygone species; birds falling from the skies, bodies slick with oil and sludge; great bears dancing raggedly in circuses; fish swollen with ingested plastics; wild apes prostituted; monkeys behind bars; insects disappearing from the skies; whales being cut, sliced, hauled and massacred off the back of ships; foxes lining coats and hats; koalas smouldering under blazing fires.

Don't look away. Because it's not over just yet.

Amidst the tragedy, a hidden world of curiosity and wonder still exists.

And in this we can find glimmers of hope.

Across the globe, an amazing orchestra of animal life is playing out in wondrous, quirky detail, revealing the resilience and spectacle of nature. In the freshwater lakes of Mexico, newly hatched axolotls are feasting on their siblings' limbs and revolutionising our understanding of nature's regeneration abilities. These small salamanders, resembling eels with stumpy legs, possess the extraordinary ability to regrow lost limbs.

Across the Atlantic, female great apes in Central Africa are delighting one another with an intimate bonding ritual, rubbing their clitorises together to strengthen their friendships and maintain peace within their communities.

In the oceans, an adult humpback whale takes on the role of 'escort' as he glides alongside a mother and her calf, protecting the vulnerable pair as they navigate the seas together for the first time.

In the air, a monarch butterfly migration paints the sky with colour and wings as millions take flight on a 4000-kilometre journey across the Americas in one of the most impressive phenomena in the animal kingdom.

On land, the earth shudders as a thundering wildebeest migration pounds across the African plains in a spectacular blur of hooves, dirt and dust. As the 1.5 million-strong herd moves in search of fresh grazing pasture, it sets the stage for a breathtaking battle of survival, with lions, leopards, hyenas, zebras, gazelles and crocodiles joining the frenzied stampede.

There is so much more to our world story than loss.

Species once thought extinct are being rediscovered. Conservation efforts driven by passionate individuals and communities are restoring habitats and bringing endangered species back from the brink.

The California condor, once nearly wiped out, is now soaring through the skies again thanks to dedicated conservation programs. The Arabian oryx, extinct in the wild by the 1970s, has been reintroduced to its native habitat, where it now thrives. The black-footed ferret, the Hawaiian crow and the Amur leopard are all testaments to what can be achieved when humanity turns its efforts towards regeneration rather than destruction.

This hidden world, filled with the strange, the tragic, the

beautiful, the profound and the delightfully odd, is shaping life on Earth and defining our individual lives.

Finding these moments of wonder, while increasingly difficult, is not impossible.

One of the pioneers of the modern environmental movement, Rachel Carson, once said, 'The more clearly we can focus our attention on the wonders and realities of the universe about us, the less taste we shall have for destruction.'[5]

This book aims to do just that. This is a story of extinction and survival, of life and death, of curiosity and perversion, of unimaginable joy and harrowing sorrow. Part celebration, part tribute to a vanishing world, this book aims to radically alter how you think about nature and extinction. It celebrates the bizarre courtships, whimsical dances and strange rituals of life and love that fill nature's playbook with humour and charm. It takes us through hunting grounds, into the deep sea, through the jungles, inside communities and into the heart of the solutions that offer a brighter future, all while shining a light on the most spectacular and devastating show on Earth: nature's dance.

The question motivating this book is: against the backdrop of extinction and destruction, can tales of hope, joy, absurdity and scientific marvel provide the fuel for humanity to confront and reverse the extinction crisis? At the very least, it aims to radically change the way in which humanity thinks about nature and considers the tangled fates of 'us and them'.

To reverse the nature crisis that we have created, a few ingredients are required. We must make the moral, economic and scientific case for action. We need to bridge social and political divides to connect with a diversity of people, each shaped by their own unique experiences, contexts and perspectives. We must invoke the art of storytelling to capture imaginations, blending the beauty and tragedy of nature to highlight both the urgency of the crisis and the potential for recovery. Our understanding of the crisis must be holistic, encompassing historical, political, economic and social dimensions. We need a visionary roadmap that outlines clear, actionable steps to address the nature crisis while inspiring hope and determination. And most importantly, we need to cultivate a sense of childlike wonder, curiosity and imagination for the natural world.

It's a tall order. And one that cannot be accomplished alone. *Nature's Last Dance* simply strives to add another thread to the rich tapestry of efforts already dedicated to safeguarding the environment. Within these chapters are countless voices – giants who have dedicated their lives, minds and souls to the protection of nature. Their stories and contributions are reflected through these pages.

This book is your invitation to join us.

Today, a five-foot-tall bronze statue stands solitary, overlooking the towering sea cliffs of the Reykjanes peninsula in Iceland. The statue is of the now extinct great auk, a memorial to a species that was hunted out of existence. This lone figure preserves the legacy of one of humanity's most defining acts on the planet: the destruction of life on Earth. In the silence of stone and bronze, the loudest truths are spoken. And the great auk – frozen in time – reminds us not just of what we have destroyed, but what we still have left to protect.

It is true that the past has borne witness to disturbing and widespread exterminations of species, the great auk being just one among myriad fallen. But while the last breath of the great auk was snuffed out on Eldey island, a movement was underway. This was an era in which nature captured hearts and minds, an era in which the magic of the natural world was nourishing the work of prolific thinkers, uncovering infinite possibilities in natural, literary, philosophical and scientific inquiry. It was a time that saw the invention of the word 'scientist', the rise of the modern conservation movement and the early stirrings of widespread public awareness of humanity's impact on, and relationship with, nature.

It was the age of Charles Darwin and *The Origin of Species*, of

Ralph Waldo Emerson and the appreciation of nature, of Eunice Foote and the discovery of the warming potential of carbon dioxide. This was the age in which people gave language to curiosity, in which the seeds of ecology, evolution, conservation and environmental justice were taking root.

So you see, for every tale of destruction, we can weave a tale of hope, regeneration and joy. Let this century be the one in which regeneration reigns over destruction and a new legacy is created for humanity.

PART 1

Origins: Extinction, Seduction and Survival

Chapter 1

Courtship and Seduction: A Twinkle in the Eye of a Kākāpō

All nature's creatures join to express nature's purpose. Somewhere in their mounting and mating, rutting and butting is the very secret of nature itself.

Graham Swift[1]

It is curious, what happens in the forested islands of Aotearoa New Zealand. Conservationists pad their heads with 'ejaculation helmets' and brace themselves for what will surely be one of the most peculiar sexual encounters of their lives. Because standing before them, displaying a dazzling plumage of green and gold, is the infamous kākāpō.

With a distinctive waddle and a friendly – if not amorous – disposition, the kākāpō is the world's heaviest parrot, and flightless at that. The kākāpō faces seemingly insurmountable reproductive challenges, not least of which is that they have revealed a penchant for mating with human heads and inanimate objects.

With just 244 kākāpōs alive today, quite a bit of pressure has been placed on this charismatic bird to improve its performance.

Once abundant throughout New Zealand, kākāpō numbers swiftly declined with the arrival and expansion of humans. In

1894, conservation efforts began to build around the kākāpō, but by the mid-1900s, it teetered on the edge of extinction, with just fifty-one left in the wild.

The kākāpō boasts a range of unique traits that make it both endearing and spectacularly ill-equipped for survival in the modern world. Firstly, it is the only living parrot on the planet that cannot fly. It has also evolved to freeze when threatened, a trait that is useful when hiding from birds of prey, but decidedly less useful when facing a human hunter. Add to this a strong, musky odour, a tendency to stomp loudly through the undergrowth and a courtship ritual that involves sitting in a hole and broadcasting your precise location into the landscape, and you begin to see how this bird has struggled to stay alive.

For much of its early history, the kākāpō evolved in a landscape free from mammalian predators – a relatively safe and rather forgiving ecological niche that allowed these more flamboyant traits to flourish. That safety vanished with the arrival of humans, and especially with European settlers who brought with them stoats, cats, rats and an enthusiastic appetite for hunting and land-clearing. Soon, the kākāpō was plummeting towards extinction.

Today, efforts to protect and restore kākāpō populations are spearheaded by the Kākāpō Recovery Programme, an initiative of New Zealand's Department of Conservation.

Under the guidance of conservationists, kākāpōs are managed on island sanctuaries, where their numbers have slowly but steadily increased. But even in these carefully curated pockets of safety, the kākāpō is still contending with a stubborn adversary: the evolutionary baggage of a world that no longer exists.

Because of its small population size, the kākāpō is plagued with the challenge of low genetic diversity and elevated levels of inbreeding. Kākāpō populations have lost up to 80 per cent of their genetic diversity since the 19th century,[2] making it difficult for them to survive, reproduce and adapt to environmental changes. Think of genetic diversity as a well-stocked toolbox, with

COURTSHIP AND SEDUCTION: A TWINKLE IN THE EYE OF A KĀKĀPŌ

each tool representing a different gene. In the game of survival, having a wide variety of tools allows for more creative solutions to unexpected challenges. Conversely, a sparse toolbox leaves you with fewer options when things get complicated.

As a result of the numerous challenges that bedevil the kākāpō, it grapples with poor sperm quality and low hatching success – problems that are exacerbated when combined with their infrequent breeding cycles, which occur only every three to four years. Since 1985, 40 per cent of all kākāpō eggs laid have been infertile, and another 20 per cent of embryos have died early in development.[3] All in all, the kākāpō is struggling with a tragically sparse toolbox.

As for the kākāpō's approach to romance, it is – like much else about the bird – endearing, inefficient and mildly catastrophic. Which is why, if you venture into the protected islands of New Zealand today, you may witness – or unwittingly be the recipient of – the kākāpō's misguided attempts to mate with virtually anything.

Male kākāpōs attempt to attract the ladies in an elaborate courtship display called 'lekking', in which they waddle up a hill or cliff together, dig a hole and start bellowing into it. The hope is that females will be irresistibly drawn to the allure of the male kākāpō shouting from its dirt pit. Unfortunately, female kākāpōs aren't interested in this particular brand of male romance unless they are well-fed: they will only mate when the fruit of their favourite tree, the rimu, produces a lot of fruit. But the rimu tree only fruits every three to four years. To make matters worse, well-fed and heftier kākāpō mums are more likely to give birth to male chicks, so *too much* food results in more male chicks born and even fewer females to carry on the species (and respond to the aroused squalling of the male kākāpō sitting atop its hole).[4] The result is that sometimes a male kākāpō, full of hope and unmet sexual desires with nary a female in sight, will engage in an unusual display of lovemaking with inanimate objects or, occasionally, humans.

In 2012, a kākāpō named Sirocco soared to international fame after being caught on camera displaying his unusual sexual proclivities. In a YouTube video that has been viewed over 27 million times, the BBC posted incriminating footage of Sirocco. Titled 'Shagged by a rare parrot', the description reads, 'Imagine being the first ever guy in the world to be filmed being sexually harassed by an endangered bird.' The video is, indeed, damning. Sirocco, charged with sexual energy and wearing the weight of the world's reproductive challenges on his shoulders, can be seen engaging in a vigorous act of coitus with a BBC presenter's head.

Sirocco is perhaps the most peculiar of an already peculiar species. Hand-raised by rangers due to respiratory issues, he imprinted on humans at an early age and swore off mating with his own kind.

Scientists, with their endless curiosity and creativity, decided to make the most of Sirocco's unique sexual preferences. And so, the ejaculation helmet was born. The maxim seemed to be: if Sirocco desires a human head, then a human head he shall have.

Armed with determination and ingenuity, Department of Conservation veterinarian Dr Kate McInnes designed a dimpled semen-collecting latex helmet. The rubber headgear is, for all intents and purposes, a hat of condoms designed to collect the sperm of the kākāpō.

Dr McInnes was inspired to make the helmet after being informed of a similar helmet that was made for a species of kestrel, an endangered bird of prey found in the forests of Mauritius. According to McInnes, conservationists there would don the helmet and 'the boy would bonk the head and they could collect the sperm'.[5] Simple, she thought.

However, there was one small problem. The Mauritius kestrel weighs just 250 grams, compared with the kākāpō, which weighs in at a hefty 4 kilograms. 'I wasn't prepared to have a 4-kilo kākāpō sitting on a little hat on my head,' said McInnes.[6]

Being a New Zealander, McInnes decided that a rugby helmet would be the most suitable option to withstand the heft of a thrusting kākāpō. With steadfast resolve, she sourced a rugby helmet from a local store and quickly began working on her master project. And so it was in the backyard of her Berhampore house in Wellington one sunny afternoon that an ejaculation helmet was remodelled from a rugby helmet. Then, McInnes, with her homemade rugby ejaculation helmet in tow, set off to the forest to commence her experiment in the name of conservation. 'We took [the helmet] down to the island and we went and visited Sirocco, and he got very excited by the whole business. And so, for about three nights in a row, I was out there in the evening with him bonking my head. He's quite heavy. He goes on for a very long time. He grunts the whole time he's doing it,' she said.[7]

Unfortunately, the helmet didn't work. You see, while the kākāpō approaches intercourse with vigour and intensity – engaging in the act for close to an hour (unlike most other bird species that only require a few seconds) – enthusiasm alone is not a reliable indicator of success. McInnes is unsure whether the helmet was a conceptual failure or not, but after seemingly endless hours of having her head mounted by the world's heaviest parrot, not a single drop of semen was produced.

The fact that the ejaculation helmet failed is neither here nor there; it's the courage to persevere that speaks volumes.[8] Today, the helmet resides in Wellington's Te Papa museum, next to 'Chloe', a motorised decoy female kākāpō, which was another failed breeding aid.

The kākāpō is not just a curious bird with peculiar mating habits; it has become a symbol of conservation and an icon of the fight to save endangered species in New Zealand. In 2010, former prime minister John Key even went so far as to dub the kākāpō the nation's 'official spokesbird for conservation'.[9]

The animal kingdom is full of extravagant courtship displays and strange sexual adaptations that offer fascinating insights into the evolution of species and behaviours. The very concept of courtship puzzled and intrigued one of history's greatest naturalists, Charles Darwin. It was Darwin, after all, who was among the first to unearth the mysteries of courtship while he was, unexpectedly, quarrelling with a peacock. Just as the kākāpōs' unique mating challenges captivate conservationists today, Darwin found himself captivated by another seemingly paradoxical trait in nature: the peacock's tail.

In 1860, Darwin wrote a letter to a friend, exclaiming, 'The sight of a feather in a peacock's tail, whenever I gaze at it, makes me sick!' It had been a year since he had published his seminal work, *On the Origin of Species*, but Darwin was now tormented by the peacock and its colourful plumage'.[10]

In *On the Origin of Species*, Darwin suggested that all living species share common ancestry, descending from a few early life forms and evolving over time through natural selection.* He noted that organisms that were better suited to their environment were more likely to survive and reproduce. The giraffe, for example, evolved a long neck to feed on leaves that others can't reach, giving it a competitive advantage.

But the peacock, with its extravagant tail feathers, was downright baffling to Darwin and represented an unsolved puzzle piece in his theory of evolution. Why would evolution favour such ostentatious and seemingly impractical displays? The peacock's tail seemed to serve no functional purpose in day-to-day survival. In fact, it was often a hindrance. It wasn't just the peacock, though; Darwin's frustration highlighted a larger conundrum in the animal kingdom.

From the flamboyant tail feathers of the peacock to the elaborate courtship dances of birds of paradise, nature seemed to be full of bewildering but cumbersome beauty that defied simple explanation.

---

*   Alfred Russel Wallace independently conceived the theory of evolution by natural selection, which is the sort of thing that would usually earn someone more than a footnote in a book. But Wallace gets the short end of the stick here because, while he helped develop the theory of natural selection, he wasn't particularly sold on Darwin's idea of *sexual* selection. To Wallace, the purpose of the peacock's flashy plumage was practical – like camouflage. And so, Wallace lurks here in the footnotes, overlooked in the same way he once overlooked the peacock's tail.

COURTSHIP AND SEDUCTION: A TWINKLE IN THE EYE OF A KĀKĀPŌ

The intricate and dazzling displays of peacocks would later become a cornerstone of Darwin's theory of sexual selection. In 1871, Darwin published his book *The Descent of Man, and Selection in Relation to Sex*, which laid the foundation for all modern work on sexual selection – and solved his peacock problem.

The peacock's tail, it turns out, was for mating – survival was, seemingly, a secondary concern. Though it made him considerably more visible to everything that might eat him, it made him more attractive to peahens and that, to the peacock, was of the utmost importance.

Sexual selection, in a nutshell, is nature's dating game. It explains the evolution of traits that improve reproductive success, even if those traits may be a hindrance to an individual's survival.

'Both sexual selection and natural selection boil down to the same thing – more of your genes in the next generation's gene pool – but they work by different means,' says evolutionary biologist and self-proclaimed dissector of snail genitals, Menno Schilthuizen.[11]

Natural selection operates through the survival of the fittest, favouring traits that enhance an individual's ability to survive and reproduce in their environment. This could mean anything from better camouflage to more efficient hunting skills, or even resistance to disease. Sexual selection, on the other hand, favours traits that increase an individual's chances of mating and producing offspring – even if those traits don't necessarily enhance survival and, in some cases, may reduce it.

This process of sexual selection, according to Darwin, occurs through two main mechanisms: intersexual selection (known as 'mate choice') and intrasexual competition.

Mate choice involves selecting mates based on specific desirable traits. It helped Darwin explain why some animals evolved seemingly impractical traits and behaviours that defied earlier understandings of survival of the fittest, the classic example being the peacock's vibrant tail feathers. Although the peacock's tail requires more energy and increases its visibility to predators, it is seemingly worth the risk – ostensibly because being alive isn't much use if no one wants to mate with you. In this case, the advantages of increased mating opportunities outweigh the dangers of greater visibility. So, peacocks with the most resplendent tails have better luck in attracting mates, ensuring that their flashy feathers are passed down to future generations. Mate choice is also observed – somewhat clumsily – amongst kākāpō, as we saw earlier. The male kākāpō, hollering from his hole in the ground, is demonstrating an acoustic and visual display that conveys to females the quality of his genes and overall biological fitness, hoping they will choose him as a mate. His deep, resonant call, combined with his physical condition, signals to females that he is a prime candidate for fathering their offspring.

Intrasexual competition, on the other hand, involves competition between members of the same sex for access to mates. For example, a male deer (known as a buck) with larger and more robust antlers may be better equipped to win battles and establish dominance over other males, which would then give him greater odds of mating with a female.

~

The mating rituals of the wild often involve elaborate dances, melodic songs, extravagant gifts and displays of physical prowess, intended to signal biological fitness, health and genetic superiority to potential mates. But they also offer clues to better understanding and protecting life on Earth.

Indeed, the Earth has music for those who listen. And if you listen closely while standing on the shores of California, you will hear the guttural roar of northern elephant seals in the throes of intercourse. This takes place during a mass gathering of seals that is so large, it often closes the beach to visitors. Meanwhile, in Florida you will hear the throaty bellowing of alligators engaged in a courtship. During alligator mating season, Florida goes into high alert, activating its Statewide Nuisance Alligator Program (SNAP), which has a dedicated hotline for locals who may get caught in the

COURTSHIP AND SEDUCTION: A TWINKLE IN THE EYE OF A KĀKĀPŌ

crosshairs of amorous alligators. In Botswana, grunts, wheezes and honks punctuate the night as hippos engage in a courtship under the star-speckled skies of the Okavango Delta. Some hippo vocalisations have been measured at 115 decibels, roughly the same volume as a motorcycle or bulldozer. While in Australia, as the golden light of dusk filters through the canopy, the forest comes alive with a medley of lyrebird mating calls. The lyrebird has a unique talent – it imitates a variety of sounds, including chainsaws, camera shutters, car alarms and even human voices. In Sydney's Taronga Zoo, a lyrebird named Echo has mastered a rendition of the zoo's fire alarm with such accuracy that he completes his performance with an 'evacuate now' announcement.[12]

Wildlife produces an array of wonderful sights and sounds when in the throes of lust, and these courtship rituals are becoming increasingly useful for ecologists. By diving into the world of wildlife courtship, we gain insights into biodiversity, species interactions and the overall health of ecosystems, all of which are crucial to maintaining life on Earth.

One way in which animal courtship is supporting the health of ecosystems is by capturing the grunts and groans of frisky koalas. In Australia, the guttural mating call of the koala is helping to guide decisions about habitat and populations.

An AI software trial using smart sensors has started in New South Wales, with the technology built to activate when it detects the sound of a koala's mating call. Sensors are fixed to trees to record the sounds of male koalas as they bellow to their intended paramour. Created by the International Centre for Neuromorphic Systems at Western Sydney University, the technology is designed to help improve tracking of the threatened species. This device allows ecologists to assess koala occupation across a landscape and then facilitate informed decision-making to protect important habitat and improve koala populations in New South Wales and beyond. The data from this technology is now being analysed to protect other iconic Australian species, such as gliders and forest owls, many of which are listed as threatened.

Sometimes, a beguiling mating call isn't enough, though. Animals employ various tactics to enhance their chances of reproductive success and demonstrate genetic fitness. From gift-giving to penis-flinging to trickery and deceit, there is great diversity in the courtships of the wild – perhaps nowhere more so than under the sea.

The male argonaut octopus has a detachable penis-arm which he throws at the female so that she may fertilise her eggs when she chooses. After the male flings his lone penis at the female, he promptly dies. The female, who is much bigger than the male – about eight times larger and 600 times heavier – keeps the penis-arm inside her shell and uses it to fertilise her eggs when she chooses.[13] While not much is known about this enigmatic creature, it is thought that the argonaut's penis-arm – which is actually a specialised arm called a hectocotylus that is used to store and transfer sperm – is an adaptation to ensure that fertilisation can occur even if the male is no longer present, enhancing reproductive success under risky mating conditions.

The anglerfish employs a drastically different strategy for reproduction. When anglerfish want to mate, the male eats a hole in the side of the much larger female and embeds himself. Their skin, blood vessels and digestive systems fuse together and his head dissolves, while his eyes, fins and organs drop off. He effectively becomes a parasite living on the female; his only role is to occasionally provide sperm for breeding. Scientists believe that this form of sexual parasitism is an adaptation that enhances the reproductive success of anglerfish in deep-sea environments.

Meanwhile, the cuttlefish resorts to trickery. Being a master of deception, the male cuttlefish will often disguise himself as a female to sneak past dominant males and gain access to actual females. The male will change colour, shape and even behaviour to closely resemble a female. This tactic, known as female mimicry, allows these 'sneaker' males to mate without engaging in direct confrontation with larger, more dominant males. In cuttlefish societies, larger males typically monopolise access to females through aggressive displays and physical confrontations. In environments where deception and camouflage enhance survival, mimicry can evolve as a favoured trait passed down through populations via natural selection.

The soap opera-like drama of animal courtship isn't just entertainment – it is shedding light on human social dynamics and even sparking new insights into feminist evolutionary biology.

Historically, studies of sexual selection have heavily emphasised male behaviours and traits, sometimes distorting or overlooking the crucial role of females in mate choice and reproductive success.

For much of history, science has fallen victim to outdated, non-scientific and patriarchal stereotypes about sex and gender roles. It was Darwin who, arguably, laid the foundations for this bias. For all his groundbreaking theories on evolution and sexual selection, Darwin was influenced by the prevailing prejudices of his era and was astonishingly blinkered when it came to understanding a female's place in the evolutionary story.

Darwin's theory and commentary on sexual selection sparked moral, scientific and social debates that continue today. Though he was one of the chief architects of evolutionary theory, Darwin constructed a worldview based on his preconceived notion that women were inferior. His biases around gender and race, in particular, were largely a product of the Victorian society that he lived in and had a profound impact on the way he theorised science.[14]

'Man is more courageous, pugnacious, and energetic than woman, and has a more inventive genius,' said Darwin in *The Descent of Man, and Selection in Relation to Sex*.[15] He claimed that there was a 'chief distinction in the intellectual powers of the two sexes', with man having attained 'a higher eminence'.[16] Similarly, Darwin likened female animals to 'wives' and believed that 'man has ultimately become superior to woman'.[17]

Darwin appears to have overlooked the fact that in mid-19th century Victorian society, almost all scientists were white Christian men, and women of this time were actively denied access to higher education, were excluded from the great scientific academies and, of course, were banned from voting, owning property, having bank accounts or holding public office. Such an unequal environment certainly makes it difficult to construct an impartial theory on the biological differences of the sexes.

As women's rights activist Caroline Kennard wrote in a letter to Darwin, 'Let the "environment" of women be similar to that of men and with his opportunities, before she be fairly judged, intellectually his inferior, please.'[18]

Unfortunately, while society has moved past many outdated Victorian-era mindsets and prejudices, the legacy of sexism in science and evolutionary theory remains. This has, in turn, given us a contorted understanding of evolutionary processes and social behaviours across species. The prevailing narrative in evolutionary biology has been that males are courageous, competitive and dominant, while females are passive and coy. Though this has its roots in observations of some species, it is a large generalisation shaped by skewed historical (and current) biases around gender and race.

Agustín Fuentes, professor of anthropology at Princeton University, accused Darwin of letting racism, sexism and misogyny 'warp' the scientific process and influence his findings. In the journal *Science*, Fuentes says Darwin's 'adamant assertions about the centrality of male agency and the passivity of the female in evolutionary processes, for humans and across the animal world, resonate with both Victorian and contemporary misogyny'.[19] Reflecting on Darwin's *Descent of Man*, Fuentes says that one merely has to look at the data to see that 'there is no biological coherence to "male" and "female" brains or any simplicity in biological patterns related to gender and sex within the theory'.

Similarly, Gil G Rosenthal and Michael J Ryan observe that Darwin's work was shaped by 'misogyny and sexual prudery'.[20] By focusing on male-centric traits, they note that many of Darwin's influential followers have followed the same narrow path – a path that largely neglects females.

Consequently, the sex differences that are emphasised today among species often stem from discrimination and social prejudices that shaped Darwin's view of the world in the 1850s, rather than from females' alleged biological inferiority.[21]

As evolutionary biologist Ernst Mayr says, 'No biologist has been responsible for more – and for more drastic – modifications of the average person's worldview than Charles Darwin.'[22] And while Darwin's theory of evolution by natural selection has been firmly established, the progression of some of his original ideas on sexual selection is difficult to evolve because of anchor bias. Anchor bias is a psychological phenomenon in which an individual tends to rely too heavily on the first piece of information they receive on a topic. This makes it challenging to accept new information on evolutionary theory that doesn't play by Darwin's original rules.

This bias has also persisted through to modern times, and, as a result, has shaped the direction of contemporary research on evolutionary biology and sexual selection, which remains overwhelmingly male-centric. Or, to be precise, penis-centric. In a 2014 journal article, researchers analysed 364 studies and publications on the topic of genital evolution, comparing the volume of publications focusing on male versus female genitalia. This revealed a startling discrepancy: male genitals dominate evolution research. Between 1989 and 2013, 48.6 per cent of studies were exclusively male-focused, while just 7.7 per cent of studies were exclusively female-focused.[23] Evolutionary theory, evidently, is hampered by outdated, single-sex bias.

Thankfully, the bonobo burst onto the scene with its bulging clitoris as a veritable feminist icon to disrupt humanity's traditional understanding of the role of females in social and evolutionary processes.

The bonobo is an endangered great ape that bears a remarkable resemblance to a chimpanzee and shares 98.7 per cent of its DNA with humans. Found in the forests south of the Congo River in the Democratic Republic of the Congo, bonobos have developed fascinating social structures. Among these structures, bonobos demonstrate that brandishing one's clitoris may just be the secret sauce of the sisterhood.

Unlike chimps and other non-human great apes, bonobos live in a largely egalitarian society that emphasises cooperation and sharing over conflict and territorialism. A key factor contributing to this harmony is the role that females play in a troop and their unique approach to sex. Bonobos are one of the few known species that can separate sex from reproduction. Sex is frequently treated as a pleasurable activity that is foundational to their relationships and communities, a social glue that strengthens cooperation and reduces tension within groups. As Professor Meredith F Small notes in *Discover* magazine, 'Bonobos seem to have sex more often and in more combinations than the average person in any culture, and most of the time bonobo sex has nothing to do with making babies.'[24]

The female bonobo's commitment to solidarity amongst women is inspiring a new generation of scientific inquiry. In a striking display of friendship, female bonobos enhance their bonds with one another by rubbing their clitorises together. This unique behaviour, known as genito-genital (GG) rubbing, can serve both social and sexual functions – fostering unity, pleasure and mutual support, and helping females strengthen alliances and gain influence within bonobo society.*

A female bonobo is approximately half the weight of a human teenager, yet her clitoris is three times as big as the human equivalent. So enlarged is the bonobo clitoris that it is 'visible enough to waggle unmistakably as she walks'.[25]

There may be a reason for this. It is thought that this

---

* Across the sea in Australia, four women – heroes, if you will – discovered that snakes have clitorises. The salient point here is that despite men making up the majority of snake researchers, they still were not able to find the clitoris. Or, didn't think to look for it. These women were Megan J Folwell, Kate L Sanders, Patricia LR Brennan and Jenna M Crowe-Riddell.

clitoral-rubbing ritual is in fact so beneficial to the species that the female bonobo has evolved to have a larger clitoris. Because clitoral rubbing among females offers an advantage to the species – by strengthening social cohesion – researchers believe that the clitoris has been selected as a favourable trait. Over generations, this results in the favourable trait (the clitoris) being passed on and becoming more exaggerated and common in the population.

Bonobos are not only expanding Darwin's original male-centric theories on sexual selection but can provide rich insights into the roles of creativity, social bonding and conflict resolution among humans. These findings could offer valuable lessons and help drive further improvements across various aspects of human society, from education to innovation to workplace productivity and wellbeing. This growing body of female-integrated research is not only reshaping our understanding of evolution but is providing a more complete understanding of the extraordinary diversity and complexity of life on Earth. Crucially, this may just help uncover the true architect of 'slut-shaming' – the age-old pastime of disparaging women for perceived sexual indiscretions.

Remarkably, the fruit fly might be the unlikely origin of slut-shaming of women. And fittingly, it is the kākāpō that can serve as a beacon of sexual liberation.

In 1948, English geneticist AJ Bateman published a study on fruit flies suggesting that male promiscuity enhanced reproductive success.[26] His assumptions became known as 'Bateman's principles' and heavily influenced human understanding of reproductive behaviour, often in harmful and misguided ways. While Darwin certainly launched this idea into mainstream, Bateman's principles reinforced the idea that men are biologically wired to 'spread their seed', and that women are naturally chaste.

In this way, a sexual double standard was allowed to blossom: in many cultures, men's sexuality was celebrated while women's became a social taboo. Science journalist Angela Saini says this idea of the chaste, monogamous woman was 'constructed by men' in a way that 'oppresses women'.[27] While Australian sociologist Michael Flood found that even today the label 'slut' carries stronger 'moral and disciplinary weight ... when applied to women'.[28]

Since Bateman first proposed his principles, many of their core assumptions have unravelled. In fact, studies across multiple species including humans show that females can gain both genetic and resource benefits from multiple mating.[29] This has led scientists to rethink traditional ideas about male and female behaviour. Yet, in humans, this research has been slower to penetrate social settings as long-held beliefs about gender roles and social norms continue to restrict female sexual freedom.

Back in New Zealand, the kākāpō joins a growing list of species that challenge the idea that males are hardwired for sexual adventure while females are naturally chaste and monogamous. According to Dr Lydia Uddstrom, veterinary advisor at the Department of Conservation's Kākāpō Recovery Programme, reproductive success is more likely when a female kākāpō is promiscuous. 'If a female mates more than once, either with the same male or ideally with different males, we see a massive increase in fertility of the eggs,' says Lydia. 'Having multiple mates is really important for female kākāpōs'.

Nature isn't just up-ending outdated notions of female sexuality but is challenging wide-ranging societal norms and gender expectations. Traits once deemed biologically male, such as promiscuity, aggression, risk-taking and dominance, look quite different when viewed through the lens of female animals.

Few studies, if any, draw a link between Hillary Clinton, bonobo clitorises and hyena hunting strategies, but Professor Jennifer Smith and her colleagues aren't like most researchers. In 2020, Smith and her team published a seminal paper that took a non-traditional approach to smashing the glass ceiling – they studied female non-human mammals.[30]

The glass ceiling – an invisible barrier keeping women from

leadership roles – was the focus of these researchers, who turned to non-human mammals for insights to address gender imbalances in human leadership roles. These findings, they offer, may just add another crack in the glass ceiling that Hillary Clinton came so close to smashing in the 2016 US presidential election.

Female-biased leadership, they revealed, is common in orcas, lions, spotted hyenas, bonobos, lemurs and elephants.

Naturally, the authors highlight the clitoris-wielding bonobos as exemplars of female leadership. In bonobo groups led by females, conflict is noticeably lower than in their male-led chimpanzee cousins. Similarly, in spotted hyena clans, the clitoris is an indicator of female leadership. Female hyenas are physically larger and stronger than males and wield the most power in their societies. The females bear a 'fully erectile, penis-sized clitoris through which they mate and give birth'. Researchers note that 'females often engage in elaborate greeting gestures involving the mutual investigation of their erectile pseudo-penises to reinforce social bonds and promote collective action among adult females'.[31]

In orca societies, females frequently lead efforts to cooperatively search and hunt for prey. Not only this, they also act as bearers of wisdom, promoting the survival of their young, who lack local ecological knowledge.

Meanwhile, in elephant societies, the female matriarch generally leads the collective movement of the herd. And it is also female elephants who intervene in inter-herd conflict and protect the herd, forming a protective circle around younger elephants when a threat emerges. Matriarch elephants, the authors note, 'serve as long-lived repositories of knowledge, sharing social and ecological information with less experienced group members and leading them away from potential threats'.[32]

Black-and-white ruffed lemurs and ring-tailed lemurs are also led by females, particularly in group conflicts, and females spearhead most collective attacks directed towards intruders.

In lion prides, lionesses virtually always lead group movements.

They engage in most of the cooperative hunting and regularly share prey within prides, acting as leaders and 'joining forces with each other to defend their territory against other prides as well as against infanticide by nonresident males'.[33]

These findings highlight factors relevant to human leadership: our traditional definitions of leadership may indeed be overly narrow, ineffective and outdated. Across these female-led societies, we see a version of leadership that thrives on balance, cooperation, conflict resolution, community-building, wisdom-sharing and adaptability – traits often categorised as 'feminine' and frequently overlooked and devalued in traditional models of power.

But this narrow view of leadership doesn't just disadvantage women; it limits men too. Both men and women are constrained by outdated ideas of what 'real' leadership should look like, where 'strength' is equated with dominance and 'empathy' with weakness. Such a rigid definition leaves little room for qualities that foster resilience, adaptability and community – all critical for modern leadership.

In nature, strong female leaders – from elder orcas to elephant matriarchs to clitoris-rubbing bonobos – embody a full-spectrum approach to leadership, drawing from wisdom, cooperation and emotional intelligence.

As one of the study's co-authors, Professor Mark van Vugt, says, it is important not to confuse leadership with dominance. 'Leadership is something that happens because there is a problem that needs to be solved by some kind of coordinated action,' he says.

By fostering coalitions and valuing knowledge-sharing, human societies can leverage this female leadership advantage for more balanced and resilient organisations, proving that nature may just offer a blueprint for how human societies could function more effectively.

And sometimes, we may just uncover the clues to this blueprint by examining the clitoris of the bonobo and the sexual delights of the kākāpō.

Chapter 2

A Legacy of Extinction: The Sound of Silence

> Homo sapiens is poised to become the greatest catastrophic agent since a giant asteroid collided with the Earth sixty-five million years ago, wiping out half the world's species in a geological instant.
>
> Richard E Leakey[1]

Can you feel it? The quiet creep of extinction. The fading of life on Earth. The way the insects no longer lash your windscreen and the last wild animal you saw was a battered body on the roadside – swollen and mangled, tyre tracks etched on its skin, blood congealed on its face. The quiet of the trees, the waning of the birdsong.

The world is fading around us. Like an old photograph, the colours are being leached. Until we are a world washed in monochrome.

This is mass extinction. Stone by stone, we sculpt the world into a grand mausoleum. And in it, we will surely bury ourselves.

In 1997, a Danish ecologist named Anders Pape Møller began counting dead bugs. Every summer for twenty years, Møller and his fellow researchers drove the same cars on the same route through the Danish countryside. With painstaking detail, they counted the dead insects on their windscreens after each ride. They used four different methods to collect data on dead insects, one of which was attaching adhesive tape to their windscreens. In 2019, Møller published the results of his decades-long research. Things were bad – really bad. The study revealed an 80 per cent decline in insect populations.[2] And with them, bird populations – which rely on insects as a food source – were collapsing.

It's not just birds and insects. Between 1970 and 2018, there has been a 69 per cent average decline in wildlife populations around the world.[3] This, according to scientists, constitutes a 'biological annihilation', one in which hundreds of species and populations are being eradicated.[4]

And in this annihilation, humanity has distinguished itself as the only species in the history of the planet to have single-handedly instigated a mass extinction of life on Earth.

A mass extinction event occurs when species disappear at a rate faster than new species can emerge. It is usually defined as the vanishing of about 75 per cent of all living species in existence within a short period in Earth's history.[5] A short period, given Earth's long history, is considered anything less than 2.8 million years.

There have been five mass extinction events in the Earth's history, all caused by natural phenomena.

The fifth mass extinction – the Cretaceous-Paleogene extinction – is the most recent and arguably the most famous. Occurring 66 million years ago, it marked the end of the dinosaurs.

The next mass extinction to follow would be unlike anything the Earth has ever experienced. We call this the sixth mass extinction, and, according to some scientists, it's already here.[6] It is marked by the dominance of one single species: Homo sapiens. Never before has one single species translocated so many animal and plant species on a global scale like this. Never before has one single species dominated and devoured life on Earth to such a scale that evolution is being redirected.

As the famed paleoanthropologist Richard E Leakey said, we have *theories* on what caused each of the 'Big Five' extinctions, some of them compelling but none definitively proven. 'For the sixth mass extinction, however, we do know the culprit. We are,' he said.[7]

The geological time period we currently live in is known as the Holocene epoch. Think of an epoch as a chapter in the long story of Earth's history. The Holocene epoch began around 11,700 years ago and marked the end of the ice age and the development of human civilisation. Yet, so drastic is the influence of humanity over the past two centuries, that this period is now widely – albeit unofficially – known as the Anthropocene epoch ('Anthro' meaning 'human') to describe the immensity and domination of humanity's impact on life on the Earth.

Great beasts roamed the Earth 11,700 years ago: towering woolly mammoths reaching 3 metres high, giant ground sloths weighing almost 4 tonnes and sabre-toothed tigers with razor-sharp canines stretching up to 18 centimetres long.[8] Then Earth underwent a period of profound transformation, emerging from the ice age into a new epoch: the Holocene. The Holocene brought fertile lands and a stable climate, enabling the rise of agriculture and sowing the seeds of the modern age. In the Fertile Crescent that stretched across the Middle East, human communities underwent drastic cultural and technological transformations that spread the world over. Crops and animals were farmed to provide a reliable food source, transforming nomadic hunter-gatherer lifestyles into more predictable agricultural societies. Around the world, people began to settle, farm and build communities. Great cities, empires and civilisations rose. Populations boomed.

By the 15th century, advancements in shipbuilding saw ocean-traversing ships travel vast distances with increasing frequency, heralding the Age of Exploration. British and European powers embarked on voyages to discover new trade routes, riches, and what they regarded as new land – often already inhabited. Farming practices evolved, paving the way for new production and manufacturing methods that would mark the dawn of the Industrial Revolution in the 18th century.

This was thought to be the golden era for humanity. At the centre of this revolution stood Great Britain. Steam engines roared to life and factories sprouted across the land, spreading throughout continental Europe and eventually across much of the world. Coal became the lifeblood of the new age, powering factories, cars, ships and trains, while iron forged into stoves and pipes helped deliver warmth to homes. Wool and cotton production surged, fuelling the textile industry. Cities expanded, life expectancy increased and human populations boomed as empires found new ways to enrich their economies.

But this was a gilded era with a dark underbelly. This period of expansion and pursuit of growth led to widespread environmental and social upheaval. Forests were razed and wetlands drained to make way for agriculture, urban developments, railways, roads and factories; vast swathes of natural landscapes were destroyed. Factories spewed smog and soot into the air and rivers and airways became choked with pollution as industrial waste was dumped indiscriminately.

The social fabric among some societies frayed. While some enjoyed immense wealth and benefitted from scientific and medical advancements, inequality widened in other areas. Many toiled in harsh conditions in factories and mines, facing exploitation and poverty. The rise of urban centres brought about overcrowded living conditions, poor sanitation and the spread of diseases.

European powers led brutal campaigns to seize vast territories across Africa, Asia and the Americas. In their pursuit of gold, silver, spices, sugar and cotton, they established plantations and mines, securing labour through the enslavement of people, particularly African peoples. The transatlantic slave trade became the cornerstone of the colonial economy, forcibly transporting millions of Africans to work in plantations in the Americas, particularly in the Caribbean. Colonial powers brought violence and displacement to Indigenous communities, who faced armed conquest, forced relocations and the destruction of their traditional ways of life.

As the exchange of goods increased across lands, so too did the spread of pathogens and invasive species. Deadly viruses and bacteria such as smallpox, measles, typhus and cholera were exchanged across continents. European domesticated animals, such as cows, pigs, chickens, goats and sheep, were transported across the seas, finding new homes in foreign lands and decimating local ecosystems. Rats and insects hitched rides on ships, spreading lethal pathogens and waging war on native species.

The Earth's ecosystems strained under such pressure. Firearms evolved, each new iteration more accurate and powerful than the one before. Hunters could kill animals from greater distances, in greater numbers, with more precision. Commercial hunting operations intensified as demand for animal products like fur, feathers, meat and ivory rose. Animal populations plummeted. The fur trade, particularly in North America, saw the near extinction of species like the sea otter and beaver. Demand for ivory led to the extensive hunting of elephants in Africa. Passenger pigeons, once numbering in the billions, were hunted to extinction. American bison populations plunged from millions to just a few hundred.

In the 19th century, fishing techniques industrialised, and large nets trawled the oceans, hauling thousands of animals in a single catch. Steam-powered whaling ships endowed with exploding harpoons took to the seas, slaughtering whales by the hundreds of thousands. Whale bodies were broken down into parts, their blubber lighting homes and lining soaps, candles, paint, cosmetics and grease for machinery. Whale bones were moulded into corsets, hair combs, tongue scrapers, brooms, chimney brushes, umbrella spines and fishing rods, while their teeth were fashioned into chess pieces, jewellery and piano keys.[9]

In the backdrop of this frenzy, carbon dioxide levels were rising. As coal, oil and gas were burned, vast quantities of greenhouse gases blanketed the Earth, trapping heat in the atmosphere.

Climate patterns began to shift as sea levels rose, while extreme weather events surged in frequency and intensity. The invisible hand of climate change was beginning to reshape the world. Prior to the Industrial Revolution, carbon dioxide levels had sat at around 280 parts per million for the preceding 6000 years. That is, for every million parts of air, 280 were made up of carbon dioxide. By 1950, carbon dioxide in the atmosphere had reached 312 parts per million. In 2023, it was 419 parts per million.[10]

By the 20th century, alarm bells were ringing. Up until this point, outside of scientific circles, the concepts of environmental decline, climate change and species extinction were largely regarded by the general public with 'curiosity and mild humour'.[11] That was, until disaster struck the city at the heart of the Industrial Revolution.

On Friday 5 December 1952, a great smog descended upon London, blanketing the city in a shroud of pollution and bringing the city to its knees. Up to 12,000 people died and hundreds of thousands suffered respiratory illness. So thick was this fog that it is said to have choked cows to death in the fields.[12] Residents were unable to see their feet as they walked. Driving was nearly impossible, so abandoned vehicles clogged the roads. The smog seeped inside buildings as well. A greasy filth covered exposed surfaces, and movie theatres closed as the yellow haze made it impossible for ticket-holders to see the screen. Birds got lost in the fog and crashed into buildings. Undertakers ran out of coffins and florists out of bouquets as the deadly impact of the Great Smog took hold.

In the weeks preceding this smog, a wintry cold had besieged the city of London. To keep warm, residents burned large quantities of coal in their homes. Smoke poured from chimneys across the city, mixing with the pollution being spewed from London's factories and vehicles. At the same time, a high-pressure weather phenomenon known as a temperature inversion was settling over southern England, trapping cold, polluted air close to the ground. This meant that the sulphurous smoke produced by coal couldn't rise and disperse. With no wind to clear it away, a toxic yellow-black smog that stank of rotten eggs smothered the city.

Five days of apocalyptic conditions were enough to shift public opinion on environmental issues, albeit slowly. In 1956, following a government investigation, the British parliament passed the *Clean Air Act*, which restricted the burning of coal in urban areas, authorised local councils to establish smoke-free zones and provided grants for homeowners to convert from coal to alternative heating systems.

From rising pollution levels and massive deforestation to devastating oil spills, the world was forced to confront the environmental crisis. The environment had well and truly punctured public consciousness.

With the dawn of the 21st century, evidence of environmental crisis mounted with undeniable force and clarity. Reports documented environmental harm and species loss in vivid, horrifying detail: ancient forests decimated, polar ice caps melting, marine life smothered by oil, waterways choked with plastic, greenhouse gases poisoning the atmosphere.

The new millennium had arrived, and Earth was fracturing under an onslaught of violence.

In 2017, a powerful declaration echoed across the globe as over 15,000 scientists from 184 countries signed a statement titled, 'World Scientists' Warning to Humanity: A Second Notice'.* They asserted, among other things, that 'we have unleashed a mass extinction event, the sixth in roughly 540 million years, wherein many current life forms could be annihilated or at least committed to extinction by the end of this century'.[13]

The evidence was undeniable: humanity's impact on the Earth was accelerating and, as a result, species were disappearing at an alarming rate.

---

* The second notice has more scientist signatories and formal supporters than any other journal article ever published.

Despite this, disagreement over whether this amounted to a sixth mass extinction was drawing considerable attention.

While a majority of scientists argue that the speed and scale of species loss compels us to accept the arrival of a sixth mass extinction event, a smaller faction challenges this view. This group argues that we lack conclusive evidence to suggest we have reached the catastrophic 75 per cent species loss benchmark that defines previous mass extinctions.

In response, the majority counters that only a fraction of the world's species has been evaluated, and there is a lack of data on bugs and many other species, which skews the overall picture of biodiversity loss.

Then the latter, more optimistic group insists that we still haven't quite reached the catastrophic benchmark set by the previous five mass extinctions, though we're enthusiastically hurtling towards it. What they are saying, in essence, is, 'We are in agreement that the house is on fire, but a few still argue it might not have completely burned down – yet.'

In the midst of this scientific battleground, another drama was unfolding, where intellectual might was being wielded like a sharpened sword – only with less bloodshed and more rock samples.

For decades, a debate has been raging in the science community that is equal parts urgent and meticulously pedantic. The question at the heart of the debate is: does the devastating impact of human activities deserve to be formally recognised as a new geological epoch?

Proponents of the idea, led by the Anthropocene Working Group (AWG), wanted to formalise the undeniable and irreversible changes that human activity has wreaked on the planet. According to them, the Holocene – a relatively calm 11,700-year stretch of environmental stability during which humans invented agriculture, cities and the concept of brunch – is officially over. We have now moved into the Anthropocene epoch – that is, the age of humans.

The answer could have profound implications for our understanding of Earth's history and our place within it; it would formalise a new chapter in Earth's geological timeline and cement the fact that human influence is now the dominant and irreversible force shaping the planet. Such a declaration would guide scientists in their research, shape terminology in textbooks, research articles and museums worldwide – and influence how future generations understand our evolving planet.

Reaching a conclusion on this matter, unfortunately, has all the urgency of a house fire and the administrative delight of a parliamentary procedures committee.

To put the matter to rest once and for all, a group of formidable experts was called upon. Enter the geologists.

Armed with metaphorical rock hammers and an unyielding determination to unearth things like the truth, these geologists were tasked with deciding whether humanity has truly left the Holocene behind and officially entered the Anthropocene.

As the guardians of the Earth's timeline, these geologists study the Earth's crust and carve up the planet's 4.6 billion years of history into phases, placing them in chronological order on a timescale.

On 4 February 2024, after fifteen years of debate, the International Commission on Stratigraphy (ICS) assembled to cast their votes either 'for' or 'against' a proposal to give formal geological recognition of the 'Age of the Humans'.

By 4 March, the results were in. Four voted in favour, twelve voted against and three abstained from voting.[14] The outcome was a decisive rejection: we were not officially, geologically, in the Anthropocene epoch.

'There are no outstanding issues to be resolved. Case closed,' declared Philip Gibbard, secretary-general of the ICS.

And that is when all hell broke loose.

Some members cried foul, claiming 'grave violations' of procedural rules and calling for an inquiry. Others accused those members of having 'sour grapes' and being 'sore losers'.

News outlets blasted headlines like 'Geologists Make It Official: We're Not in an "Anthropocene Epoch"'.[15]

To the casual observer, it would seem that humanity's impact on the planet had been exaggerated. But this is far from the truth.

The voting members were not disputing humanity's impact on the planet. What they were actually disputing largely came down to dates, definitions and tradition.

To qualify as a new geological epoch, specific criteria must be met, including having a clear starting date that can be pinpointed in rock and soil layers. But as it turns out, nailing down this starting date for humanity's impact isn't as straightforward as it sounds. In this case, some experts claimed it was impossible to attribute one specific date to the start of humanity's broad planetary influence.

But there were other objections too. Stanley Finney, fondly known as 'the big phallus' of geology, disputed the recognition in its entirety, claiming that the Anthropocene is a political idea rather than a scientific one.[16]

Others thought the epoch was just too young to be recognised – at least by the standards of the Earth's 4.6 billion-year history. As former chair of the AWG Jan Zalasiewicz says, geologists 'work in geological time'.[17]

What is important to note here is that the enormity of humanity's impact on Earth was not in question. Even the International Union of Geological Sciences (IUGS), the voting committee's parent body, acknowledged that the Anthropocene 'will remain an invaluable descriptor in human-environment interactions'.[18]

Despite the vote, the AWG stands fully behind its proposal, which 'demonstrated beyond reasonable doubt' that the planet has moved well beyond the stable conditions of the previous era, and that these changes are irreversible.[19]

What this means, in a nutshell, is that everyone agrees humanity is wreaking havoc on the planet but there is strict protocol for announcing it.

While geologists debate whether humanity's impact warrants a new epoch, biologists are counting the casualties. But both are

A LEGACY OF EXTINCTION: THE SOUND OF SILENCE

converging on one blunt truth: the planet is no longer operating within the relatively stable boundaries of the Holocene, and the fallout is being measured in extinctions.

For decades, discourse surrounding extinction has been mired in misinformation, perpetuating enduring myths that distort public understanding of the extinction crisis.

One of the most persistent misconceptions is the belief that extinction is a natural and expected part of Earth's history, and therefore the current rate of extinction of species is normal.[20]

Phrases like 'extinction is the engine of evolution' and 'species constantly go extinct' are missing vitally important pieces of nuance.[21] These generalisations are appealing fodder for the continuing denial of the extinction crisis we are weaving. But the devil is always in the detail, and detail doesn't make for a great headline. There is simply no headline or soundbite that can capture the complexity and catastrophe of humanity's relationship with the natural world.

While it is true that extinctions are a normal part of evolutionary processes, this argument misses fundamental details, one of which is the background extinction rate, which describes the natural rate at which species go extinct without human interference.

The background extinction rate is estimated to be around one extinction per million species per year. Which means that under typical circumstances, in one single year, roughly one species will become extinct out of every million species.

Yet over the past 200 years, extinction rates have increased sharply, coinciding with the rise of industrial society. Today, current extinction rates are 1000 times higher than natural background rates of extinction and, if current trends persist, future extinction rates could be up to 10,000 times higher than the natural background levels.[22]

Another common misconception about extinction is rooted in a misinterpretation of evolution and natural selection. If you've ever heard the term 'survival of the fittest' bellowed at your Saturday morning CrossFit class, then you have likely been exposed to one of the ways this term has misled discussions about extinction and evolution.

When people talk about 'survival of the fittest', they tend to imagine raw strength, speed or some kind of bloodthirsty dominance – traits that might win you an athletics competition, or at the very least a bar fight. Used in this way, humanity's ability to dominate others – or nature as a whole – is a source of pride, celebrated as an honour that marks evolutionary triumph. It can be used to justify anything from ruthless competition to human (or social) superiority to extinction of species.

In this flawed interpretation, humanity is the ultimate apex predator, uniquely exempt from, and superior to, nature's rules.

While there is no question of humanity's dominance as a species, this narrative is a deeply distorted and dangerous misunderstanding of Charles Darwin's theory of evolution by natural selection, particularly when it leads some to view the mass extinction (or control) of other species or social groups as a natural outcome of our ascendancy.

The concept of survival of the fittest was made famous by Charles Darwin's evolutionary theory, though it was a term he had borrowed from English sociologist and philosopher Herbert Spencer.

In Darwinian terms, 'fitness' doesn't refer to physical strength, speed or dominance, as it is often (mis)understood today. Instead, it refers to the ability of an organism to survive and reproduce in a specific environment. The 'fittest' individuals are those best adapted to their ecological niche, meaning they are more likely to survive and pass on their genes to the next generation. Think of a puzzle piece that snugly fits into its unique place rather than a chiselled bodybuilder.

This distinction is important. Throughout history, the ideology of human supremacy has been wielded as a weapon to justify mastery over others, often under the banner of survival of the fittest.

In the late 19th century, the significance of Darwin's thoughts extended beyond the realm of biology, and the concept of survival of the fittest began to take on a much more sinister life of its own. Philosophers, economists and political scientists applied Darwin's concepts to their own fields, and before long, a distorted version of Darwin's theory of natural selection was being used to fuel supremacist views and justify atrocities.

One way in which this supremacist ideology manifested was in the form of social Darwinism, which influenced and propelled scientific racism and, later, Nazism.

Social Darwinists held that the life of humans in society was a struggle for existence ruled by survival of the fittest. They warped and twisted theories of natural selection to argue that social and economic inequalities were a natural result of the 'fittest' humans rising to the top.

Among many other flaws in their arguments, proponents of social Darwinism fundamentally misunderstood the notion of 'fitness' and erroneously equated it with wealth, power and social status.

As social Darwinism evolved, it gave rise to the scientific racism and the eugenics movement, which advocated for compulsory sterilisation, euthanasia and, ultimately, ethnic cleansing and genocide. These twisted beliefs found their darkest expression in the policies of Adolf Hitler, who brandished these ideas to legitimise the horrors of the Holocaust.

'Since the inferior always outnumber the superior, the former would always increase more rapidly ... Therefore a corrective measure in favour of better quality must intervene. Nature supplies this by establishing rigorous conditions of life to which the weaker will have to submit and will thereby be numerically restricted,' said Hitler.[23]

To be clear, Darwin's theory of natural selection explains how species adapt to their environments over time, favouring traits that enhance survival in specific conditions. It doesn't suggest that wiping out other social groups or species en masse, or even dominating them, is a mark of success.

While social Darwinism has been comprehensively debunked and dismissed by the scientific community – and the global community at large – echoes of this sentiment have subtly shaped narratives around humanity's relationship with nature.

Among some fringe groups, the idea of humanity's supremacy – the belief that we have the 'power to change nature in our interest' and hold 'dominion over nature' – is not only accepted but celebrated.[24] In this view, our role as masters of our environment is seen as a natural and laudable achievement, fulfilling what some perceive as humanity's ultimate purpose: to shape and dominate the natural world.

This narrative is not just scientifically inaccurate – it's also dangerous. It can lead to a view that human-induced wildlife extinction is natural or even necessary. It suggests that we are somehow above nature, when in reality we are deeply intertwined with it. Moreover, it absolves us of responsibility for an environmental crisis of our own making.

When we drive species to extinction, we're not demonstrating strength; we're destabilising the very systems that support us and threatening the delicate balance of ecosystems. The more species we lose, the more precarious our own future becomes. Far from being a triumph of evolution, human-induced extinction is a sign of our failure to honour and understand the evolutionary balance of the natural world. We're not proving our strength by driving other species to extinction; we're eroding the foundation of life on Earth.

We are playing a dangerous game with nature that we will most certainly lose. And in doing so, we set the stage for our own demise.

~

Legend has it that in 1872, a German scientist placed a live frog in a pot of tepid water.* Then, he lit the flame and watched as the

---

* The 19th century saw a series of experiments where scientists sought to uncover the mysteries of life and death through the hapless frog.

water boiled, the frog along with it.[25] It was with this gruesome experiment that the term 'boiling frog syndrome'[26] was born. It is a grim metaphor that aptly describes our march to extinction, where gradual changes go unnoticed until it's too late.

For the frog, the temperature rises ever so slowly. At first, the changes are barely perceptible, just a gradual warming that the frog doesn't heed. The heat starts to intensify and the frog remains oblivious to danger, adjusting to the rising temperature. Until it's too late. The water boils, and with it, the frog's fate is sealed.

Extinction is a silent storm. First, the insects vanish, and birdsong falls silent. Life drains from the oceans, coastlines erode and forests wither. Then fall the reptiles and amphibians. The mammals soon follow. And then it is too late. This storm will take us down.

There is a relentless pulse of tiny moments that pave the way for mass extinction. A building is erected atop the last of a bird breeding ground. A ship collides with a whale. A wild animal is crushed by a car. A forest is razed. An oil tanker ruptures in the sea.

It's a place none of us know until we are already there. But the signs are there, like a creeping shadow. It comes to us in the night, weakens us at the knees, knocks us off our feet, blinds and deafens us. It swallows us whole, drowns us. It's when the glow of the moon is swallowed by the bright city lights. It's the tsunami that rips through the town and the sound of sirens as a city is engulfed in flames. It's the collapse of an ice shelf and the last breath of a species lost forever. It's the parched throats in the midst of a drought and the poisoned air that fills a child's lungs.

This is extinction, weaving its way into our lives.

And it all seems normal – that is, until we start counting the dead bugs on our windscreen.

Chapter 3

A Spectacle of Survival: A Humbled Army

> The enemy is the tough, prolific, gangling marauder of the sand plains whose species, ever since the beginning of agriculture in the State, has invaded, in a frenzy of hunger, some of the finest fields.
> *The Sydney Morning Herald*[1]

The year was 1932 and the Australian military had just lost a war to a flightless bird. This wasn't the outcome the government had anticipated, of course. But in a struggle for survival, the emu proved fittest.

The Great Emu War was a peculiar chapter in history for a country that's already perfected the art of the absurd. Australia, ever the overachiever in eccentricity, has a unique talent for turning its quirks into national lore. This is, after all, the country that responded to the mysterious vanishing of its prime minister, who drowned at sea, with the most fitting tribute they could muster – naming a public swimming pool after him. The Harold Holt Memorial Swimming Centre stands proudly in Glen Iris, Victoria, known fondly by locals as 'Dead Harry's'.[2]

The interwar years in Australia were marked by severe economic hardship. The devastation of World War I had left deep scars, and the onset of the Great Depression only intensified the struggle for Australians. The Australian economy, heavily dependent on agriculture, was suffering, and nowhere was this more acutely felt than in Western Australia, a key agricultural hub with vast tracts of land devoted to wheat farming.

In an effort to boost agricultural output and honour returning servicemen, the government had allocated parcels of land in Western Australia to many veterans after the war. But by the 1930s, farmers and veterans found their lands under siege – not by human enemies, but by emus.

Emus numbering in the tens of thousands descended upon this crucial farmland in Western Australia, trampling crops and wreaking havoc on the agricultural sector.[3]

Emus are endemic to Australia. Featuring prominently in Indigenous stories and culture, they are also the second largest bird in the world, after the ostrich. Each year, they migrate across Western Australia in search of food and suitable breeding grounds. Their migration in 1932 was not unusual in itself, but the sheer scale of the movement was unprecedented, driven by drought, habitat disruption and the lure of abundant wheat crops in newly cleared farmland.

The farmers, faced with destruction of their livelihoods, appealed to the government for help. The situation was dire: their crops were being decimated, and the economic pressures of the Great Depression meant they could ill afford such losses.

In response, the Australian Government took a drastic and unprecedented step: they declared war against the emu.

The operation was placed under the command of Major GPW Meredith of the 7th Heavy Battery of the Royal Australian Artillery. The goal was clear: reduce the enemy emu population and protect farmers' crops.

What followed was a series of comical and humiliating skirmishes that would cost Australia a great deal of ammunition, as well as dignity.

Soldiers arrived in Campion on 2 November 1932. On the first day of battle, just twenty emus were killed. Two days later, the military launched an ambush on the birds. But the mission went awry. Just twelve of the 1000 assembled birds were killed.

'They can face machine guns with the invulnerability of tanks,' said Major Meredith.[4]

After six days of battle, the military, having fired 2500 rounds of ammunition and killing less than 500 birds, withdrew from the war.

'The emus have won every round so far,' cried former minister of defence Albert Ernest Green, according to official parliamentary records, which were, indeed, documented.[5]

Unwilling to be defeated by birds, the Australian military effort was embarrassingly reinstated and reinforcements were called in on 12 November.

This attempt proved marginally more successful. But after almost one month of war, it was becoming painfully obvious that the Australian Army was no match for the emu. The ratio of bullets expended to casualties was simply too high to justify continuing.

On 10 December 1932, the army, reaching an underwhelming climax, withdrew its troops for good. The emus were declared victors of the war.*

Today, Australia boasts the rare distinction of being the only country in history to lose a war to a flightless bird. The Great Emu War is, perhaps, the most quintessentially Australian illustration of survival of the fittest.

So, how did a species of flightless birds manage to defeat a modern military? The answer lies in the emus' unique and advantageous survival traits. Survival traits are physical or behavioural characteristics that give an organism an advantage in surviving and reproducing in its environment. Over generations, these advantageous traits – such as speed, size, migratory behaviours

---

*   After the soldiers returned home, federal Labor parliamentarian AE Green was asked if the troops would receive a medal.

or sharp beaks – get passed down, gradually becoming common in the population.

This process is known as adaptation. Adaptation doesn't happen overnight; it evolves through natural selection, one of the core mechanisms of evolutionary change.

Throughout this war, the emus demonstrated a spectacular array of traits that allowed them to outmanoeuvre and ultimately defeat the Australian military.

The birds didn't achieve victory through direct combat; instead, they leveraged their own unique suite of evolutionary advantages, employing a blend of speed, group coordination and savvy environmental knowledge. They simply outsmarted the Australian military and then wore them down with frustration and humiliation.

When faced with military firepower, the emus would employ their remarkable speed and agility to escape from areas where they were being targeted. Emus can run at speeds of up to 50 kilometres per hour. Their ability to outrun their pursuers meant that by the time the soldiers repositioned to target another group, the emus had already regrouped elsewhere, rendering their efforts largely ineffective.

Emus also demonstrated a tactical advantage through sophisticated group behaviour. When confronted by the might of the Australian military (which, in this case, involved soldiers chasing flightless birds with machine guns), the emus would scatter and then regroup as a unit, making it impossible for the army to target and manage them effectively. This strategy of breaking into smaller groups rendered the entire military's efforts futile, much like a toddler trying to catch confetti in the wind.

The emus' adaptability to various environments also played a crucial role in their survival. While the military struggled with the harsh terrain and weather of the Western Australian Wheatbelt, emus navigated the landscape with ease. When the military attempted to flush emus out of one area, the birds would simply migrate to less accessible regions or find refuge in areas where their presence was less detectable.

Finally, the emus managed to achieve a level of psychological warfare that few enemies had before them. Their constant disappearance and reappearance – combined with their ability to carry on destroying crops as if nothing was amiss – turned the battlefield into an absurd theatre. Such casual defiance from the emu, of course, further demoralised the soldiers.

An army never expects to lose a war to a flightless bird – or any bird, really. But the emu simply boasted survival traits that even the Australian Army couldn't contend with.

~

Nature is brimming with remarkable adaptations. From colour-changing camouflage that allows predators to vanish in plain sight to bioluminescent bodies that light up the deep sea and turtles that breathe through their genitals, adaptations are survival tools honed over millennia to help species survive and reproduce in their environments.

But in the face of widespread human interference, how do species adapt? And can they?

The answer to this is anything but straightforward, involving a complex interplay of genetic, social, behavioural and ecological factors.

Perhaps the most famous case of adaptation is witnessed in a species of insect that arouses almost universal indifference among the general public: the moth.

Rarely considered unless it is circling a lightbulb, the peppered moth is actually one of the clearest illustrations both of the effects of human-induced environmental change and of Charles Darwin's famous theory of evolution by natural selection.

The peppered moth's existence could be described as a long, tedious exercise in survival punctuated by occasional moments of profound annoyance. Typically found in the northern hemisphere, the peppered moth had traditionally boasted white wings speckled with black. For millennia, this moth had been painstakingly honing its adaptation skills to blend in seamlessly with its habitat, its wings allowing it to rest camouflaged on its favourite lichen-covered tree trunks. Here, it could while away the days in peace, largely ignored by predators and, indeed, everyone else.

In this environment, the white wings of the peppered moth were an evolutionary advantage because they rendered the moth nearly invisible to predators. Then, just to make things interesting, a genetic mutation kicked in that produced a black variety of moth. Initially, these black moths were at a disadvantage: their dark wings contrasted sharply with the lichen-covered bark, making them prime targets for predators. As a result, fewer black moths survived to reproduce, keeping their numbers low compared to the better-camouflaged white peppered moths.

This situation persisted until the Industrial Revolution swept through the British Isles, belching out clouds of soot and smoke that blanketed entire landscapes in grime. Pollution killed the lichens and darkened the tree trunks, turning the previously advantageous light colouration of the peppered moth into a liability. In this new, soot-darkened landscape, the black moths, once at a disadvantage, found themselves better camouflaged. Their survival rates increased and their population surged as they became less visible to predators. Suddenly, the once ingenious and carefully crafted white wings became a death sentence; the dark-winged moth was now the new darling of the moth world.

Since moths are short-lived, this evolution by natural selection happened quite quickly. The first black peppered moth was recorded in Manchester, England, in 1848 and by 1895, 98 per cent of peppered moths in the city were black.[6]

Then the mid-20th century saw the introduction of pollution controls through the *Clean Air Act*, which improved air quality. As tree trunks lightened with cleaner air, lichen made a comeback, and the light-coloured peppered moths reclaimed their place in the spotlight. The black moths fell out of favour once more. The balance of natural selection swung back, proving that in the game of survival, adaptation is less about a fixed advantage and more about keeping pace with an ever-changing environment.

Sadly, having so successfully weathered the soot-stained chaos of early industrialisation, the peppered moth is now in decline. Between 1968 and 2002, its numbers in Britain fell by almost two-thirds.[7] The causes of this decline are still unknown, though one might wonder if, after all the upheavals and sudden shifts, the moths have simply grown weary. After all, there's only so much adapting one species can do before it decides it's had quite enough.

If species like the emu and peppered moth can find ways to survive and adapt to both war and industrial revolution, what does that mean for the broader spectrum of life facing human-induced environmental changes? This raises another important question: if species can just adapt so well, why should we worry about protecting the environment?

Australia might be able to help answer this.

You'd be forgiven for assuming that Australia's war with an emu and shameless tribute to a drowned prime minister were isolated events. But, of course, you would be mistaken. In Australia, irreverence is a national sport played with gleeful enthusiasm. And in this game, no one is off limits. From spreading a rumour that their own prime minister had soiled himself in a McDonald's to trying to sell the country of New Zealand on eBay, insolence and mockery is woven into Australia's cultural narrative.

Continuing this tradition, Australia unsurprisingly took to naming one of its native birds, the Australian white ibis, with the finesse of a blunt axe. Widely known as the 'bin chicken', the ibis is a long-legged wetland bird that has, in recent years, acquired a reputation as a raider of bins and a 'tormenter of children's playgrounds'.[8] Its expansion into the city rose as rapidly as its reputation; such a stir has this bird caused that it is generally met with either revulsion or glee by the Australian public, depending on whom you ask. In 2016, 8000 people registered for International Glare at Ibises Day in Sydney, where attendees were asked to gather at their local park and 'glare and show your general distaste toward ibises'.[9] But in 2017, the ibis was voted by the public as Australia's second most popular bird.[10]

The ibis lives by the mantra that any publicity is good publicity. Originally inhabiting inland wetlands in Australia, the birds were driven out of their natural habitat, which was degrading due to extensive agricultural development, drainage of wetlands and urban expansion. The ibis quickly moved into urban centres in search of food and habitat. Today, paying a fitting tribute to its moniker, the ibis can be found merrily rummaging through landfill and waste, or sitting atop a bin.

To the casual observer, it would look as though life had given the ibis lemons and then spat in its face for good measure. But it would be a grave mistake to underestimate this particular dumpster diver. Since moving to cities, its populations have flourished, so much so that the native bird is considered a pest in some areas. The ibis has, according to experts, 'greatly benefited from human settlement'.[11]

Yet the list of formal complaints submitted to the government against the ibis is long, specific and strident. Ibises have allegedly committed the following crimes: damaging vegetation, competing with local wildlife, transmitting disease, colliding with aircraft, threatening passenger safety, scavenging from bins, dispersing litter, harassing people and stealing food. Though it may be useful to note that the ibis likely has a similarly long – and arguably more compelling – list of complaints against humans.

Dr Dominique Potvin, senior lecturer in animal ecology at the University of the Sunshine Coast, points out a curious paradox facing urban species. While certain adaptable species like the ibis can thrive in cities, their success can often backfire. 'That success comes with the risk of them becoming viewed by people as a "pest"', she says. 'Their adaptability and resilience has allowed them to persist after their original habitats have been degraded,' she says, but now their close interactions with humans have meant our 'perspective on them has shifted'.

Such urban species may appear abundant in urban settings, she notes, but they are often vanishing in their native habitats. The risk is that human perception of an urban population may not reflect the reality of that species, which means policies, priorities and management 'are not necessarily based on data or evidence, but rather opinion'. This mismatch between perception and reality, she argues, complicates efforts to protect species where it's needed most.

Irrespective of public sentiment, the ibis makes an interesting case study in adaptation. In its new urban setting, the ibis has become a symbol of adaptability. Instead of retreating or declining, the ibis has thrived in the urban sprawl, finding sustenance among a vast buffet of human waste.

~

Yet in stark contrast to the success story of the ibis is the plight of many other species, such as the koala, the regent honeyeater and the Leadbeater's possum – all of which are critically endangered.

While the ibis provides quite a remarkable story of resilience, it doesn't reflect broader species trends across Australia. You see, Australia has a rather long and illustrious tradition of making species go extinct. The country holds the record for most mammal extinctions of any continent on Earth and has accomplished a rather unique achievement of driving its most iconic species – the koala – to the brink of extinction.

Why can an ibis thrive in cities amidst trash while species like the koala, the Leadbeater's possum and the regent honeyeater perish? There are many reasons to explain this, not the least of which is that few species can survive a relentless onslaught of habitat destruction. But one reason to explain the difference between species like the ibis and the koala could be the role of 'specialist' and 'generalist' species. The ibis is a generalist species, meaning it can eat a wide variety of foods and thrive in a range of habitats. While the regent honeyeater, Leadbeater's possum and koala are all specialist species, meaning they have a limited diet and stricter habitat requirements. Species with highly specialised diets or habitats may find it challenging to shift to a new environment if their specific needs are not met.

The divergent fates of the ibis and the koala highlight a critical aspect of adaptability: generalists tend to fare better in rapidly changing environments than specialists. The ibis, with its remarkable ability to exploit everything from wetlands to city bins, has flourished in urban environments. The koala, by contrast, is a specialist – reliant on eucalyptus forests and, more precisely, on the leaves of a limited number of eucalyptus species for both food and water. As these habitats are fragmented or cleared, the koala's options diminish. In this way, specialist species like the koala are more likely to be endangered than generalists, surviving only in a narrow range of conditions and/or relying on specific food sources, while generalist species are able to survive in a wider range of conditions and on a wider range of food sources.

This comparison underscores a broader principle in ecology: the adaptability of a species is often linked to its ecological niche.

Dr Potvin says that while scientists do think there is 'an advantage to being a "generalist" in terms of habitat and diet,' this isn't necessarily a prerequisite for survival, 'and nor do all generalists tend to do well in urban environments'. She says there must be more to the story. 'It's likely a combination of a lot of different features about urban habitats as well as the species themselves.'

The bin chicken's rise as an urban survivor is both a story of resilience and a cautionary tale. It shows us that while some species can adapt to the chaotic environments we've created, many others – like the koala, the regent honeyeater and the Leadbeater's possum – are left behind, unable to cope with the rapid changes brought about by human activity. And often, the presence of native animals in cities – like the ibis – often masks their decline in native habitats, leading to misplaced conservation priorities.

As we see with the koala, not all species can adapt with the speed and adeptness of the moth or thrive in urban environments like the bin chicken. While there are many species that are revealing a remarkable ability to adapt to human-induced change, the overwhelming majority are collapsing under the force of humanity.

So great is the impact of humanity that scientists are confident that humans are 'worse for nature than the world's worst nuclear accident'.[12]

On 26 April 1986 at 1.23am, Reactor 4 of the Chernobyl Nuclear Power Plant, located in what was then the Soviet Union (now Ukraine), exploded during a late-night safety test that went catastrophically wrong. The explosion ripped through the reactor's roof, releasing a massive plume of radioactive material into the atmosphere. The initial blast killed two plant workers instantly, and the ensuing fires sent more radioactive particles into the air, turning a disaster into an environmental apocalypse.

In the hours that followed, emergency workers, firefighters and plant staff – most unaware of the full scale of the danger – rushed to the site to contain the fires and limit the spread of radiation. They fought flames and tried to cool the reactor, many without protective gear, directly exposing themselves to lethal doses of radiation. The graphite used in the reactor's core ignited, causing fires that raged for days. The released radiation – about 400 times more than the atomic bomb dropped on Hiroshima in World War II – spread across Europe, contaminating vast areas.

Pripyat, the nearby city home to plant workers and their families, continued its daily routines for several hours, even as radiation levels spiked. It wasn't until the afternoon of 27 April, more than twenty-four hours after the explosion, that the 49,000 residents were hastily evacuated, told only to pack for a short absence. They never returned. Over the subsequent years, 350,000 people would end up being evacuated from surrounding areas.[13]

In the weeks that followed the incident, the Soviet government attempted to downplay the disaster, but the truth couldn't be contained. Helicopters dropped sand, clay and lead into the burning reactor in a desperate attempt to smother the radiation, while

A SPECTACLE OF SURVIVAL: A HUMBLED ARMY

thousands of 'liquidators' – conscripts, volunteers and workers – were sent to clean up the site, many receiving fatal doses of radiation in the process.

After the Soviet Union eventually admitted to the disaster, a 30-kilometre exclusion zone was established around the reactor. But the damage was already spreading.

Twenty-eight operators and firefighters, plant workers and rescue personnel died within a few weeks of the disaster, their bodies buried in lead coffins as a precaution against contamination.[14] Thousands more would suffer from radiation-related illnesses in the years that followed, their lives marked by pain, illness and premature death.

Immediately after the Chernobyl explosion, the environment became a nightmarish tableau of destruction. Trees in the now infamous 'Red Forest' surrounding the plant turned a sickly hue and died en masse as poison saturated the land. Animals, too, suffered. High doses of radiation caused a grim litany of problems: birds with deformed beaks, rodents with strange tumours and insects that were either dead or mutated beyond recognition.[15] It was a place where nature was fighting a losing battle against the lingering effects of humanity.

The radiation caused genetic mutations in many species. These mutations led to a range of abnormalities, including physical deformities and reproductive issues. For example, some birds and insects were born with malformed wings, and many animals showed signs of reduced fertility.[16]

But then something curious happened.

With humans gone, nature started to reclaim the land, radioactive or not. At first, it was subtle – a deer spotted here, a boar there. Then it became clear that wildlife wasn't just returning; it was thriving. Chernobyl became home to a burgeoning wildlife population. Wolves, boars, deer, elk and even the elusive lynx began to appear in numbers that baffled scientists.

Animals appeared to be adapting to the radiation.

Some birds in the area, for example, have developed physical adaptations such as higher levels of antioxidants, a sort of biological shield against the DNA-damaging effects of radiation.[17] This suggests a possible evolutionary response to the extreme conditions. Additionally, there is evidence that some animals may be developing higher resistance to radiation over generations.

There have also been observations of behavioural adaptations, which may be the result of the absence of human interference. Some species of birds, for example, have learned to avoid highly contaminated areas, which could be an evolved behaviour that reduces their radiation exposure. Boars and deer have taken to the abandoned fields and forests, feasting on the overgrown vegetation that now covers the ruins of human settlements. Even the rivers, once feared to be too contaminated for life, are teeming with fish, which are promptly snapped up by otters and herons.

The world's worst nuclear disaster has given rise to one of the most unexpected wildlife sanctuaries on the planet, and in this, the Chernobyl exclusion zone has become a living laboratory for studying these complex interactions, offering unique insights into how life responds to extreme environmental stress. In 2016, the Ukrainian part of the exclusion zone was declared a 'radiological and environmental biosphere reserve' by the national government.[18]

While the long-term effects of radiation on wildlife are still not fully understood, studies of Chernobyl have led some scientists to conclude that the negative effects from radiation are 'not as large as the negative effects of having people there'.[19]

In short, wildlife adapting to human destruction is the exception, not the rule. Most species and ecosystems are remarkably resilient, but that resilience has its limits and is not enough to withstand a continued onslaught from humanity or the relentless scale and speed of modern habitat destruction, pollution and climate change. And as Chernobyl teaches us, there is still time to change this trajectory, but not much.

Chapter 4

Unravelling Ecosystems: Hippos from Hell

> In nature, nothing exists alone.
> Rachel Carson[1]

Upon meeting an amateur exotic pet owner, you will likely discover that one of three things has befallen them in their life: they have been hospitalised after their beloved tiger named Nibbles 'accidentally' mauled them; their African grey parrot is inadvertently crippled and/or neglected and promptly dies, or at the very least hobbles around its cage with a bow leg and a crooked wing; or their exotic iguana is 'freed' into the wild, sparking a widespread biological plague or invasive species outbreak.

And that is precisely how Pablo Escobar came to be the creator of Colombia's most unusual environmental problem.

In the 1980s, Pablo was a busy man. He ran the world's largest drug-trafficking syndicate, was feverishly ordering the deaths of thousands of people and was almost single-handedly nurturing and scaling the cocaine industry. But like all successful entrepreneurs, he understood the importance of maintaining hobbies to ensure a strong work-life balance. Pablo's hobby was collecting hippos.

Pablo transformed his private estate, Hacienda Nápoles, into a menagerie. He imported an assortment of llamas, cheetahs, lions, kangaroos, tigers, ostriches and other exotic fauna to populate his personal zoo. Yet the pièce de résistance of his collection was the quartet of hippos he had imported from a zoo in Texas. Regarded as the deadliest land mammals on the planet, the hippos were a testament to Pablo's unique brand of opulence and questionable judgement.

The hallmark of a truly memorable dictator is, ostensibly, their ability to remain a source of irritation long after they are gone. And that is indeed what Pablo accomplished. In 1993, Colombian police shot Pablo dead, and his menagerie of exotic species suddenly found themselves orphaned. While veterinarians were able to remove some of the exotic animals, others unfortunately died. But not the hippos. Bereft of their dictator and deemed too dangerous to approach by authorities, the hippos eventually absconded to the nearby Magdalena River. Then they multiplied. And continued to multiply. What started as four hippos swelled into a herd of roughly 200. This has regrettably presented a different sort of burden in Colombia.

The smallest change can upset an ecosystem, depending on the type of ecosystem and its resilience. And very often, when humans meddle in ecosystems – like shipping a hippo from one continent to another – it can set off a chain reaction of environmental, social, economic and political consequences.

Dubbed 'the hippos from hell', Colombia's imported hippos have somehow become both a beloved pop culture phenomenon and an ecological and political nightmare.[2] Statues have been mounted in their honour and a tourism industry has emerged around them. Yet their presence has sparked fierce debate and, more worryingly, is threatening Colombia's delicate ecosystems.

In their homeland of Sub-Saharan Africa, hippos are considered ecosystem engineers. Like beavers that build dams or elephants whose footprints create tiny ponds for frogs, ecosystem engineers are organisms that play a dramatic – and vitally important – role in shaping and altering their environment.

UNRAVELLING ECOSYSTEMS: HIPPOS FROM HELL

The hippo's engineering gift to the world arrives unwrapped and unceremoniously dumped in the water in the form of excrement. The hippo feeds its environment with this gift, depositing its nutrient-rich dung into the water, which acts as a fertiliser, stimulating aquatic food chains and altering water chemistry and quality. Usually, hippo excrement is highly beneficial to aquatic systems, forming the backbone of entire ecosystems, transforming rivers into thriving habitats and allowing life to bloom.

However, under unusual circumstances, hippo excrement can have the opposite effect. An article in the journal *Nature Communications* describes this effect as, 'Organic matter loading by hippopotami causes subsidy overload resulting in downstream hypoxia and fish kills,'[3] which, if crudely translated, means, 'Hippo poo can suffocate the life out of marine ecosystems and fish.'

Pablo Escobar's 'cocaine hippos' may unfortunately be doing just that. Far from their native habitat, these hippos are upsetting Colombia's delicate ecosystems, predominantly through their bowels. So great is the threat that the Colombian government declared them a toxic invasive species, making them the largest invasive animal in the world.

By excreting waste into lakes and rivers, the hippos can trigger harmful algal blooms and change the composition of the water.[4] This water is a habitat for animals including manatees and capybaras, and researchers fear that Colombia's vulnerable species could become displaced or endangered as a result.*

Colombian biologists recently predicted that by 2040, if no intervention is put in place to control the hippos' breeding, the population will grow to over 1000.[5] That would roughly equate to up to 12,600 kilograms of suffocation-inducing hippo excrement per day.

---

* Of course, scientists debate the degree to which this non-native species is a problem. Scientists like to disagree. After all, critical thinking and open debate compel new ways of thinking and expand our understanding of the world. Scientists may disagree on this too, though.

Colombia's ecological battle with the cocaine hippos and their prolific waste morphed into a social and political crisis in 2009 when the government sanctioned a hippo hunt to address the burgeoning hippo population.

The plan backfired spectacularly when the hunting party not only killed a male hippo named Pepe but chose to commemorate the kill with a photo shoot where they posed around Pepe's corpse like trophy hunters. Unsurprisingly, this sparked a swift and furious public outcry, and Pepe became a martyr and the face for a petition to end the hippo cull. Realising they had misread the room, the government promptly suspended the hunt and pivoted to a new strategy.

In 2023, the Colombian government unveiled a three-pronged strategy to address the hippo crisis. The centrepiece of the plan is to sterilise the hippos, but it also includes plans to transfer hippos to other facilities or sanctuaries and in some cases activate an 'ethical euthanasia protocol'. The sterilisation program aims to manage the population more humanely and sustainably, reducing the number of new births while allowing the existing hippos to live out their lives without further exacerbating the ecological impact. The only hurdle is that sterilising a hippo is markedly different to giving your cat the snip.

The Colombian government will now dedicate efforts to mobilising teams of veterinarians, conservationists and experts to take on the task of trudging through the dark to track down the world's most deadly land mammal that weighs three tonnes and was previously owned by a notorious drug lord. They must then knock the hippo out with a tranquiliser that is shot from a specialised gun that requires a highly trained operator with a firearms license, before performing an incredibly delicate and invasive surgery on it. Simple.

Invasive or alien species – that is, a species that causes harm in a new environment where it is not native – is one of the major drivers of biodiversity decline.[6] As is the case with the cocaine hippo, these new arrivals can disrupt food chains, outcompete native species or even become apex predators where none existed before.

Invasive species have contributed to 60 per cent of known global extinctions in the modern era, fundamentally disrupting the way ecosystems work.[7] As ecologist Guillaume Latombe says, 'The relentless pace at which alien species are being transported and introduced globally will alter ecosystems for centuries, if not millennia.'[8]

Invasive species can be introduced in various ways – from being transported on the bottom of your shoe to hitching a ride on a boat – but the rise in exotic pet ownership has notably increased the number of these 'pets' being released into the wild. The result of this is very often an invasive species outbreak.

Japan is one country that learned this the hard way.

One of the most wonderful things about children is that they have a unique and profound ability to amplify the folly of their parents, then present it onstage with all the flair of a high school talent show. In the 1970s, the children of Japan earnestly embraced their role as pint-sized harbingers of parental ruin when they set off one of the largest biological invasions in Japanese history. And it all started with a children's book.

In 1963, a children's book entitled *Rascal: A Memoir of a Better Era* told the story of a young boy named Sterling and his adventures with his pet raccoon, Rascal. The story soon reached Japan, where it was translated into a fifty-two-episode cartoon TV series which uterly capivated the nation's children.

Soon, every child in Japan wanted a pet raccoon just like Rascal. Unfortunately, raccoons are not native to Japan, so Japanese families started importing raccoons from North America to own as exotic pets. They did this with gusto, importing 1500 raccoons each month to the delight of their children.

Things took a dramatic and unforeseen twist with the airing of the series finale of *Rascal the Raccoon*. You see, the main character, Sterling, came to the realisation that his pet raccoon would be much happier if he were 'free' in the wild. The show, reaching a heartwarming conclusion, revealed Sterling releasing his lifelong friend back into the forest.

Naturally, the children of Japan agreed that their pet raccoons should absolutely also experience the taste of freedom and promptly released their pet raccoons into the wilds of Japan – the wilds, of course, being Japanese cities.

Almost fifty years later, Japan is still grappling with the costly invasive species outbreak. As raccoons have few natural predators in Japan, they were able to quickly establish a population and are now found in forty-four of the country's forty-seven prefectures.[9] So great is the problem that a raccoon hotline has been established to report sightings of the animal. Eventually, the Japanese government banned importing the animals, but it was too late – the curse of Rascal the raccoon had taken hold.

Raccoons have embraced their role as an invasive species in Japan with gleeful abandon. They have taken to raiding homes, entering air vents beneath floorboards, nesting in roofs, up-ending gardens, stealing goldfish from ponds, destroying temples and rummaging through rubbish.[10] They disrupt ecosystems and have had a huge impact on the ecology of Japan, feasting on rare and endangered native species, including salamander and crayfish.[11] They invade cattle farms, infiltrate fish farms and damage crops, and they caused more than 450 million yen worth of agricultural damage in 2022 alone.[12] Overall, invasive species in Japan have cost the economy US$728 million.[13]

As with Pablo Escobar's cocaine hippos and Japan's children's cartoon turned raccoon invasion, the trade in species doesn't necessarily have to be illegal to be harmful. Wildlife trade encompasses both illegal and legal trade. It covers a wide range of 'transactions' – a veritable smorgasbord of buying and selling organisms, from purchasing a bit of timber, to owning a pet raccoon, to wearing an alligator-leather belt, to shooting a rhino for its horn. There is, of course, a spectrum of harm. But the wildlife trade, aside from very often being a threat to animal welfare, human health and biosecurity, is indeed a major driver of extinction. And most of it is done both legally and virtually unregulated.[14] Throughout human history, we have built a vast library of case studies that showcase our proficiency in causing widespread social, environmental, economic and political harm through the trading of species.

During the 1980s, the exotic pet trade was booming in the US, and Miami, Florida, was (and still is) a haven for wildlife trade. Over the next three decades, tens of thousands of Burmese pythons were legally imported from Southeast Asia to be kept as pets. Pet owners, soon realising that owning a 5-metre python may have been a bad idea, released them into the wild. To make matters worse, a hurricane in Florida struck a python breeding facility, sending more pythons into the wild. The pythons quickly made their way to the Everglades National Park – an area rich in biodiversity – and wreaked havoc upon it. Female pythons can lay fifty to one hundred eggs per year and the snakes have few natural predators in the region, so their population expanded rapidly. This has had catastrophic impacts on ecosystems in the region, with pythons responsible for at least a 90 per cent decline in native mammal populations and a rapid decline in bird species.[15] The importation of Burmese pythons was banned in the US in January 2012 and, as of 2021, it is illegal to own, buy or sell one of these pythons. But by this stage, the legal trade had already fuelled demand for the python and established transport routes and supply-chain networks, which effectively paved the way for a seamless handover from legal trade to illegal trade.

In its native environment, the Burmese python is listed as 'vulnerable to extinction'. The reasons? Habitat loss and over-exploitation via wildlife trade. The irony of this is not lost on many. Florida wants to kill the Burmese python. In Southeast Asia, they want to save it. And both regions are investing significant time and effort to either kill the species that was introduced or save the species that was exploited.

In the US, there is an annual ten-day 'Florida Python Challenge', in which participants hunt down pythons. The winners with the most python kills take home a cash prize. In 2025, Florida spent

US$805 million of their budget on Burmese python removal and Everglades restoration projects.[16] Between 1960 and 2020, invasive species cost the US at least $1.22 trillion.[17] This does not include the billions of dollars state and federal governments have spent on restoring the Everglades, which is rapidly deteriorating as a result of the python.

While these incidents might seem like isolated mishaps by people who are most certainly compensating for something with their pet apex predators, they're actually symptomatic of something much larger and infinitely more problematic: humanity's longstanding tradition of driving extinction by destabilising ecosystems. And very often, there are profound implications for nature and humanity alike.

Nature operates on a system of checks and balances. When those checks fail – when, for example, predators disappear or invasive species outcompete native species – we can trigger 'trophic cascades'. It's a scientific term that, when used in this context, translates loosely to 'everything across the food chain starts building or unravelling in ways we didn't foresee'.

In order to truly decimate an ecosystem – its structure, function and the species that sustain it – a few critical ingredients are required. You see, ecosystems are often fragile in their interconnectedness but equally quite resilient – it almost takes a dedicated and concerted effort to destroy one. First, you must relentlessly take from it, removing animals and plants right across the food chain, from predators and prey to organisms in the dirt. You could do this through hunting or to buy and sell them on a legal or illegal marketplace. Then you could raze their trees and poison or empty any nearby streams. You could erect towers, mining sites and buildings on top of them, while also ensuring that you are powering the towers, mining sites and buildings with fossil fuels to make certain greenhouse gases become trapped in the atmosphere. Those heat-trapping gases will change the structure of – and often impair – the ecosystem. If you continue that persistently, you will have weakened the ecosystem enough that it does not have time to recover. While it is at its weakest point, you could then release an invasive species into that ecosystem. This should most certainly be a species that you have taken from another faraway ecosystem, preferably one that has frequent breeding cycles so that they proliferate quickly. It is even better if that faraway species can become the new top predator in town.

These actions, in essence, represent the five major threats to biodiversity and species survival. They are, in summary: habitat destruction, pollution, climate change, over-exploitation of organisms (such as wildlife trade and hunting) and invasive species. We are doing them all exceedingly well.

Each of these threats is deeply intertwined and mutually reinforcing. Habitat destruction strips away forests – the lungs of the planet – allowing greenhouse gases and pollutants to accumulate. Pollution, often a byproduct of industrial activities, seeps into waterways, poisoning organisms at every level of the food chain. Over-exploitation of a species, through the wildlife trade, for example, can have a ripple effect and lead to invasions like the Burmese python's takeover of the Everglades. Essentially, these threats can accelerate the breakdown of ecosystems and drive further extinctions.

Yet the reverse can also be true. A disturbed ecosystem can be restored. When done carefully, a little spark of life can set off a chain reaction that sets recovery processes in motion. Where we've caused collapse, we can – slowly, carefully – build back resilience. Human intervention – or, in some cases like Chernobyl, the absence of human destruction – can help set these recovery processes in motion. Not always, but sometimes.

Restoration isn't just about repairing what's broken. It's about understanding that ecosystems are dynamic, ever-changing systems. They don't return to an 'original' state but instead evolve towards a new balance.

In the case of the Everglades, there is no clear-cut solution to overcoming the python invasion – at least, not yet. But while some ecological catastrophes may seem beyond repair, others have shown us that recovery, though difficult, is possible. For every tragedy like the Burmese python invasion, there are examples of recovery where human intervention, guided by science and patience, has helped ecosystems thrive.

Regrettably, one of the ways in which we came to better understand ecosystem restoration was through the great saga of the sea otter.

When sea otters sleep, they float on their backs and hold hands with one another to keep from drifting apart. Nestled together, they find sanctuary in shallow underweter havens along the Pacific Coast, curling themselves around kelp. So busy are they foraging, playing, cuddling and slumbering, these otters don't hear the canoes stealthily gliding through the water towards them. Until it's too late. The sound of clubs and spears pierce the night, and soon the water runs red. This was 'The Great Hunt', and it marked an era in which the fur trade was at an all-time high. And the most prized creature of this trade was none other than the sea otter.

In the 18th and 19th centuries, the fur trade was a fiercely competitive global enterprise, opening a vast international trade network. At the centre of this network was the sea otter – namely, sea otters from the North Pacific Ocean. Sea otter fur was considered to be the most desirable and valuable fur in the world, used primarily to line winter clothing. The sea otter's fur was, indeed, of the 'most exquisite fineness and richness in both colour and texture'.[18] This is because, unlike other marine mammals, sea otters lack a layer of blubber and rely entirely on their thick, water-resistant fur to maintain body heat in cold waters. Unfortunately, this furry adaptation became a death sentence once it was noticed by fur traders.

Traders from North America, Russia, Britain and other countries competed fiercely for sea otter pelts, driven by demand across China and Europe, where the fur was used to make coats, hats and other luxury items. Russia spearheaded this trade and was the world's largest supplier of fur, acquiring pelts of sea otters and other animals from the Indigenous peoples of the Pacific Northwest and Alaska. So powerful was the fur industry that it played a pivotal role in Russia's expansion into Siberia, the Russian Far East and its colonisation of the Americas, where it traded in fur-bearing animals such as Arctic foxes, lynxes, sables and sea otters.

The hunt for the sea otter was frenzied and demand was high – everybody from nobility to royalty wanted a sea otter pelt. By the 17th century, a set of four luxurious, silver-tipped sea otter pelts could fetch a sum equivalent to a house in Victoria, British Columbia.[19]

But hunting an otter was not for the faint-hearted. 'No other kind of hunting is so exhausting and involves such exposure as this method of chasing the sea otter,' reported the *Hartford Weekly Times* on 21 July 1892.[20] While sable, marten, mink and even ermine could be trapped or shot without extraordinary trouble, the sea otter proved more elusive and had to be sought 'as diligently as the diamond, for three centuries of experience has made him wise'.[21]

Hunters would use clubs or spears to kill the otters as they were the least likely to damage the fur. Sometimes a stray otter may be shot from the land as he played in the surf, but the chief methods of his capture were 'the surround' and clubbing.

The *Carroll Herald* newspaper reported the methods in painstaking detail on 15 August 1883:

> They attack only at night. Several canoes, with three to paddle and one to spear in each, advance quietly as may be to a spot where otters have been seen during the day. The first who discovers an otter darts his spear and generally with successful aim. The animal dives, and all the boats at once form a circle sufficiently large to be sure that they have surrounded him and that he must rise within, and then they wait patiently. The otter remains under water from fifteen to thirty minutes if he is not wounded: if he is dragging the spear-pole behind him he must

come up much sooner ... If within reach, soon another spear is thrown; if out of reach, a great shouting and splashing of paddles is kept up to compel the timid animal to dive again without delay. This continues time after time until worn out and exhausted, he rises where some one can give him a fatal blow.[22]

After killing the otter, hunters would skin the animal immediately to prevent the fur from deteriorating. The skinning process involved carefully removing the pelt, which includes both the fur and a thin layer of the underlying skin. The pelts were then cleaned, stretched and dried. They were often treated with salt to prevent decay during transportation, especially when the furs had to be shipped over long distances to markets in Europe, Russia or China. They were typically auctioned or sold directly to furriers, who would then prepare them for use in high-end garments.

The allure of the sea otter's fur set off a global trading frenzy, and hunters descended upon them with reckless abandon.

Unsurprisingly, sea otter populations crashed. Once numbering in the hundreds of thousands, by the mid-19th century their numbers plummeted to mere thousands with some groups wiped out entirely. The near extinction of otters also unleashed a destructive trophic cascade, dramatically altering the marine environments they once ruled. You see, the sea otter is a keystone species and a top predator in its ecosystem, meaning it plays a major role in keeping its ecosystem from descending into chaos. It performs this role largely by dining, with considerable enthusiasm, on sea urchins – prickly little gluttons that, when left unchecked, graze kelp forests into oblivion. As ecologist Kristy Kroeker says, 'Where you see otters, you are more likely to see a kelp forest.'[23] As otters disappeared, urchin populations exploded with no natural predators to keep them in check. These urchins ravaged vast underwater forests of kelp, which serve as vital refuges for countless other species. And when kelp forests fall, so too do the many species that depend on them.

Efforts to save the sea otter began in 1911 with the North Pacific Fur Seal Treaty, which created an international prohibition on hunting fur seals and sea otters at sea. But by then, the unravelling had already begun. More than 99 per cent of sea otters had been killed.[24] Hardly a sea otter was to be found in all of the North American Pacific Coast.

In the 1970s, sea otters and other marine mammals received additional protections under the *Marine Mammal Protection Act 1972*, which prohibited the killing of sea otters in US waters and enforced stricter regulations on commercial fur hunting. During this time, non-government organisations (NGOs), conservationists, scientists, Indigenous peoples and academics were hard at work. They began documenting not just the devastating impact of the fur trade but the vital ecological role of the sea otter, working to reintroduce them to their historical ranges and spur ecological recovery. The efforts of these groups would mark a critical turning point for the sea otter and other marine mammals.

The resurgence of sea otters has had a remarkable impact on the surrounding ecosystems. 'With sea otter recovery, kelp forests regained diversity and balance,' says Andrew Johnson, California representative of conservation organisation Defenders of Wildlife. He points to studies by the Monterey Bay Aquarium, which revealed that kelp cover expanded or persisted in areas with otter populations, with otters even helping buffer the effects of climate change.[25]

By controlling sea urchin populations, otters allowed kelp forests to regenerate. These forests matter, not just because they are vital habitat for marine life, but because they act as nature's coastal defence, softening the blow from storms and erosion. They're also tremendously good at tackling climate change. Just like trees, they absorb carbon dioxide, but on a scale that's hard to ignore: a kelp forest can absorb twenty times more carbon dioxide than the same area of land forest.[26]

In recent decades, sea otter reintroduction programs have helped repair ecosystems disrupted by their decline. Defenders of Wildlife has been instrumental in coordinating efforts for potential reintroduction of otters across California, collaborating with Indigenous tribes like the Yurok Tribe. Andrew highlights notable successes, like that of the southern sea otter population, which has grown from just of a few dozen individuals along the remote Big Sur Coast in California to about 3000, now spread from the north and west of Santa Barbara to south of San Francisco. 'We hope to help sea otters return to a 900-mile section of the West Coast where they used to roam before near-extirpation in the fur trade,' he says.

The recovery of the otter is more than a charming comeback story; it's a hard-won demonstration that human intervention can sometimes repair what we've broken. The lesson here is not that every ecosystem can be restored to its former glory, but that given the chance, nature has a remarkable ability to find its feet again and offer a renewed path forward.

In contrast to the Burmese pythons' destructive ripple effect, the otters' return shows how thoughtful intervention can bring life back to an ecosystem, even in the wake of devastation.

Humans have an extraordinary talent for looking at the intricate, millennia-old balance of nature and deciding that what it really needs is a hippo shipped from another continent. Or, better yet, heavy machinery to remodel a forest.

But it's never just one python, one otter or one raccoon that pays the price. Behind every story of a single species lies a vast and complex web of intricate relationships that has the potential to set off ecological, social, political and economic chain reactions. In nature, no species exists in isolation. Each is a thread in a vast, intricate web of life. Tug at one thread and the entire web will inevitably wobble. It only takes one. Just one. Then everything teeters.

But here's the thing: sometimes, with care – or just the absence of interference – nature returns.

PART 2

A High-Stakes Game of Extinction and Survival

*This chapter contains graphic descriptions of animal suffering and the impacts of human activity on wildlife which some readers may find distressing.
Reader discretion is advised.*

Chapter 5

Desperation:
The Loneliest Rhino in the World

One last surviving member lived on and on, reminding us of our past sins to his race and beckoning us to try one more time to save them.

Craig Stanford[1]

To slice off a rhino's horn, you must move swiftly. There is no time for a clean cut now that armed rangers roam the plains. It's best done at night, under the cover of darkness. A fuller moon will provide more visibility, but a new moon offers superior concealment. The rhino's sight is poor; she won't see you coming.

First, you must shoot the rhino. Use a tranquiliser – a quick shot of etorphine (M99) or carfentanil to the shoulder or hindquarters. The dart will penetrate her thickened skin. The anaesthetic will seep into her bloodstream and quickly depress the central nervous system. A high-powered rifle does the job too. A modified .303 bolt-action or .375 H&H Magnum.* The bullet will tear through muscle, bone and flesh, and likely vital organs. One shot should

---

* Often, maiming occurs when shots are poorly placed or the weapon lacks the necessary stopping power for an immediate kill. Weapons that maim are often modified.

maim, not kill. If her calf is nearby, ignore it. If it cries, shoot it. Both mother and calf will wake up and bleed out eventually.

Once the rhino is immobilised, you must hack away at her snout. Use a panga machete – or, for the impatient, a chainsaw. You'll cut through connective tissue, arteries and flesh. You will split open the cavity in her nasal passage right between the eyes. There is a network of blood vessels braiding through her face. Blood will spray. It will coat your skin, fill your mouth, linger on your tongue. The horn will detach, and the rhino will be left with a gaping, bloody wound on her face, skin flayed open, raw tissue exposed.

The earth will be soaked in blood on this night.

That is how you kill a rhino. That is how almost 10,000 rhinos have been killed in the last decade.[2] That is one rhino killed every nine hours. A rhino's horn has a black-market value of US$22,257 per kilogram.[3] It is more valuable than gold, diamonds and cocaine.* It is one of the most sought-after substances on the black market. And it is made up of keratin, the very same substance as your fingernails.

And it is for this reason, that on 19 March 2018, conservationists watched in torment as the last male northern white rhinoceros on the planet took his final breath and signalled the extinction of yet another species.[4] Sudan, as he was known, died in Ol Pejeta Conservancy in Kenya. He had spent the last decade of his life under twenty-four-hour armed surveillance.

Countless species have been claimed by extinction. But every so often, there is one that is so profound, so devastating, so obliterative, so preventable that it punctures public consciousness. And in those few moments – those few brief seconds – the world experiences just a microcosm of the pain that conservationists, scientists, rangers and volunteers deal in every single day.

This all-consuming, harrowing ache watching the suffocation

---

* Rhino horn is more valuable by weight. The value of raw rhino horn has fluctuated over the past ten years.

of a species they have dedicated their lives to; knowing they are losing, knowing they are alone, knowing they are outmatched and outgunned, and doing it anyway. Because the alternative is unfathomable, criminal even. And for a few short days, every once in a while, as the world reads a headline of a fallen icon, they feel a glimmer of hope that maybe, just maybe, things will change. But then the world moves on, and these conservationists, scientists, rangers and volunteers keep going. Wake up, beg for funding, fix the fence, plant the tree, care for the injured, petition the government, survey the population, sample the DNA, assess the habitat, tag the animal, analyse the data. Wrap the corpse. Rinse and repeat.

And still, species fall. And still, people look away. And still, there is no money.

When you are engulfed in flames, the only thing you can do is surrender or fight. These are the people who choose to fight every day a battle that they know they likely won't win. But they cling to fragile hope, praying for just one moment – one single moment – that may bend the odds in their favour. They wade through the depths of inhumanity and depravity and apathy and cruelty and never sag beneath the weight of such injustice. And in this, they carry a burden for us.

It is so arbitrary how we send species to extinction. So casual. The scales of a pangolin, the tusk of an elephant, the fur of a seal, the blubber of a whale, the flesh of a dolphin, the wings of a bat, the beak of a bird, the fin of a shark, the skin of a snake, the bones of a tiger, the bile of a bear, the shell of a turtle, the fur of a fox. We come for their body parts, flatten their homes, find them in forests, pull them from trees, dig them from burrows, haul them from seas, catch them in rivers, tear them from deserts; we bludgeon them on ice, slaughter them in lakes, poison them in reefs; we rip babies from mothers, tear fetuses from wombs; we carve up packs and shoot down flocks, and etch our violence upon the bodies around us then sit atop the ruins and think them to be jewels.

The end of a species is often a silent vanishing. But sometimes, every once in a while, it's a tempest.

There is a word to describe the last surviving member of a species or subspecies. It's 'endling'. And once an endling dies, that species is declared extinct.

Martha the passenger pigeon was an endling. In 1914, Martha was found dead on the floor of her cage at Cincinnati Zoo. With her death, the passenger pigeon was declared extinct.

In the late 1870s, there were up to five billion passenger pigeons – the most abundant bird in North America. So large was their population that their flocks would darken the skies for three days straight. The passenger pigeon fell victim to the fallacy that no level of exploitation could threaten such an abundant species.

In 1857, a bill was brought forth to the Ohio state legislature seeking protection for the passenger pigeon. Senators scoffed at the idea, proclaiming, 'The passenger pigeon needs no protection. Wonderfully prolific, having the vast forests of the North as its breeding grounds, traveling hundreds of miles in search of food, it is here today and elsewhere tomorrow, and no ordinary destruction can lessen them'.[5] The passenger pigeon was slaughtered mercilessly in droves, at the same time that their habitat was being destroyed. In just a few short decades, they went from a population of five billion to zero.

When Martha died, she was roughly twenty-nine years old, with a palsy that made her tremble. Martha's corpse was extracted from her cage. She was dangled upside down, held by her feet, lowered into a tank of water and frozen upside down in a 136-kilogram block of ice.[6] She was then shipped to the Smithsonian Institution in Washington, DC, where she was skinned, dissected, photographed and mounted. A memorial statue of Martha stands on the grounds of the Cincinnati Zoo.[7]

On 7 September 1936, 'Benjamin' the Tasmanian tiger was found dead on a slab of concrete in a cramped enclosure after being accidentally left outside in the cold. He froze to death. So misunderstood was the Tasmanian tiger – known to science as the thylacine – no one even realised that 'he' was actually a female – a mature, but still relatively young, adult female. Fewer were aware that her name was never actually Benjamin. The last nameless female spent her final days pacing the chain-link walls of her enclosure at Beaumaris Zoo in Hobart, Australia, where she died, marking the end of her species. Hunting, extensive habitat destruction and the introduction of competitive species and invasive diseases exacted a heavy toll on the thylacine. For decades, the Tasmanian government had placed bounties on them, and at least 3500 were killed by hunters between 1830 and the 1920s.[8]

The government granted official protection to the species on 10 July 1936, which meant that the sole surviving Tasmanian tiger was protected for the last fifty-nine days of her life while inside a cage. They were officially declared extinct in 1986 after years of numerous purported sightings.

In 2000, a tree fell on Celia, the last surviving Pyrenean ibex, a species of wild goat. It is thought that extensive hunting and competition with livestock signalled the downfall of this species. They had existed for tens of thousands of years before humans came along. After her death, scientists cloned Celia in a Spanish laboratory. The clone was born and then died shortly after, marking the second extinction of that species. Celia is the only animal to have become 'unextinct', and also to become extinct twice.

In 2012, it was Lonesome George. George was a giant male Pinta Island tortoise and the last known individual of the subspecies. The subspecies had existed for thousands of years but was wiped out by the introduction of goats to the island, which decimated their food supply. Lonesome George's caretaker was a park ranger named Fausto Llerena, who rode his bike to visit George every day. The tortoise would waddle to meet him at the gate of his breeding centre on Santa Cruz Island in the Galápagos. On the morning of Sunday 24 June 2012, George didn't come to the gate to meet

Llerena. George had died in the night at over one hundred years old. He had no surviving descendants. With his death, the Pinta Island tortoise went extinct. To Llerena, the tragedy was much simpler. 'He was my best friend,' he said.[9]

The golden toad went extinct. So did the western black rhinoceros. So did the Caribbean monk seal. And the desert bandicoot. And the Chinese paddlefish. The Hawaiian black-faced honeycreeper. The little Marianas fruit bat. The Steller's sea cow.

Scientists count one million species threatened with extinction.[10] But the truth is, we don't really know how many species have already gone extinct – or how many are at risk. The vast majority of Earth's species remain undiscovered. Even among the species we do know, we have data on only a tiny fraction of them.[11]

What we do know is that many, many species are teetering on the brink of extinction. And for some of them, so dire are their circumstances that we are merely holding our breath and hoping for a miracle.

Dr Chloe Buiting is a woman who takes a saw to the face of a rhino and slices off a horn. It would be easy to think her the villain. But in this story, she is the hero. She is one of a small band of veterinarians who are operating in crisis mode to save a species on the brink and do what few others will: cut the horn off the rhino before someone else does, because that 'someone else' will not do it kindly.

So dire is the situation for rhinos throughout Africa that conservationists are responding with DEFCON 1–level urgency: keeping insurance populations in Australia; zoo exchanges and captive breeding programs; 3D-printed synthetic rhino horn; sustainable rhino-horn production; resurrection; injecting poison and pink dye into rhino horn to reduce its value on the black market; attaching GPS trackers; installing twenty-four-hour armed guards; and, in desperate cases, slicing the horn clean off the rhino.

Buiting is abused for the work she does; it's confronting for some. But if you're a surgeon in a war zone, you do whatever you can to

DESPERATION: THE LONELIEST RHINO IN THE WORLD

save a life. And make no mistake: for the rhino and this team of vets, this is a war zone. Desperate times have called for desperate measures, and rhinos are utterly, profoundly desperate.

Rhino-horn amputation – or 'dehorning' – is a conservation strategy used to protect rhinos from poaching. It's a painless procedure. Veterinarians sedate the rhinos, blindfold them and insert earplugs, then use a chainsaw to cut off the top of the rhino's horn, avoiding the nerves. The entire procedure takes about twenty minutes, after which dehorned rhinos are monitored closely. Rhino horns, just like fingernails, grow back quickly. Often, the horn will return within twelve to twenty-four months, requiring another dehorning procedure.

The practice is not without controversy. While it has proven effective in reducing the risk of poaching, it's a short-term solution that requires continuous monitoring and repetition. One study reported that the black rhino – the more aggressive relative of the white rhino – exhibits behavioural changes as a result of being dehorned.[12] The black rhino, being 'stripped of their main armament',[13] seems to feel more vulnerable, which may disrupt their social behaviour and networks. Though one would imagine that being dead would also disrupt their social networks.

It is important research that has raised equally important questions that may help conservationists consider whether these behaviour changes may impact the black rhino's breeding potential. Despite this, the author of the study, Vanessa Duthé, says, 'At the end of the day, a dehorned rhino is better than a dead rhino.'[14]

While the sight of a dehorned rhino might feel jarring for some, the procedure is giving the rhino a fighting chance at survival.

Conservation was traditionally a field of science that chiefly focused on preserving, protecting and maintaining wild habitats. A pocket of land would typically be ring-fenced to ensure the ecosystem was responsibly managed and species would be protected. In this way, conservation used to focus on protecting nature by keeping humans out.

But it wasn't enough.

The rate and scale of habitat destruction, climate change and extinctions have risen sharply. In today's increasingly globalised world, the expansion of trade and travel has created unprecedented connectivity, allowing goods, people and ideas to move across borders with ease. However, this connectivity has also introduced significant challenges for conservationists. Globalisation and advanced technology have intensified habitat loss, illegal trade and resource exploitation. Demand for natural resources is now global and unparalleled. Timber, beef, fish and energy (like coal and gas) move seamlessly across the world, each product having been extracted from nature in some way. This has led to the overexploitation of wildlife and their habitats as overfishing, hunting, logging and mining driven by global markets often deplete resources faster than they can be replenished.

Species can also be transported and traded across the world with greater ease, making it easier for traffickers to move wildlife products across borders. As borders become more porous, pathogens, including zoonotic diseases, and invasive species become harder to contain, disrupting local ecosystems and threatening native wildlife. Added to this, climate change, fuelled by industrial activity, exacerbates these issues by altering habitats and threatening species' survival.

Traditional conservation methods are, quite simply, buckling under the relentless onslaught of habitat destruction, pollution, climate change and invasive species. In response, conservation strategies are evolving from solely focusing on conservation to a broader approach that includes regeneration. This shift emphasises the need to view humanity's relationship with nature as part of a broad, complex and interconnected system. Yet despite this evolving mindset, the staggering rate of species decline is demanding bolder and, in some cases, more radical solutions. Many conservationists are merely operating in survival mode: trying to keep remaining populations alive while fervently hoping for systemic change. They're forced to treat the symptom, not the cause, patching a haemorrhage with a bandaid, fully aware that true change hinges on sweeping overhauls across political, economic and social systems.

On the ground, the stakes have never been higher, and conservationists are getting desperate. While Dr Chloe Buiting is sawing off horns to save the rhino and Dr Kate McInnes is donning an ejaculation helmet to save the kākāpō, conservationists around the world are increasingly employing imaginative and often radical and risky tactics to save species on the brink.

In Australia, scientists deploy toad sausages via helicopter to save the northern quoll.[15] In Africa, rhinos are blindfolded and dangled upside down by their ankles from helicopters in order to relocate them to safer areas.[16] Scientists in New Zealand developed a 'kiwi deodorant' to mask the bird's strong scent and protect them from invasive predators,[17] while in England, zookeepers encourage flamingos to mate by installing mirrors and playing tape recordings of flamingo calls.[18] To support breeding efforts, scientists around the world are inserting probes into the anuses of cheetahs, flying foxes, Siberian tigers, black-footed ferrets, African elephants, giant pandas and many other species.[19] These probes deliver electrical pulses to the nerves responsible for erection and ejaculation, allowing scientists to collect semen for artificial insemination or genetic testing.

Fish are being shot through canons; snot is being collected from whales via remote-controlled helicopter; wildlife carers are dressing up in moose, panda, crocodile and whooping crane costumes in the name of conservation; robot crocodiles and lizards are being built; artificial habitats are being constructed; genes are being modified, de-extinction efforts explored and species cloned. To protect species, nothing is off limits.

But for many, the odds are stacked against them. And sometimes even the most well-intentioned efforts can spectacularly backfire.

In the 1980s, everyone knew the Sumatran rhinoceros was heading towards extinction. There are five species of rhino in the world, three of which are critically endangered. But the Sumatran rhino has the dubious distinction of being considered the most endangered due to its rapid rate of decline.

In 1984, a bold and unprecedented international agreement was forged between zoos, conservationists and governments, setting in motion a plan to save the world's rarest rhino species from the brink of extinction.[20] The program would, among other things, capture forty Sumatran rhinos from the wild and transport them to zoos and reserves around the world in order to establish captive breeding programs and ensure 'the long-term survival of the Sumatran rhino'.[21] Seven of these rhinos were sent to the US, three to the UK and one to Thailand, while seven remained in Indonesia and twenty-two in Malaysia.

Sadly, these efforts went disastrously, tragically awry. Within eleven years, the program collapsed without any rhinos being successfully bred. And more than half of the captured rhinos were dead.

The mortality records for Sumatran rhinos 'rescued' between 1984 and 1995 paint a gruesome picture. While conservationists could, in some cases, catch the rhinos, they had a much harder time keeping them alive. The captive breeding program in Sabah, East Malaysia, captured thirteen rhinos between 1987 and 2014. One died in the pit trap. One died within a day of being captured. One escaped through an electric fence. Another died overdosing on the sedative given to him. And another died when a tree fell on him in his enclosure, breaking his back. A few died from unknown causes and some of tetanus, salmonella and incurable cancer. By 2019, all thirteen rhinos were dead. None had successfully bred.[22]

Peninsular Malaysia's program had a total of fourteen Sumatran rhinos, captured between 1984 and 1994. By 2003, all fourteen rhinos were dead. None had successfully bred. One male died within a month because he was fed full-cream milk by one of the keepers. Another died after she got her head caught in the iron bars of her cage and broke her neck trying to free herself. Another died after getting caught in a wire snare trap and had a wound that became infected. But the worst twenty days for Sumatran rhino conservation came in October and November of 2003. Five rhinos died: four females and one male. A female died on 28 October. Ten days later, a male died. The day after that, another female died. One week after that, another female. Then the day after that, another female. No one knows for certain what happened – the post-mortem records state bacterial infection as the cause. But, whatever the case, with them went the entire captive population in Peninsular Malaysia.

The US captive breeding program did not fare much better; mortality rates were high. As journalist Elizabeth Kolbert notes, 'The zoos were feeding the animals hay, but, it turns out, Sumatran rhinos cannot live off hay; they require fresh leaves and branches. By the time anyone figured this out, only three of the seven animals that had been sent to America were still living, each in a different city.'[23]

It took the program seventeen years and dozens of rhinos to finally produce a baby, in 2001. A female rhino named Emi gave birth at the Cincinnati Zoo after six years of unsuccessful attempts.

Today, there is only one survivor left of the original rhinos captured for this program. Her name is Bina, and she resides at the Sumatran Rhino Sanctuary (SRS) in Indonesia. She's been in captivity for thirty-four years. Conservationists are still trying to get her to breed.

The rhino-saving program was deemed a disaster. In 1995, conservationist Alan Rabinowitz wrote an article titled 'Helping a Species Go Extinct', which was published in the journal *Conservation Biology*.[24] The scathing report ignited a flurry of criticism towards the program.

So where did it all go wrong?

Firstly, many of the rhinos caught for the program were 'doomed' rhinos, meaning that they were often older or already experiencing a decline in health.

The rhinos were also fed the wrong diet from the beginning. Very little was known about the Sumatran rhino and conservationists were ignorant of the needs of this rare species of rhino. Zoologists and conservationists were accustomed to caring for black and white rhinos and mistakenly assumed that the Sumatran rhino would have the same dietary needs as the others. 'I think there likely was a sentiment that a rhino was a rhino was a rhino,' said Susie Ellis, former executive director of the International Rhino Foundation.[25]

There was also the issue of hygiene and improper veterinary care. In Malaysia, rhinos were 'kept for long periods in facilities that lacked basic hygiene protocols and biosecurity measures, leading to bacterial infections and eventually, deaths'.[26] Added to this, many veterinarians were not aware of how to treat the Sumatran rhino.

To make matters worse, Sumatran rhinos are incredibly difficult to breed, and conservationists at the time knew very little about their complex reproductive biology. Not only were infertility rates high, but pregnancy failures were alarmingly common. One rhino, Rosa, suffered eight miscarriages. Another, Emi from Cincinnati Zoo, had five miscarriages in five years.[27]

Despite this, an immense amount of research and investment has been made over the years, and numerous lessons have been learned along the way. Today, scientists and conservationists have a much better understanding of the complex reproductive lives of Sumatran rhinos. As of 2024, six Sumatran rhinos have been born in captivity, bringing the total captive rhino population globally to eleven. In 2023, Cincinnati Zoo sent its last remaining Sumatran rhino back to Indonesia. His arrival meant that the entire population of Sumatran rhinos is now in Indonesia. This captive population may just be all that is left of the Sumatran rhino. Their wild populations have continued to plummet, and today there are just thirty-four to forty-seven Sumatran rhinos left in the wild. As Elizabeth Kolbert says, 'In an ironic twist, humans have brought the species so low that it seems only heroic human efforts can save it.'[28]

DESPERATION: THE LONELIEST RHINO IN THE WORLD

The story of the Sumatran rhino is one of heartbreak, tragic missteps and a last-ditch fight to save a species on the brink of extinction. At its centre is a group of people who have been on a thankless mission for decades, often the subject of widespread criticism.

It may be easy to blame conservationists and scientists for the blunders made in the race to save a species. But the salient point here is: they aren't the root cause of the extinction crisis. And it takes a special sort of hero to build an ejaculation helmet, saw off a horn or sprint across a room with rhino semen in order to save a species. If the Sumatran rhino has a future, it's owing to these dedicated conservationists, scientists, veterinarians and caretakers who relentlessly persevered to save the Sumatran rhino despite being pitted against seemingly impossible odds.

With six Sumatran rhino births in captivity over the past twenty-three years, there remains a glimmer of hope. And for the conservationists in this program, that glimmer of hope is what they hold on to.

Chapter 6

A Call to Arms: The Greatest Rescue in History

> Hope begins in the dark, the stubborn hope that if you just show up and try to do the right thing, the dawn will come. You wait and watch and work: you don't give up.
>
> Anne Lamott[1]

In the early hours of 23 June 2000, a cargo ship called the MV *Treasure* sank off the coast of South Africa. The vessel broke apart between Robben and Dassen islands – home to the largest colonies of vulnerable African penguins. It was carrying 1300 tonnes of fuel oil.

Within hours, thick black oil was gushing from the ship, spreading for kilometres across the water. It wasn't long before thousands of oil-soaked penguins began flooding the shores of nearby islands and beaches.

This wasn't a huge amount of oil compared to other disasters. Approximately 2680 barrels of oil spilled into the sea from the MV *Treasure*. By comparison, the 2010 British Petroleum (BP) oil spill from drilling rig *Deepwater Horizon* spewed 4.9 million barrels of oil. But the spill in South Africa was in the feeding and breeding ground of nearly half the world's total population of African penguins. And the clumping effect of the oil was rendering their wings useless. They weren't going to survive.

A CALL TO ARMS: THE GREATEST RESCUE IN HISTORY

News of the oil spill began to trickle through nearby communities. The future of Africa's only penguin species was in danger. Seventy-five thousand African penguins faced immediate threat from the spill.[2] What happened next was one of the most remarkable wildlife rescue missions in history.

It started with one phone call. And then another. Then right across the world, phones started ringing. Messages chimed between communities, countries and continents. Emails were sent. Supplies were gathered. Cars were fuelled. Flights were booked. Clean-up crews were dispatched. Zoo and aquarium staff woke at all hours of the night, alerting one another – Brazil, Ireland, New Zealand, Australia, France, Germany, Poland, Singapore, Canada, Hungary, the Netherlands, the UK and the US all heeded the call and all rallied their teams.[3] Networks of volunteers mobilised. Charities assembled. The South African Foundation for the Conservation of Coastal Birds led the charge. The Department of Environmental Affairs and Tourism arrived. The International Fund for Animal Welfare (IFAW) arrived. The World Wide Fund for Nature (WWF) arrived. International Bird Rescue arrived. The Red Cross arrived. A call to arms had been made, and right across the globe, people roused from their beds and they came. They were readying to save the African penguin.

And then came the volunteers. They descended upon South Africa. They came en masse, from every corner of the planet. Cape Town was ground zero, and the world was coming to help. Mums and dads, grandmas and grandpas, children of all ages. Students. Locals. Tourists. Doctors. Nurses. The young and the old. The experienced and inexperienced. They came alone and they came in groups. They came in sneakers. They came in boots. They came in designer clothes and in wetsuits and in business suits and in dresses. The fishermen showed up. The medics showed up. The media showed up. The chefs showed up. The schoolkids showed up. They arrived on foot and in cars and on bikes and in planes and on boats and in helicopters. They travelled through the night, across all time zones. They brought food and they brought drinks. They brought spray bottles and toothbrushes, wash tubs and garden hoses. They came with towels and heat lamps, with syringes and needles and catheters and ointments and vitamins. They came with wheelbarrows and forklifts and pallets of frozen fish. They came with rubber gloves and tables and chairs and clipboards and two-way radios. They came. And they kept coming.

Local fishermen took to cleaner waters to catch fresh fish for the oiled penguins. The Red Cross bandaged and stitched fingers shredded from force-feeding penguins and gave tetanus shots to droves of injured volunteers. The South African Air Force installed wiring, lighting and ventilation systems. Ferries, which usually shuttled tourists, carried boxes of oil-soaked penguins to safety. People in ski boats surfed breakers with boxes of birds under their arms. Locals set up hotlines for potential volunteers. In July, six people answered 28,346 phone calls from those wanting to help; they spoke with people as young as six and as old as ninety-six.[4] A seven-year-old girl donated the contents of her piggy bank to the penguins. Australians knitted sweaters for birds and shipped them across the ocean. This was the largest volunteer workforce ever assembled.[5]

And the volunteers kept coming. Every day, for months and months, they kept coming. Some were experienced, but most had never handled birds. They fed penguins, they washed them with toothbrushes, they scrubbed oil off thrashing bodies, they swam penguins in makeshift pools, they cleaned pens and set up tents and worked through all hours of the day and night. Three hundred and sixty-three tonnes of fish were fed to the birds, 6350 tonnes of beach sand were brought in for bird pens and 7,550 litres of detergent was used to clean birds.[6] All in all, 45,000 people showed up, donating more than half a million hours of their time to save the penguins.[7]

And it worked.

Forty thousand African penguins were saved. Ninety per cent of oiled birds that were captured were rehabilitated and released back into the wild. The call for help was answered. The world had showed up for one little penguin.

It happens, you know. These people, they are everywhere. These incredible, beautiful, kind people who show up. Who give everything. All over the world, people are showing up for wild species and environments.

On the frontlines of the environmental crisis, a hidden army marches. Quietly gathering in every corner of the planet, they are a force that spans continents, an unstoppable movement of ordinary individuals doing extraordinary things – mostly as volunteers.

Among them is Gracie.

Gracie stands in the forest and gazes in wonder at a Tasmanian masked owl. This owl is cryptic, mysterious and rarely seen during the day, if at all. It's also endangered – critically endangered. The owl is perched in the hollow of an old tree; a female. She stares back at Gracie with deep obsidian eyes. Her face is framed by a heart-shaped mask of pale white, her plumage a mottled blend of chestnut, grey and brown. For Gracie, this is an unbelievably special moment.

The forest she stands in is the Snow Hill Forest Reserve in Tasmania. This forest – home to the rare masked owl – will soon be logged by Sustainable Timber Tasmania. Gracie says this is unacceptable. We have no right to take away their home, she says. 'I promise we will protect you,' Gracie whispers to the owl. Then she walks away. Back to work. Back to trekking through the forest, sifting through soil, finding roosts, collecting clues in habitats, recording bird song. Monitoring. Cataloguing. Photographing. Teaching.

Gracie is a citizen scientist. She spends her days scouring the forests for endangered species. She carries binoculars. Her face is scrunched and serious, tilted upwards, gaze on the treetops. She is finding the homes of some of Australia's most endangered birds, species like the swift parrot and the Tasmanian masked owl. Her findings are then reported to local scientists and the Bob Brown Foundation and are used to help Australia better understand the location and behaviour of its critically endangered species.

Some weekends, Gracie guides people through threatened forests to help them understand the things most important, the things that need protecting. The trees and the soil and the birds and the bugs and all of the wonderful life in the forests. Other days, Gracie speaks to audiences and shares her knowledge. She tells people how the forests of Tasmania are being destroyed. How these forests are beautiful. How they are joyful. How every day there is less and less habitat for species. How it makes her sad. How it makes her angry. How these animals are dying. How these trees – these ancient trees – are being cut down and shipped off on the backs of trucks. She tells them she is going to stop this. She will fix this. She is working every day to fix this. After all, Gracie made a promise to an owl. And she intends to keep it.

Gracie is twelve years old.

There are people in this world whose tiny footprints leave the most profound mark. Where there is fear and despair, they bring courage and passion. Where there is apathy and discontent, they bring beauty and joy. Where there is confusion and complexity, they bring simplicity and clarity. It is simple, they know. We are nature, and nature is us. So, we live amongst it with care, we open our eyes and our hearts and when we take from the world, we give back to the world. A full circle. Because we know what the right thing to do is. But somewhere along the way, we tell ourselves that doing the right thing is hard. That doing the right thing will have to wait until later.

But these tiny people know more than us. They know it is not hard. They know, in fact, that it is the easiest thing in the world. To be kind, to walk gently, to have wonder in your heart and fire in your eyes. That the softness inside you is a strength. That curiosity is a gift. They know we have taken too much, that the Earth is sick. But in the same way that we have made it sick, they know we can help it get better. So, they do that. They help it get better.

It was Christmas Eve 2015 and Lisa Palma was being dangled upside down by her feet inside a stormwater drain. She was saving a bird, of course. But this was all in a day's work for Lisa. After all, she was the CEO of wildlife emergency response organisation Wildlife Victoria. But long before her time as CEO, she was a volunteer wildlife rescuer and carer – a job she continues to this day.

This particular Christmas Eve, Lisa had received a call about a rainbow lorikeet trapped inside a drain. It was getting dark, and Lisa was expected home to meet her partner. She called him. Yes, a bird was trapped, she said. Yes, she wasn't coming home. Again. She wasn't leaving this lorikeet, she said. She knew the bird needed help. It would die in that drain without her.

She peered through the metal grate of the drain and watched the helpless lorikeet. The streets were quiet now and darkness had fallen. 'I knew the bird would die down there, so I just sat there. I just sat there quietly, by myself as it got dark and it started to get cold, working out what to do,' she said.

But then something curious happened. One by one, people started emerging from their homes. Then people started stopping their cars on the street and approaching Lisa. What was she doing, they wanted to know. 'Is everything okay?' they asked. They were ordinary members of the public, and they were coming to help.

A vet nurse drove past the street on her way home from work and saw Lisa sitting alone. She gave Lisa her jumper. An elderly couple came out and offered her a coffee and some sandwiches. Then a man walked over, crowbar in tow, and pried the metal grate off the drain. Soon, more people began to gather. A crowd had formed, there to show support for Lisa and this little lorikeet.

With the grate removed, Lisa lowered herself headfirst into the pitch-black drain. Lo and behold, the bird was alive and well. A Christmas miracle. 'I looked down, and here's this recalcitrant lorikeet – who are incredibly cheeky birds, by the way – uninjured,' she said. She then knotted some towels together and carefully lowered them into the cavity. The lorikeet grabbed the towel with its beak, and Lisa gently pulled it out, caught it and put it in a rescue basket. As the lid of the basket closed, the whole street erupted in cheers. One woman approached Lisa shortly afterwards and pulled her into a hug. Then she looked at Lisa and said, 'I hope something good happens to you.'

Lisa represents one of the thousands of volunteers who dedicate their entire existence to the service of wild species: wildlife carers and rescuers. They are the unsung heroes of the world, working tirelessly to protect and save the world's most vulnerable creatures. They are the first responders when a car collides with a wombat. They are the people who dangle themselves upside down in a drain when a bird is trapped. They are the resuscitators when a koala is scorched by a blazing bushfire. They are the healers when a kangaroo is caught in a barbed-wire fence. They are a lifeline to species and provide an immense service to the countries they operate in. They use their own cars, they buy their own equipment and they always, always show up for native animals. Women make up 86 per cent of this industry in Australia, and 70 per cent of all rescuers and carers are over the age of forty-six.[8]

Sleepless nights, broken marriages, weekends nursing sick wildlife: this is the life of those in one of the world's most under-recognised, under-appreciated and predominantly voluntary workforces. Lisa says that every single one of her staff and volunteers, are people who care. They care deeply. 'They don't turn up at work just to do a nine-to-five job. They are caring, honest, hardworking people who are all mobilised around wildlife,' she says.

Lisa is both a wildlife carer and rescuer. Wildlife rescuers are typically the first to respond when a native animal is found in danger or distress. Their primary role is to safely capture and retrieve the injured or stranded animal from hazardous situations – whether it's from a road accident, a trapped environment or a climate disaster. They assess the immediate condition of the animal and take the necessary steps to stabilise it, often transporting it to a wildlife carer or vet for further treatment. Rescuers operate in the field, often facing unpredictable conditions, and can handle frightened or injured wildlife with care and skill.

A CALL TO ARMS: THE GREATEST RESCUE IN HISTORY

'You are heading out at all times of the day and night to rescue native animals in distress,' Lisa says. 'The sad fact of being a wildlife rescuer is you are not dealing with healthy animals. You are dealing with sick, injured, orphaned animals, and the majority of the cases end in euthanasia. It is incredibly confronting, incredibly challenging,' she says.

Wildlife rescuers spend countless hours in their cars, racing from one emergency to the next, with long waits at vet clinics in between. But you can always spot one by the state of their car. 'I have a whole range of nets, three different ladders, crates of different sizes, different sized gloves, a spare set of clothes and gumboots in my car. You need to be prepared for every species, in every condition,' says Lisa.

Lisa has been everywhere and seen everything. 'I have had to hold animals as they have taken their last breath. I have had someone holding my legs upside down into a tunnel off a freeway to rescue ducklings. I've climbed trees, been on roofs, climbed ladders, pulled animals out of chimneys,' she says.

She talks of the 2014 heatwave in Victoria, Australia. Most people wouldn't remember this day, she says. But for people working in wildlife 'we will never forget it'. She speaks of the twenty-seven heat-stressed possums lying listlessly in the front room of her house. Of thousands of animals suffering extreme heat stress. Of possums with horrific burns on their feet from hot tin roofs. Of animals dying of thirst. 'It was awful,' she says. 'I am going to get emotional.' And she does.

That day, as temperatures soared above 45 degrees Celsius in Victoria, Lisa and her fellow volunteers were on their feet for twenty-two hours, rescuing wildlife from 5am until 3am the following day.

'It is an incredible privilege, the feeling you get when you release an animal – that otherwise would have died – to its home in nature. It is indescribable.' Lisa says that if she were taken from the world tomorrow, she would be okay with that, knowing that she has saved thousands of lives. 'Honestly, if I left the world tomorrow, I would go in peace,' she says.

It's incredibly rewarding, she says. Every day, wildlife rescuers are dealing with caring members of the public. 'People don't ring Wildlife Victoria if they don't care. The type of person that will ring us is one who has seen an animal in distress and wants to help,' Lisa says. It is the little interactions with caring members of the community that help her keep going: a simple thank you, a cold glass of water after a tough rescue or a homemade jar of chilli oil from a grateful member of the public. 'Those moments stick in your memory,' she says. 'You're not alone.'

While wildlife rescuers focus on the immediate, often urgent task of saving animals, wildlife carers take over the longer-term process of healing them and preparing them for release. And, yes, it is largely voluntary. These people convert their homes into wildlife sanctuaries, spend hours foraging in the bush for native food for animals and wake at all hours of the night to care for the sick, injured and orphaned. They build makeshift pens, enclosures and sheds in their backyards. They fill their cupboards with incubators, formulas, blankets, towels, medications, heat pads and beds for sick animals. They might have baby bats in their lounge room; kangaroos, possums and flying foxes in separate pens in their backyard; wombats in their study; gliders in pens in the hallway. Wildlife carers are the rehabilitators, the nurses, the carers of the sick. They work with the young, the old, the orphaned and the unwell, and they nurture them over weeks or months until they can survive independently.

'The mornings are always cleaning the cages,' Lisa says. 'You are stripping the cages, you are putting piles of soiled items in the wash – usually four to six loads of washing a day at least.' There is also food collection and preparation, where carers painstakingly forage through the bush to gather the right variety of food and enrichment for each of the species in their care. Eucalyptus, acacia, flowers and fruits for the possums; nectar, pollen, tree sap and insects for the gliders; raw meat for the tawny frogmouths and kookaburras; snails for the lizards. They ensure nocturnal animals are in bed and turn their attention to the diurnal animals. They treat injured, sick and orphaned animals throughout the day and night: bandaging wounds, applying dressings, treating infections, bottle feeding babies.

'You might have fractures, wounds to bandage. Lots of adult birds – kookaburras, tawny frogmouths, parrots – they might have concussion. I have had so many kookaburras tucked under my arm as I have wedged their beaks open to put insectivore balls in their mouth and tried not to get my fingers bitten,' Lisa says. They fix pens and enclosures. They train volunteers. They coordinate with vets. They administer medication. And that's just before lunchtime, of course.

Like many wildlife carers, Lisa wakes every three hours to feed the animals in her care. 'Often, you've got animals that are orphaned that need feeding on three- to four-hourly intervals. I am so used to the midnight, 3am and 6am feeds. Because you know if you don't feed them, they could die,' she says.

Right across the world, wildlife carers and rescuers are being pushed to the brink. They operate in the shadows, propping up a vital system on shoestring budgets, paying out of their own pockets to rescue, feed and care for native animals. But in Australia, the situation has become dire. They're running out of time, money and people to do the work that nobody else will.

'We are out of business in two and a half to three years. We are out of money,' says Lisa. It's a heavy weight to carry for the CEO of Victoria's primary wildlife emergency response organisation. Just last year, the Victorian government did not commit any more funding to Wildlife Victoria, and Lisa says her organisation was already stretched to breaking point well before that. This constant financial strain is tearing at the very fabric of the wildlife rescue and care community. It creates division, competition, factions and individualism.

'The whole sector is chronically underfunded. Chronically,' says

Lisa. Wildlife Victoria has operating costs of $7 million per annum. It is experiencing at least a 20 to 25 per cent year-on-year increase in demand on their services. It is at the frontline of climate change, habitat destruction and urbanisation, and native animals are being impacted at an extraordinary rate. But it gets just $500,000 from the Victorian state government, with no commitment of any funding after this year. 'I don't want to be the CEO that causes our organisation to end,' she says.

Lisa says the community is marked by fragmented, struggling groups of volunteers trying to deal with the crisis that is facing wildlife. 'It is not okay. It is not good enough,' she says. 'We need more funding, more insightful and proactive legislation. We need more recognition for the vital public service we all provide.'

Across the volunteer workforce, she sees compassion fatigue, relationship breakdowns and financial ruin. Added to that, they are grappling with an ageing volunteer workforce. 'I genuinely worry that we are not attracting enough enthusiastic, capable young people into wildlife rescue and care. In an environment where we have an ageing volunteer workforce and ongoing demand for our service, we are really facing into a tipping point,' she says.

Despite the challenges, Lisa is incredibly proud of the work that she and her network do. 'I believe we are doing great work. Do I want to do more? Yes. Is it challenging? Yes. But we work as a team. Our workplace feels like a home. It is friendly, it is welcoming. One thing I have really noticed about people in the environmental sector is their high intellect. People are incredibly passionate and intelligent,' she says.

Would Lisa do anything else with her time? Not a chance.

'This is the most challenging job I have ever done in my life,' she says. 'But what else would I do? There is nothing else I would do. This is it.'

More often than not, change starts on the ground. It's grassroots, community-led and voluntary. It's the one person who finds an injured bird and calls for help. It's the man that dreams of filling his neighbourhood with trees. It's the schoolchildren taking photos of butterflies to support scientific research. It's the person who strikes up a conversation with a neighbour and starts a campaign to pick up litter in their local park. It's the women who knit sweaters for penguins and send them to South Africa during an oil spill. It's the person who builds a community garden on their rooftop. It's the young woman who signs up to become a wildlife rescuer and tells her friends. It's the retiree who petitions the local council for speed humps to protect wildlife from vehicle collisions. It's the young boy who collects blankets from his friends to donate them to the local wildlife shelter. It's the farmer who finds an injured possum and nurses it back to health. It's the people who notice the wonder and tragedy around them and think, 'I can do something.'

These small actions, they ripple outward. They start movements, you see. These people inspire others to follow their lead, and they create change in their communities.

Throughout the world, these seemingly small acts have inspired powerful, enduring change.

Wangarĩ Maathai was a woman who laughed heartily and with her whole body. As a child, she listened to birds and memorised their names. She perched around the fire with children and elders. They told stories spun from ancient threads and woven with wisdom. The kind of stories that passed through generations. The stories that shape values and lives and histories.

She said the experiences of childhood are what mould us, that they make us who we are. 'How you translate the life you see, feel, smell, and touch as you grow up – the water you drink, the air you breathe, and the food you eat – are what you become,' she said. As a child, her surroundings were 'alive, dynamic and inspiring'. She lived in a world where 'there were no books to read, where children were told living stories about the world around them, and where you cultivated the soil and the imagination in equal measure'.[9]

She watched her mother plant seeds and till the soil. As she grew, she saw changes in the lands and waters. She noticed rivers would rush down the hillsides and along paths and roads when it rained, and that they were muddy with silt. That was soil erosion, she knew. 'We must do something about that,' she said to herself. And so, she did.

Wangarĩ Maathai was born in Nyeri, Kenya, in 1940. She was the first African woman and the first environmentalist to win a Nobel Peace Prize. She was the first woman in East and Central Africa to earn a doctorate degree. And she planted trees.

In 1977, Maathai founded the Green Belt Movement, mobilising Kenyan women and children to plant trees. Under her leadership, more than 35 million trees were planted, which then inspired the United Nations (UN) to launch a campaign that has led to the planting of a further 11 billion trees worldwide.[10]

She said, 'The trees would offer shade for humans and animals, protect watersheds and bind the soil, and, if they were fruit trees, provide food. They would also heal the land by bringing back birds and small animals and regenerate the vitality of the Earth. That is how the Green Belt Movement began.'[11]

She said, 'Now, it is one thing to understand the issues. It is quite another to do something about them.'

She thought of what could be done, rather than worrying about what could not be done. 'I didn't sit down and ask myself, "Now let me see, what shall I do?" It just came to me: "Why not plant trees?"'[12]

Her mission was to plant trees across Kenya to heal the environment, alleviate poverty and end conflict. She saw, before most people, the connection between environmental degradation and human suffering. 'Poor people will cut the last tree to cook the last meal,' she once said. 'The more you degrade the environment, the more you dig deeper into poverty.'[13]

Along the journey, Maathai never walked alone. She said the success of the Green Belt Movement lay in 'the work of the tens of thousands of women and men who toiled over the millions of tree seedlings they plant, water, and nurture to maturity'.[14]

'None of this journey would have taken place without the spirit, courage, and dedication of the people, especially the thousands of women who believed in me and started to plant trees alongside me in Kenya, and stood up for the environment when it became necessary to do so,' she said.[15]

Maathai's campaign started with the planting of a simple tree. And she persevered, until millions more were planted. Her work was rooted in community; she recognised the complexity and dilemmas that communities faced, and she 'connected complex environmental issues with their impact on ordinary lives'.[16] She was fearless: 'a thorn in the side of the male dominated Kenyan authorities'.[17] She spoke truth to power. She was an activist. A feminist. A healer. A mother. A woman of the trees. The former Kenyan president called her a 'mad woman' and 'subversive'.[18] She was beaten unconscious by thugs and state police. Her husband divorced her for being 'too strong-minded for a woman'.[19] She challenged the divorce in court and lost. Then she called the judge 'incompetent and corrupt'.[20] These remarks landed her in jail for six months. And through it all, she never shrank. Through her acts and through her deeds, her community followed her. Then her country. Then the world.

Maathai died in 2011. Her legacy lives on through her daughter, Wanjira Mathai. The Green Belt Movement continues today. Following Maathai's death, then UN secretary-general Ban Ki-moon called her 'a pioneer in articulating the links between human rights, poverty, environmental protection and security'.[21] Former US vice-president Al Gore said she 'devoted her service to her children, to her constituents, to the women, all people of Kenya – and to the world'.[22] Former US secretary of state Hillary Clinton said, 'Her death has left a gaping hole among the ranks of women leaders.'[23]

From the tree-dotted landscapes of Kenya where the Green Belt Movement took root; to the shores of New Zealand, where communities guard sea turtle hatchlings; to Mexico's Sierra Gorda, where a biosphere once scarred by logging was revived by a grassroots campaign led by music teacher Martha Isabel 'Pati' Ruiz Corzo – life on the frontlines of the environmental movement is pulsing with hope. These are not isolated endeavours. They are pieces of a larger puzzle, contributing to global environmental health, one small action at a time.

At the heart of many grassroots movements is citizen science. People like twelve-year-old Gracie who contribute to real-world research despite not having formal scientific training. From rural villages to bustling cities, citizens are collecting data on everything from pollution levels to bird populations, playing a key role in shaping environmental policy and scientific understanding. Around the world, citizen science projects have surged, gathering invaluable data on pollution, biodiversity, climate change and species health. These data points, collected by thousands of individuals across continents, feed into global databases that help scientists monitor species health, track the effects of climate change and advocate for stronger environmental protections.

Charles Darwin was one of the original citizen scientists. Driven by a fascination for the natural world, he collected data on the plants, animals and environments he encountered. Like a detective piecing clues together, he meticulously recorded his observations in a diary. He would spend days on end tracking species around the world, observing their intricacies with remarkable precision. This combination of curiosity, determination and observation would, of course, fuel one of the most groundbreaking theories in scientific history: the theory of natural selection.

Citizen science allows anyone – regardless of age, background or expertise – to contribute meaningfully to the sciences. And that they do.

In Kenya, Maasai warrior communities in Narok County collect, identify and record indigenous plants, documenting the medicinal properties of each and preserving traditional plant knowledge while addressing issues related to deforestation.[24]

A CALL TO ARMS: THE GREATEST RESCUE IN HISTORY

In Cameroon, Baka hunter-gatherers collaborate with farmer communities and NGOs to monitor animals and collect data on illegal wildlife crime, gathering evidence to inform effective forest management legislation.[25]

In Australia, school students, mostly in Year 4, named seventeen new insect species and documented 5000 species using DNA. The students were credited in a journal article published in *Austral Entomology*.[26] In Brunei, citizen scientists discovered a new species of beetle.[27] In Montenegro, they discovered a new species of giant slug.[28] In Hawaii, citizen scientists captured the first ever observation of humpback whale sex, and there was a twist – both whales were male.[29] In Germany, citizen scientists scoured 2.2 tonnes of sand to identify the amount of plastic on the coastline; this will feed into plastic policy decisions.[30] On iNaturalist, one of the most popular citizen science platforms in the world, 3.4 million citizen scientists have recorded 214 million observations across approximately 491,000 species.[31]

Citizen scientists report shark sightings to one another, they mobilise to monitor and clean up ocean plastic, they track butterfly and bird populations and migrations, they identify planets, they document and report wildlife and environmental crimes, they find platypus and search for insects and scour the oceans for life. They document coral bleaching, they track invasive species, they observe and record the presence of bees, butterflies and other pollinators in local parks and gardens, and they make discoveries – world-changing discoveries.

Citizen scientists don't just gather data. They change the world by contributing their time and passion. In doing so, they help ensure the survival of species and ecosystems under threat. These are the people on the frontlines of the environmental crisis. They are guided by passion and hope and desperation and anger. These people, they see magic because they look for it. For them, it took just one spark: the desire to plant a tree or save a penguin, or just to be a good neighbour. And once you stoke the spark, there is no stopping it. It is an addiction, a duty, a calling.

True change takes root in the steely determination of everyday people. People who do small things. Things that become big. These people come in all shapes and sizes and varieties. And sometimes, they come in the form of a twelve-year-old girl who gazes upon an owl and believes with all her heart that she can protect it.

PART 3

The Tangled Fates of Us and Them

Chapter 7

The Miracle Makers: A Drunk Moose and an Economist

> No human technology can fully replace 'nature's technology' perfected over hundreds of millions of years in delivering key services to sustain life on Earth.
>
> Marco Lambertini[1]

When China is mad at another country, it takes back its pandas.

Throughout its history, the People's Republic of China has used giant pandas as a gesture of goodwill towards its international allies and competitors, loaning pandas to other nations as diplomatic gifts.* During the 1950s, Chairman Mao Zedong was known to send pandas as gifts to his communist allies, which included North Korea and the Soviet Union. In 1972, following US president Richard Nixon's historic visit to China, the Chinese government gifted two pandas to the US, signalling an easing of Cold War tensions between the two countries. Gifted pandas are sprinkled throughout zoos around the world as a form of soft power, a strategic diplomatic tool to shore up alliances. But when that goodwill dries up, China's response is swift and

---

* This is under a lease model in which pandas are loaned for around US$1 million per year.

furious. To signal its displeasure, it demands a rapid return of its beloved pandas.

In 2010, then US president Barack Obama greatly insulted China by meeting with the Dalai Lama. The US was accused of 'hurting the feelings of the Chinese people',[2] and two pandas were promptly pulled from their homes in US zoos and sent back to China via a dedicated FedEx plane (known as the FedEx Panda Express).

Indeed, nature is a force with the power to both unite and divide humanity. It nourishes and it destroys. It paints sunsets and unleashes tempests. Nature is the arena where humanity's greatest triumphs and tragedies unfold. Nature is a miracle maker.

Millennia ago, ancient civilisations forged alliances with nature, training falcons as hunting partners to survive the harsh desert. Today, the falcon is an Arabian icon, so esteemed that they are issued passports in the United Arab Emirates (UAE). Qatar Airways even allows passengers to travel with a falcon in the cabin with them.

In the US, the bald eagle has reigned as a national emblem since 1782. It is a majestic symbol of pride and strength that earned its enduring place on the nation's seal.

In the walled Ethiopian city of Harar, a unique partnership exists between humans and hyenas. Hyenas are fed by humans in exchange for cleaning the streets and keeping malevolent spirits at bay.

In the Pacific Northwest, Indigenous Coast Salish peoples identify as 'salmon people',[3] recognising salmon as a lifeline woven into their cultural heritage and survival. Festivals and ceremonies are held in honour of the salmon, paying tribute to the power it has to 'unite people' and connect and enrich communities.[4]

From the silent warhorses that carried soldiers into battle to the wolves that coevolved with humans to become their most trusted companions, nature is the foundation of humanity's story. It has shaped our identity, sculpted our societies and carved out the pathways of our politics, health, livelihoods and economies.

Every thread of civilisation, every innovation, every triumph owes its roots to the natural world. Nature isn't just our history – it is the blueprint of humanity itself.

And that blueprint is endlessly inventive. The natural world has not merely shaped us; it has schooled us. Time and again, nature's solutions have become humanity's breakthroughs, translating evolutionary marvels into technological leaps. In the most unique ways, species find their way into human society, sometimes revolutionising entire fields with far-reaching impacts on human society, from agriculture and medicine to technology and defence systems.

NASA created a gripping system inspired by the gecko's sticky feet, aiding astronauts in navigating the International Space Station. This technology is also being adapted for robots to tackle space junk. Space researcher Gareth Meirion-Griffith says, 'If nature hadn't come up with this, I don't think anyone would have ever thought of it.'[5]

The elephant's trunk inspired the development of the bionic robotic arm.[6] The whale inspired the design of wind turbines.[7] The kingfisher bird inspired Japan's high-speed bullet train.[8] Bats inspired the development of ultrasound technology for medical imaging.[9] Bird flight is the basis of aviation, as inventors from Leonardo da Vinci to the Wright brothers used soaring birds as early guides for creating lift and aerodynamic shapes.[10] Spider silk, strong enough to inspire blockbuster movies, is now being explored as the future of military armour.

Most importantly, the very functioning and survival of humanity is inextricably and directly tied to the plants, animals and ecosystems that make up the natural world.

If the idea of breathing, eating and wearing clothes appeals to you, then you will undoubtedly be supportive of nature. After all, humanity is entirely, irrevocably dependent on nature for essential, life-sustaining services. From tiny insects to vast forests, nature provides the food we eat, the air we breathe and the homes we live in.

Trees act as the Earth's lungs, filtering pollutants and producing the oxygen we breathe. Pollinators like birds, bees and bats fuel our global food system by spreading pollen across plants, allowing crops to reproduce. Healthy soil puts food on our plates, purifies our water, protects us against flooding and combats drought. Coral reefs protect coastlines from erosion while supporting fisheries and tourism. Nature provides the raw materials for our homes, the energy we use and the clothes we wear.

Nature also heals us.

The leaves, bark and sap of plants contain secrets that have fuelled human medicine (and health) for centuries. Over 80 per cent of registered medicines either come from, or have been inspired by, the natural world.[11] From aloe vera soothing burns to aspirin found in the bark of willow trees, to morphine that was synthesised from poppies to relieve pain, to anti-cancer agents used in chemotherapy drugs, nature has been the foundation of both ancient and modern medicine. Medicines used to treat conditions such as Parkinson's disease, Alzheimer's and malaria often contain compounds originally discovered in plants.[12] Seventy per cent of drugs used for cancer are natural or nature-inspired products.[13] Conservative estimates suggest that we are losing one vital drug every two years because of our relentless assault on the natural world.[14] As science communicator Peter Bickerton says, 'Right now, the cure for cancer, or COVID, could be going extinct.'[15]

Nature's influence isn't always grand or obvious – sometimes, it's as unassuming as a caterpillar. Indeed, the humble caterpillar may be the architect of global commerce; it is, after all, responsible for one of the earliest trade networks, connecting East Asia with the Middle East and Europe. Silk threads developed from the larvae of the silk moth led to the development of the ancient Silk Road. This trade network played a central role in shaping global trade, politics and cultural exchanges for centuries, facilitating not just the exchange of goods like silk, spices and precious metals but also the spread of knowledge, ideas, language and innovations.

Nature is the often-overlooked scaffolding upon which our

progress is built, influencing the formation of human societies in profound ways; everything from our most advanced technologies to our most enduring political systems and trade networks to our wealthiest economies would collapse without the gifts that nature provides.

Equally, the loss of nature robs us of our health, humanity and livelihoods. It creeps up on us, a slow unravelling. But sometimes, every now and then, this unravelling is punctuated by moments of shattering consequence and clarity.

In India, the decline of vulture populations has cost the economy US$350 billion and caused widespread fatalities.[16] Vultures are nature's very own clean-up crew, and they have a very important job: to eat the dead things. Living off the carcasses of wild and domestic animals, their cleaning service is vital to the health of humanity and the planet. By eating the remains of decaying bodies, they prevent stench, significantly curtail carbon emissions and eliminate disease and bacteria that can be released during decomposition – without them, disease can spread.

Unfortunately, this noble career path hit a snag in the 1990s. India's vultures began dying en masse due to diclofenac, an anti-inflammatory drug widely used by veterinarians to treat livestock. Birds that fed on carcasses of cows that had been treated with the drug suffered from kidney failure and died. The 50 million-strong vulture population plummeted to near zero – a population decline of more than 99 per cent. When vulture populations collapsed, the consequences were far-reaching. With their disappearance, the scavenging services they provided disappeared too, and carrion was left in the open for long periods, leading to an increase in the populations of rats and feral dogs, and subsequently in the incidence of rabies. Deadly bacteria and infections proliferated, and there was increased water pollution from carcass dumping and surface runoff.[17] Half a million people died and ecosystems were degraded as a result. In 2006, India, Pakistan and Nepal banned the veterinary use of diclofenac, and vulture populations have been slowly recovering.

Across the world, nature protects, defends and sustains ecosystems and humanity alike. Mangroves provide crucial protection from storm surges and flooding, saving coastal communities over $65 billion and safeguarding more than 15 million people.[18] The Great Barrier Reef supports tourism services that generate $5.7 billion annually,[19] drawing visitors from around the world. Beavers create wetlands that enhance biodiversity, improve water quality, mitigate flooding and provide natural services worth millions. In fact, a long-delayed government project to build a dam and restore wetlands in Czechia was quietly overtaken by beavers, who did the job themselves – saving taxpayers US$1.2 million in the process.[20] Without waiting for permits or paperwork, the beavers built their own structure and restored the wetland that had been earmarked for development.

A shade cloth at your local hardware store might cost $300, but trees provide that same cooling and shade for free, while also improving air quality and supporting wildlife. Even species like the ibis play a surprising role in ecosystem management, acting as natural pest control by preying on cane toads, one of Australia's most invasive species. These are services that nature provides to humanity – services we often overlook and for which we do not pay.

In 2011, a moose was found drunk out of its mind and stuck in a tree in Sweden. It was late in the evening, and Per Johansson heard a roar coming from his neighbour's garden. It was there that Johansson found a moose completely plastered, wedged in an apple tree. The moose appeared to be sick, drunk or 'half-stupid', observed Per.[21] He promptly called the local fire and rescue department, which arrived with a fire engine and a jeep.

'When we arrived, we used the winch to bend down the apple tree so the moose could get himself out of the tree. Once free, the moose collapsed on the ground and fell asleep. So we let him sleep it off and went back home,' said Anders Gardhagen, the Gothenburg Fire and Rescue Services spokesperson.[22]

THE MIRACLE MAKERS: A DRUNK MOOSE AND AN ECONOMIST

Two years later, news reports of intoxicated moose surfaced in Stockholm, claiming that 'a scourge of drunk moose are terrorizing the nation'.[23]

First, there was an incident in August involving a 'gang of alcoholic moose'[24] causing a disturbance.* 'Five drunken elk were threatening a resident who was barred from entering his own home,' the official police incident report said. Radio Sweden was quick to respond, alerting locals to the troublesome elk.[25] 'Get ready for the season of drunken elks,' the dispatch read.[26] 'As ripe fruit falls from the trees and ferments on the ground, it is time for some of Sweden's most majestic wild animals to act in a most un-regal manner.'[27]

Then, less than a month later, police were called after a seemingly drunk and belligerent moose got into a brawl with a swing set and then became entangled with its chains. The animal eventually managed to free himself, but not before moving the swing about 500 metres from the rest of the playground, the police stated.[28]

Residents of Alaska face similar issues with drunk moose. In Anchorage, Buzzwinkle the moose made a name for himself as the town drunk. Buzzwinkle earned his title after he stumbled through Town Square Park, pulled Christmas lights from the trees, then wandered into Bernie's Bungalow Lounge – a martini bar – and ate a pile of fermented crabapples with the Christmas lights still dangling from his antlers. According to locals, Buzzwinkle was a fixture in Anchorage, and was frequently seen drunk, 'staring off into the distance', while wearing his Christmas lights.[29]

---

* The word 'elk' can mean different things depending on where you are. In North America, elk refers to *Cervus canadensis*, a large deer also known as wapiti. In Scandinavia and the UK, elk often refers to *Alces alces* – the animal North Americans call a moose. Meanwhile, *Cervus elaphus*, or red deer, is common across Europe and is a separate species altogether. The result is a confusing overlap of names for three different animals.

News (and police) reports suggest that moose are getting drunk off fermented apples. Scientists, however, aren't quite convinced that the moose are, in fact, drunk. Volatile or 'drunk' moose behaviour can stem from a combination of environmental pressures such as shrinking forests, availability of food, human-wildlife conflict, climate change and predator-prey dynamics.[30] And perhaps occasionally from indulging in ethanol-rich fermented fruits. While this behaviour may seem like trivial moose antics, it actually offers powerful clues into the workings of entire ecosystems and ultimately – and perhaps surprisingly to many – can affect the long-term prospects and profit margins of businesses around the world.

Indeed, the behaviour and dietary habits of moose set off a cascade of ecological interactions – some beneficial, others less so. These effects ripple through the ecosystems they inhabit. And eventually, this will come to land square in the profit-and-loss statements of companies like IKEA.

Like the sea otter, moose are considered a keystone species (in certain ecosystems): they play a crucial role in determining the structure and function of the ecosystem they inhabit. Without its keystone species, the ecosystem would be dramatically different or cease to exist. In fact, through their foraging, trampling and even their excrement, moose shape entire forests and, by extension, can even influence climate systems. As they graze, they influence which plants thrive and which fade into obscurity, while each step they take changes how nutrients are distributed throughout the soil. Their mere presence even shapes predator-prey dynamics – for better or for worse. In short, moose are ecosystem engineers – but not always the good kind.

Yet, the relationships between moose and their habitat, humanity and the broader ecosystem are anything but simple. The ecological and societal benefits of moose depend and vary based on these relationships. In some areas, their grazing habits can encourage plant growth, boost plant diversity and strengthen ecosystems while in other areas, they can overgraze, leading to the erosion of soil health and the decline of plant diversity. In Norway, for example, moose have been likened to that of the forestry industry for their impact on ecosystems, where they may even be a driver of increased greenhouse gas pollution.[31]

These dynamics reveal the dual role moose play: as both architects of their habitat and agents of change or, sometimes, destruction. Their impact runs deep, and is often heavily influenced by human activities, ecosystem dynamics, and environmental conditions. Their very existence is woven into the fabric of the ecosystem, affecting everything from how much carbon a forest can store to the presence of other species and to the availability of trees and fresh water.

And here's where the corporate world should pay heed. Every company is dependent on nature's complex web of life. Whether it be an organisation using timber to build a new factory, a beverage company depending on fresh water to produce soft drinks, a supermarket relying on the existence of pollinators to keep its shelves stocked, a fashion company requiring healthy soil to ensure cotton production for their T-shirts, a cosmetics company relying on plant-based ingredients for their beauty products, a tourism company relying on pristine natural landscapes for expeditions or a furniture company depending on a moose's antics – all companies depend on and are at the mercy of nature. And when nature suffers, so too does business.

IKEA's business model is heavily dependent on nature. They are one of the largest private landowners of forests worldwide and use around 21 million cubic metres of logs in a single year.[32] That means that one tree is logged every second for an IKEA product. But without forests, and without species that sustain these ecosystems, IKEA has no business. Equally, as one of the biggest users of wood in the world, IKEA has a huge impact on the world's forests and forestry industry – including playing an outsized role in changing the trajectory of nature decline. Recognising this, IKEA has committed to no deforestation by 2026 and has worked out a sustainable strategy to become 'forest-positive', with the goal of improving forest management worldwide by 2030.[33]

It's not just IKEA. Every company and economy on the planet is directly dependent on nature. Mars Incorporated, the multi-billion-dollar purveyor of chocolates and childhood obesity, is learning this lesson firsthand. After all, Mars owes a significant chunk of its fortune to a creature so small, you've probably swallowed it while jogging: the midge. The midge is a humble, biting relative of the fly, and this irritating insect is tasked with a job of disproportionate importance. It is the near-exclusive pollinator of the cacao tree – the plant that produces cocoa beans, and with them, chocolate.

You see, cacao plants are fussy. Their flowers are small, awkwardly shaped, and largely unappealing to most pollinators. But not to the midge. The midge, in its baffling commitment, squeezes into those weird little flowers like it was built for it. Which, evolutionarily speaking, it was.

This means that the chocolate industry, valued at over $140 billion, is built almost entirely off the work of an insect smaller than a pinhead.[34]

But the midge is in strife. Like many under-appreciated workers, it is being slowly eradicated by the very system it sustains. Midges thrive in dark, damp, shaded environments typical of tropical rainforests – habitats that are rapidly disappearing due to deforestation, agricultural expansion and climate change. Nearly 70 per cent of the world's cocoa is supplied from West Africa's cocoa belt.[35] And forests across the region are being cleared at an alarming rate – driven largely by logging, mining and, ironically, the expansion of cocoa plantations to meet rising global demand for chocolate. Côte d'Ivoire alone has seen 80 per cent of its forests destroyed over the past fifty years, and with it, the shaded, humid understories and leaf-littered floors where cacao-pollinating midges breed are stripped away.[36]

In trying to produce more chocolate, we are steadily undermining the ecological systems – and insects – that make chocolate possible.

The result will be a reduction of the world's supply of chocolate and far-reaching consequences for the 5 to 6 million cocoa farmers and 40 to 50 million people worldwide whose livelihoods depend on cocoa – not to mention the soaring price of chocolate that you're likely already witnessing.[37]

For all its scale, the industry still depends on a fragile set of ecological relationships – stable rainfall, healthy forests, fertile soils, and the quiet work of an insect few people have ever heard of. In fact, some studies predict that by 2050, many of the areas currently used for cocoa farming in West Africa could become unsuitable for growing cacao trees.[38]

The midge may be nearly invisible, but without it, so is chocolate.

And with this, Mars is discovering that no matter how vast its industrial might, it remains deeply reliant on nature. After all, even billion-dollar companies are only as sweet as the ecosystems they depend on.

Mars is keenly aware of its impact on, and its dependence on, nature. Faced with falling production, the company sequenced the genome of the cacao tree in 2010, an endeavour that would allow them to access its genetic information with the aim of developing trees that resist disease and climate change impacts, as well as produce more cocoa. Since then, their efforts have only increased. They have committed $2.7 billion towards sustainability initiatives, aiming to reduce land use, greenhouse gas emissions and unsustainable water consumption.[39] They are exploring cocoa alternatives and have vowed to reduce deforestation across their product supply chain and to work with their suppliers to implement stronger nature initiatives.

Whether or not companies like Mars and IKEA realise it, their business models are inextricably tied to the dietary and behavioural habits of everything from the moose to the midge. By ensuring the health of these species and their ecosystems, these businesses are safeguarding their own supply chains and the ecological integrity of the landscapes they depend on. After all, as a drunken moose bumbles through a forest, or a midge pollinates a cacao tree, they are quietly influencing the very resources that fuel the world economy.

To destroy nature is to destroy the long-term viability of the company. Nature isn't just a resource; it's the foundation upon which entire industries rest. Clean water, pollination, fertile soil, climate regulation: these are the life-support systems of the planet and, by extension, the life-support systems of global commerce – and humanity.

In the 1970s and 1980s, scientists were growing increasingly alarmed by the rate at which humanity was driving species into extinction. Frustrated that humanity was taking nature's services for granted, they coined the term 'ecosystem services'. Nature performs valuable, practical and measurable functions, they argued; functions that underpin our health, wellbeing, economies and very existence. The aim was to illustrate the tangible benefits and importance of nature's services, with the hope that this would encourage humans to protect nature, rather than passionately dismember it.

It was an effective strategy. Just as Eminem sparked a generation of boys with bleached blonde buzz cuts and Paris Hilton turned handbag-sized dogs into the must-have accessory of the 2000s, the term 'ecosystem services' started gaining traction. First, the UN referenced it. The European Union (EU) soon followed suit. So did the World Bank. And the World Economic Forum. S&P Global joined in. Even the Obama administration got involved. The big guns in town were talking about ecosystem services.

Ecosystem services are defined as the goods and services nature provides humanity. They are, in essence, nature's gift basket to humanity – everything from timber and clean water to crop pollination and climate regulation. They stabilise our world, underpin our lives and even offer inspiration. From the air we breathe to the food we eat, these services are the invisible engines powering our existence.

Nature's services should, indeed, be obvious. Yet somehow, they often go unnoticed. We flip a light switch, fill our glass with water or furnish our house without stopping to consider that forests and wetlands are working tirelessly behind the scenes to provide these services.

In 1997, a group of environmentalists and economists published a paper that estimated the global value of ecosystem services to be a staggering US$33 trillion per year, a figure that comfortably dwarfed the global economy at the time (a mere US$18 trillion).[40] This meant that economists were looking at species like bees and bats and calculating the value of their pollination services. For example, without bees, we'd be forced to hire legions of humans armed with tiny paintbrushes to pollinate plants by hand. The cost of this? US$217 billion a year.[41]

The idea that nature had measurable value was sending economists to the metaphorical brink of orgasm. What if we could put a price tag on nature and its valuable services? Striding forth from their windowless rooms, their minds were abuzz with the ways in which nature could be quantified, priced and traded.

Economists and environmentalists alike were highlighting the clear shortcomings in our current economic system. Human economies simply cannot expand indefinitely on a planet with finite resources. Nevertheless, the modern economic system has largely been built on the comforting notion that natural resources could be extracted, consumed and disposed of ad infinitum, without any adverse consequences to a business's profit margins. Environmental destruction is simply viewed as the unfortunate but unavoidable cost of progress – less than ideal, sure, but acceptable if it keeps the gears of industry turning.

As Sylvain Vanston, former head of climate change and biodiversity for the French insurer Axa, said, 'From a business standpoint, a dead tree is worth more than an alive tree. Exploiting and destroying nature makes you reap more immediate benefits than protecting it. But when nature collapses, entire regions and entire sectors might become stranded.'[42]

The economy, it was coming to be acknowledged, is no longer

fit for purpose to create a habitable future for humans and nature. And with this, economists, clutching their spreadsheets in barely contained delight, started working with environmentalists to devise a plan. They began to re-engineer global financial and economic architecture. Their aim was simple but ambitious: integrate the value of nature into economic decisions. They worked to develop new frameworks that accounted for ecosystems as capital – essential, finite and worthy of investment and protection.

The idea was to stop treating nature as an optional extra and start treating it as essential infrastructure. Forests, wetlands, pollinators and oceans weren't just scenic backdrops; they were providing services worth trillions, usually for free. By framing ecosystems as assets – not necessarily because they could be owned or sold, but because they quite literally keep economies functioning – economists hoped to make them visible in the places that mattered: balance sheets, budgets and boardrooms.*

These frameworks, they contended, would ensure 'capital flows towards activities that protect and conserve natural ecosystems, rather than degrade and destroy them'.[43] This way, economic decisions would take into account the value of nature's services – like vultures cleaning up carcasses or moose helping to spread seeds. It would, ostensibly, reconcile a capitalist growth economy with nature protection. In this way, they asserted, the protection of nature could be expressed in a language that economists, policymakers and CEOs would understand.[44]

---

* In practice, this shows up in things like natural capital accounts, where governments track the health and value of ecosystems, or environmental profit-and-loss statements, where companies tally up their impacts on nature – water use, emissions, land degradation, that sort of thing. The point isn't (necessarily) to put a price tag on nature so it can be sold off, but to ensure that damaging it finally appears somewhere on the balance sheet, rather than being treated as a rounding error.

One of the downsides of this is that economists and businesses have a singular talent for adopting things and then making them sound both terribly important and unforgivably tedious.

As American publisher Alfred A Knopf Sr said, 'An economist is a man who states the obvious in terms of the incomprehensible.'[45] And thus, nature entered the world of economics and became utterly baffling to almost anyone without a PhD in economics or corporate jargon. 'Nature' became 'natural capital', 'ecosystems' became 'ecosystem services' and more new phases started forming like 'nature positive', 'biodiversity net gain', 'Green Wall Street', 'natural capital accounting', 'nature's balance sheet', 'natural capital assets' and 'green aggregates'.

This shift in language drew criticism from many scientists, activists and Indigenous communities, who argued that reframing living systems as 'assets' or 'capital' risked reducing nature to financial terms. But language wasn't the only thing drawing opposition. The very essence of putting a value on nature was, for many, problematic. It's a rather controversial way of thinking about nature – though it's edging its way into the mainstream. This grand exploration of the nature-economics nexus has, to date, left a large number of people in a state of reasonable bewilderment.

Putting a value on nature has neatly divided opinions into two dominant camps: those calculating the market value of a whale and those positively nauseated by the idea. The former camp believes the economy can be reconciled with nature with some tweaking. They are of the opinion that placing a value – and perhaps a price – on nature can achieve this. While the latter group sees this as a breakdown of civilisation as we know it and wonders if perhaps capitalism has overreached just a touch. Broadly, this group calls for a more radical transformation, which may include greater emphasis on scaling down ecologically destructive and unnecessary production and consumption. However, both camps tend to agree that we are hurtling towards planetary demise.

Proponents of tying economic value to nature say we can work within our current economic system and continue to enjoy economic growth while simultaneously protecting nature by valuing ecosystem services. As environmental economist Ed Barbier of Colorado State University says, 'We have put a price on nature. And that price is zero.'[46] This group argues that when something appears to have zero value in our economic system, it tends to be treated with reckless abandon. They believe that giving nature a place at the economic decision-making table is the only way to safeguard it.

Valuing and putting a price on nature, they argue, will highlight humanity's economic dependence on it. By attaching value to ecosystem services like clean water, soil health and pollination, society will be forced to recognise just how much we rely on these 'free' services. In this way, the conservation and protection of nature will be incentivised while its destruction will be disincentivised.

For example, in our current system, a chemical company may sell fertiliser to improve farm crop yields. But that fertiliser may run into waterways and harm marine life, dismantling entire ecosystems. That may reduce fish stocks and harm the livelihoods of fishers, or it may drive species extinction or contaminate the drinking water for communities. This harm is viewed, in our modern economy, as merely a side effect of doing business – it is not the problem of the company that produced the fertiliser. In theory, the company that produces the fertiliser should be responsible for damage it causes. But, in practice, much of that damage is quietly offloaded onto the public – and the natural world. If the pollution is dramatic enough – say, a large chemical spill – there might be a fine or a carefully worded apology within a press release. But most environmental harm doesn't arrive with fireworks – or press releases. It happens slowly, unevenly and just far enough away to feel like someone else's problem.

So, the clean-up costs fall to governments, which is to say, you. The health impacts are absorbed by hospital systems, and the ecosystems simply degrade politely in the background, largely unnoticed. The company keeps its profits. The public picks up the tab.

When we value nature, we can put a price on it, which could mean that the fertiliser company would pay for its negative environmental impact.* Suddenly, those external costs are brought into the company's financial equation, encouraging it to consider activities that don't harm the environment. This is, proponents say, a way to make sure businesses consider not only their profits but also the true cost of their impact on nature and public health.

Detractors, on the other hand, had a near syphilitic reaction to the notion of putting a value on nature. They are broadly of the opinion that nature's value is infinite and immeasurable, and that putting a price on it is a 'fantasy that distracts us from real efforts to save the planet'.[47] Nature, they argue, is an intricate web of interdependent species, ecosystems and cycles that support life in countless ways that cannot be fully captured in economic terms.

Conservation biologist David Ehrenfeld was slightly more strident in his view, remarking that, 'Common sense and what little we have left of the wisdom of our ancestors tells us that if we ruin the earth, we will suffer grievously.' He added, 'I am afraid that I don't see much hope for a civilization so stupid that it demands a quantitative estimate of the value of its own umbilical cord.'[48]

For others, viewing nature through the prism of its utility to humanity is exploitative and human-centric. Journalist George Monbiot calls the approach 'morally wrong, intellectually vacuous, emotionally alienating and self-defeating'.[49] He says that by integrating the environment into the world market, 'you are effectively pushing the natural world even further into the system that is eating it alive'.[50]

---

*   This might involve pollution taxes, resource-use fees or restoration obligations, but it can also create opportunities for innovation and investment in nature-regeneration practices – like paying farmers to preserve forests that protect water supplies or funding pollinator habitats.

By viewing nature as a range of services and economic values, complex webs of life may be reduced to a mere collection of resources and economic values for human consumption. This can lead to a transactional and narrow view of nature, as if the mountains, oceans, rivers and forests exist solely for human benefit.

Monbiot believes that nature is being subordinated to the human economy, noting that, 'Ecological processes are called ecosystem services because, of course, they exist only to serve us.'[51]

Added to that, assigning economic value also often prioritises the perspectives and interests of those with financial influence, such as large, multinational corporations and wealthy nations. This can marginalise local communities and Indigenous peoples, who often rely on non-monetary relationships with nature and are best placed to steward these ecosystems sustainably. When nature becomes commodified, it risks being 'owned' and controlled by powerful economic actors, undermining traditional rights and conservation by local communities.

Others argue that putting a price on nature is governments' way of shirking responsibility, outsourcing conservation to the private sector to avoid implementing effective regulation. This is often done on the reasoning that the government cannot afford to protect nature. Yet as Polly Hemming, director of the Australia Institute's Climate and Energy program, states, 'billions of dollars in subsidies and tax breaks' go to fossil fuel companies every year'.[52]

Writer Jeff Sparrow says that this approach means environmental decisions are being taken away from the public and entrusted instead to 'the market's invisible hand'.[53]

'Given the glaring relationship between profit and extinction (think of the logging firms clearing the Amazon), you might wonder at the mental gymnastics required to present financialisation as an alternative to immediate government regulation,'[54] he says.

The debate is more nuanced than simple pro-con binaries. Across the wide spectrum of views on putting a price on nature, almost everyone is in agreement that the current economic model is excelling at its unofficial mission of dismantling the natural world.

Calculating the economic value of ecosystem services is highly complex and fraught with uncertainties. Who decides what's valuable? Is a panda worth more than a turtle? A river more than a dam? And once we assign a price tag, can nature simply be bought and sold? What happens to the parts we deem less valuable? Nature isn't made up of isolated 'units' it's a complex web where small changes can have big, unpredictable effects. Yet, our ways of valuing it often fail to capture this complexity and risk, reducing nature to a neat line item on a balance sheet.

So, where does all this leave us? Perhaps somewhere in the middle. An overhaul (or dismantling) of the entire economic system, while theoretically appealing to many, may not be practical given the rate of ecosystem collapse and species extinction. In the limited time we have left to alter our destructive course, the answer will likely be, as UN secretary-general António Guterres said, doing 'everything, everywhere, all at once'.[55]

Recognising nature's value – economic and intrinsic – can help highlight the gifts we take for granted, from pollination to water purification to chocolate. If done well, it could also be key to driving more investment into nature and communities. Perhaps the most urgent step, though, is to rethink our relationship with the natural world entirely, and this means recognising that we are part of the larger community of beings, rather than a ruler of nature.

Nature is the source of our stories and symbols, our art and ingenuity. It fuels our health and feeds our bodies and builds our homes. From pandas that offer peace and diplomacy and salmon that nourish life and spirit to caterpillars that build trade networks and tiny insects that power the chocolate industry – every ounce of humanity is rooted in nature.

NATURE'S LAST DANCE

Nature is our joy, our wonder and our lifeline. It extends its miracles to us – gifts that we may call services. And while we have built economies and empires upon these gifts – and found joy in their abundance – we must remember this simple truth: nature does not belong to us. Instead, we belong to nature.

Chapter 8

Wolves of Wall Street: Cowboys and Whale Poo

> Climate change is about how we live,
> biodiversity is about whether we live.
> Dr Frauke Fischer[1]

In many cultures, cheating on your partner is frowned upon. But cheating, much like polluting, may just be a problem of equivalence.

In 2007, a now-defunct website, cheatneutral.com, offered a unique service for customers wanting to be unfaithful. It claimed to be able to offset infidelity.

'At Cheatneutral, we believe that we should all try to reduce the amount we cheat on our partners, but we also realise that fidelity isn't always possible. That's why we help you neutralise your cheating,' the website proclaimed. 'When you cheat on your partner you add to the heartbreak, pain and jealousy in the atmosphere. Cheatneutral offsets your cheating by funding someone else to be faithful and not cheat. This neutralises the pain and unhappy emotion and leaves you with a clear conscience.'[2]

'By paying Cheatneutral, you're funding monogamy-boosting offset projects – we simply invest the money you give us in monogamous, faithful or just plain single people, to encourage them to stay that way.'[3]

The site even displayed 'user testimonials' extolling the

guilt-soothing effects of offsetting one's cheating. 'Seb was so angry with me, I felt really bad about what I'd done. I came to Cheatneutral to offset the side effects of my cheating, and later on, Seb said the only reason he could forgive me was because I had offset my cheating with Cheatneutral. Thanks to Cheatneutral, we're still together, I can feel good about my cheating, and I've helped reduce global cheating as well. If I do cheat on Seb again, I'll definitely be calling Cheatneutral.'[4]

The website was, of course, a parody – a pointed, albeit amusing, critique of carbon offsets.

The analogy is crude but perhaps apt. After all, if there is one subject in the environmental sector that might just provoke a more impassioned reaction than infidelity, it is the carbon market (or, more precisely, the carbon offset).

Climate change entered the public consciousness as a scientific curiosity in the late 19th century when researchers discovered that carbon dioxide and other greenhouse gases (methane and nitrous oxide, among others) have an annoying habit of hanging around and trapping heat in the atmosphere, creating a warming effect on the planet. The predicted consequences of this are bleak: rising sea levels; melting polar ice; more frequent and extreme floods, fires and droughts; collapsing coral reefs; wildlife extinctions; disease; irreversible damage to ecosystems and industries (like agriculture and food production); and far-reaching impacts on vulnerable communities.

By 1988, the world's scientists had seen enough. They banded together, forming the Intergovernmental Panel on Climate Change (IPCC), a no-nonsense UN body tasked with assessing the climate and telling the world just how badly we were messing up the planet. The IPCC's first report stated that human activities – like the burning of fossil fuels – were substantially increasing greenhouse gas levels, driving global warming and disrupting Earth's climate stability.[5]

With these grim forecasts in hand, the UN orchestrated a diplomatic extravaganza at the 1992 Earth Summit in Rio de Janeiro. This was climate change's big debut, its step up from obscure academic circles to political prime time. After a seemingly endless labour, the summit gave birth to the UN Framework Convention on Climate Change (UNFCCC), an agreement to essentially limit climate change. Its message was clear, but also vague and abstract enough to be mildly ignored if one was so inclined.

Since that fateful 1992 summit, world leaders have dutifully assembled each year under the banner of the UNFCCC to ponder the climate conundrum. For the next twenty-three years, leaders repeatedly expressed and reviewed their commitment to combatting climate change. Things were ratified, hands were shook, arguments were had. Global greenhouse gas emissions were not reduced. You see, it is difficult to get almost every country in the world to agree on anything, particularly something that has the appealing objective of achieving 'stabilization of greenhouse gas concentrations in the atmosphere at a level that would prevent dangerous anthropogenic interference with the climate system'.[6]

Then in 2015, climate change finally had its red-carpet-at-the-Met-Gala moment. The UN Climate Change Conference (COP21) in Paris delivered a landmark climate deal with real teeth. Named the Paris Agreement, this was a legally binding agreement in which 196 countries pledged to keep global temperature rises 'well below' 2 degrees Celsius but aiming for no more than 1.5 degrees Celsius compared to pre-industrial levels.* If half a degree of warming sounds like splitting hairs over a decimal point, it's not. It translates

---

*   The Paris Agreement is legally binding in terms of process: countries are required to submit emissions targets (Nationally Determined Contributions, or NDCs), report on their progress and update their commitments every five years. However, the actual content of those targets is not legally enforceable, and there are no penalties for failing to meet them. In other words, countries are legally bound to promise – just not to deliver.

to several hundred million additional people facing floods, fires, drought, poverty and displacement as a result of climate change.[7]

Another goal of the Paris Agreement was to achieve 'net zero' by mid-century. The plan? Slash emissions with things like renewables, then mop up the leftover emissions by pulling greenhouse gases out of the atmosphere using trees and technology. The chosen mop? The carbon market.

In a nutshell, carbon (or, more broadly, environmental) markets are a trading hub where 'credits' are bought and sold. A carbon credit represents one tonne of carbon dioxide or the equivalent amount of a different greenhouse gas (like methane) that has been reduced, removed or avoided from entering the atmosphere. Think of it as a receipt proving a pollution-busting action has taken place, whether that's planting trees, protecting forests or funding renewable energy.

This is the financialisation of the environment in practice: take something messy, essential and alive, and convert it into units that can be packaged, priced and traded. Carbon credits are one such invention – an attempt to translate the work of ecosystems like forests and wetlands into economic decision-making. If nature can't be protected for what it is, the thinking goes, perhaps it can be protected for what someone decides it's worth.

The next part is where the controversy comes in. Once you have purchased a carbon credit, you can release one tonne of pollution into the atmosphere and then wave the carbon credit in the air to signal that you have, indeed, neutralised your pollution. This is called a carbon offset.

The great hope of carbon markets is that they will attract capital from the private sector, helping to fill the multi-trillion funding gap for environmental projects. Theoretically, everyone wins: companies pay to ensure they are not adding more pollution to the atmosphere and can label themselves as 'carbon neutral'; financial markets can start illustrating that there is more economic incentive for businesses to restore a forest rather than destroy it; investors profit from green projects that they've financed, which are now getting funding from the carbon market; ecosystems thrive; communities at the heart of environmental restoration projects are funded and able to thrive; and no new pollution enters the atmosphere.

In essence, carbon markets work by allowing companies or countries to fund projects that reduce emissions elsewhere if they can't fully cut their own pollution right now. This should theoretically mean that there is no overall increase in greenhouse gas emissions.

Unfortunately, there have been some snags.

Investigative journalists and researchers started uncovering widespread abuses across the carbon market – primarily the voluntary (unregulated) carbon market. Soon, reports were coming in thick and fast from major news outlets around the world. Headlines painted a chaotic scene: 'carbon cowboys'[8] were reportedly raking in millions of dollars through dubious carbon credit schemes and 'cashing in on protected Amazon forest'.[9] News outlets dubbed it 'the Wild West of the carbon market',[10] while carbon market 'certifying bodies' such as Verra (which are like the referees of carbon markets) were slammed for issuing 'junk offsets'.[11] Investigations claimed 90 per cent of Verra's carbon offsets – meant to restore rainforests – were 'worthless' with the potential to 'make global heating worse'.[12] There were claims that projects had exaggerated their emissions reduction, and of 'ineffectual' carbon offsets that were 'knowingly hindering the energy transition'.[13]

The abuses didn't stop at environmental deception. Human rights violations cast a dark shadow across the carbon market, as reports of abuse and exploitation of poor, rural and Indigenous communities emerged. In Kenya, hundreds of members of the Ogiek community reportedly had their ancestral homes pulled down with 'axes and hammers' to make way for carbon offsetting projects.[14] In Peru, residents claimed their homes were being 'cut down with chainsaws and ropes' as the carbon market boomed.[15] There were 'phantom credits'[16] that were discovered to not be reducing any emissions at all, and timber companies were claiming carbon offset credits simply for leaving trees untouched – trees they'd had no intention of cutting down in the first place.[17]

There was widespread deception and there were legal battles. One carbon offsetting executive was charged with fraud by US federal authorities after he allegedly helped to manipulate data from projects in rural Africa and Asia in order to fraudulently obtain carbon credits worth tens of millions of dollars.[18] The voluntary carbon certifier Verra was sued by communities in Brazil. The biggest companies in the world were caught in the fray after purchasing carbon credits that might have little to no impact on actual emissions reduction. Gucci, Netflix, Disney, easyJet, Boeing, Salesforce, BHP, Spotify and Shell were all being dragged into question over their carbon neutrality claims tied to these offsets.[19] Delta Airlines was sued over its 'carbon neutrality' claim.[20] EnergyAustralia was sued.[21] Danone was sued.[22]

All hell was breaking loose.

The UN weighed in, calling for an end to 'rank deception' and 'bogus net zero pledges'.[23] More than eighty nonprofits joined forces to argue that carbon offsets were undermining genuine net zero action.[24] Academics weighed in.[25] Even the Pope weighed in.[26]

It was open-season anarchy in the carbon market and across the vast networks of buyers, sellers and intermediaries, it appeared nobody was safe.

The root of these rather unfortunate outcomes was a very creative interpretation of what it meant to reduce greenhouse gases. Companies, it seemed, were over-relying on carbon offsets and using them as a free pass to delay tangible action.

In 2019, the UN Environment Programme, with the kind of patient exasperation typically reserved for explaining dinner etiquette to a toddler, reminded everyone that carbon offsets were meant to be a last resort, not a 'get-out-of-jail free card'.[27]

The idea, in theory, was that companies would use offsets sparingly, only when no viable alternatives existed to reduce their emissions in the short term. Increasingly, concerns were being raised that companies were continuing business as usual and just offsetting – sometimes using offsets that, as we saw, were having little to no positive environmental impact (or even making things worse).

Soon, the term 'greenwashing' came to enter the public vernacular. Greenwashing is the art of misleading the public to believe that a company or entity is doing more to protect the environment than it actually is. As more corporations and governments pledged to become net zero, greenwashing cases skyrocketed. In 2023 alone, 230 climate change lawsuits were filed worldwide, with forty-seven of them centred specifically on greenwashing.[28]

As lawsuits came rolling in and carbon market scandals hit the headlines, corporations started quietly retreating from their net zero claims. Investment in carbon offset projects plummeted. Public trust in carbon offsetting as a tool for climate action eroded. The gravest consequences of this were perhaps felt by communities. All carbon projects were tainted by the offset quagmire now, and legitimate projects that were actively reducing emissions through reforestation and habitat restoration projects found their funding streams in jeopardy.

Attacks on the carbon market were raining down relentlessly, coming from every conceivable direction. And that is about the time that nature (and biodiversity) made its dazzling entrance into the carbon market debate.*

Apart from the trifling matters of fraud, deception and human rights abuses, the carbon market had another problem: its premise was flawed. You see, it was narrowly focused on carbon dioxide (or greenhouse gases) and seemingly ignored the much bigger picture

---

* Nature is the broader system – the landscapes, oceans, climate and all living and non-living things not created by humans. Biodiversity refers specifically to the variety of life within that system: different species, genetic diversity and ecosystems. In short, nature is the whole show; biodiversity is the richness of its cast.

– things like biodiversity, ecosystems and the natural systems that actually make climate stability possible.

Climate change and the destruction of nature, it turns out, aren't separate crises – they're more like two sides of the same unwieldy coin, rolling towards the cliff with humanity cheering it on from the sidelines. You see, nature is critical for addressing climate change, while climate change drives nature loss. So, the idea took hold that saving one without saving the other might not be a good idea.

To illustrate this, one might look at the case of Ecuador. In the Andes, for example, carbon offset projects planted fast-growing pines in the páramo grasslands to capture carbon. While effective at storing carbon dioxide, these pines crowded out native plants, drained water and degraded the soil, ironically damaging the ecosystem they were meant to preserve.

In the race to reduce emissions and plant trees to store carbon dioxide, it turned out that we may have been overzealously erecting single-species forests around the world – what are known as monoculture plantations. And the problem with monoculture plantations is that they often create inhospitable environments for species. They don't support the richness, complexity and diversity of nature. In fact, they destroy it.

Scientists – mostly ecologists – armed with this knowledge reasoned that perhaps, just perhaps, the best way to save the planet wasn't by blanketing it in endless rows of identical, single-species tree plantations that would quietly smother the rich tapestry of life on Earth. You see, nature is a package deal. Air, water, soil, plants, animals and all the mysterious, wonderful cycles that hold them together function as an interconnected web. Climate, in this grand scheme, is one part of the whole. You can't really just solve climate change at the expense of nature and hope the other bits and pieces will sort themselves out. Focusing solely on greenhouse gases while ignoring nature as a whole is much like trying to treat a heart condition with really enthusiastic dental cleaning.

In the background of this, a group of largely under-appreciated nature experts – heroes, if you will – caught wind of the growing chatter on nature and seized their moment. And so it was with pomp and flair that the whale poo activists charged onto the scene to set the record straight on the nature-climate fiasco once and for all.

According to them, whales are extraordinary allies in the fight against climate change. When a whale does its business, its nutrient-rich poo fertilises phytoplankton – microscopic plants that absorb vast amounts of carbon dioxide from the air. That fertilised phytoplankton becomes food for krill, which in turn becomes food for – yes – whales. It's a charming little loop of life that keeps the ocean humming along. And when whales die, they sink to the ocean floor, taking the carbon in their bodies and storing it underwater for centuries. Each whale, simply by existing, captures and stores as much carbon dioxide as thousands of trees.[29] In summary, more whales mean more poo, more poo means more phytoplankton and more phytoplankton means more carbon pulled from the air. It's an exquisitely honed natural system – delicate, interconnected, essential.

Whales are not just vital to overall ocean health and wondrous beauties that enrich our lives – they are actually a climate change solution. *Nature* is a climate change solution. When we look solely at the environment through the lens of carbon dioxide or greenhouse gas emissions, we fail to see the very natural systems that already know how to do the job; the climate change solutions right in front of us – solutions that we are destroying.

So you see, paying to reduce our pollution via monoculture tree plantations that are inhospitable environments for species while simultaneously destroying species like whales and their habitats that naturally – through their very existence – protect the planet and provide pollution reduction services for free is, in fact, a singularly terrible idea.

With nature (and biodiversity) now officially important, some decision-makers worked on several ideas to protect it that were almost incredibly successful but ultimately were not. Then, after much deliberation, they concluded that the answer was indeed quite obvious. With emphatic nods and self-satisfied smiles, they revealed their solution to the nature crisis: a biodiversity market.

Soon, biodiversity offsets, which had been quietly tumbling around for a few decades, began to gain prominence. In 2023, while the world was embroiled in a ferocious debate about the role of carbon markets, the Australian Government introduced 'world-first' biodiversity market legislation, which would be a voluntary national market to incentivise private investment in nature. This legislation was able to draw on the experience of a similar state-based biodiversity scheme in New South Wales that was deemed a comprehensive failure.[30] Not to be deterred by decades of evidence of ineffectual nature policy, the federal government announced the 'Nature Repair Market', alongside a statement proclaiming that it was creating a 'Green Wall Street'.[31] This arguably meant it would either usher in a new era of investment towards biodiversity or, alternatively, create new opportunities for predatory behaviour and subprime credits, and trigger the nature equivalent of the Great Recession.

The idea of biodiversity markets sets off familiar but arguably more creative arguments. Unlike carbon markets, which are centred around one singular, quantifiable unit of measurement – greenhouse gas emissions – biodiversity markets have the unenviable task of trying to place a unit of measurement and value on the complex web of life: ecosystems and species and their interactions. One of the primary points of contention over Australia's Nature Repair Market, and biodiversity markets in general, is whether or not we can offset nature.

At first glance, a biodiversity offset could conceivably mean that we could smack a koala across the face in one forest and pay for another koala to keep its favourite tree in another forest. It's difficult to know for sure, though, because the rules for biodiversity markets are excruciatingly tedious and sufficiently vague.

Biodiversity credits (and biodiversity markets) aim to address biodiversity destruction by creating financial incentives for conservation. They function similarly to carbon credits but focus on funding the protection, restoration or sustainable management of ecosystems to preserve biodiversity. For example, a biodiversity credit might represent the preservation of a hectare of koala habitat in Australia or the restoration of coral reefs in the Caribbean. The intention is to provide an alternative funding stream for conservation, tapping into private capital that might otherwise remain unavailable.

But as with carbon, the questions are dizzying. Should the metric be species numbers, the presence of endangered species or the preservation of critical habitats? Which has more value – a forest in Australia or one in Colombia? How do you value ecosystems that hold cultural or spiritual significance? And importantly, why can't the government just directly pay to protect nature?

In Australia, the Nature Repair Market legislation finally passed both houses of government but with some modifications: no offsets allowed, and no use of the word 'market'. Whether businesses and investors will choose to voluntarily purchase biodiversity credits, or even how exactly the market will work, remains to be seen.

Poorly designed carbon and biodiversity markets – often riddled with loopholes and exploited by opportunistic third parties – have cast a long shadow over those doing meaningful work. As trust in the system erodes, the damage falls hardest on community-led projects that actually protect ecosystems, support people and reduce emissions. Broad-brush criticism ends up tarnishing all initiatives, including those delivering tangible results – like Indigenous-led conservation or local reforestation efforts. Worse, it risks cutting off funding for the very projects that should be scaled, pushing them further to the margins.

Rowan Foley and his team are part of an emerging Indigenous-led environmental market industry, one that breaks free from the carbon sector's limitations. A Wondunna man of K'gari (formerly known as Fraser Island) Badtjala people, Foley is a ranger by trade and the founding CEO of the Australian nonprofit Aboriginal Carbon Foundation (AbCF). First Nations peoples' 'new approach', as he likes to say, is 'only 65,000 years old', a nod to the rich history of Australia's first scientists and the traditional custodians of its lands and waters: Aboriginal and Torres Strait Islander peoples, the oldest continuous living cultures in the world.

According to Foley, as well as other experts, the carbon market suffers from 'carbon tunnel vision' and is a result of narrow European-centric thinking that often ignores broader cultural, social and environmental benefits. 'If the primary aim of land management continues to simply be producing carbon credits, then it's always going to fail,' he says.

Indeed, it is the narrow focus on carbon that has left many unconvinced of the benefits of projects across environmental markets. Foley says it's not enough to simply sequester carbon; you need to honour the land, understand it and, most importantly, respect the knowledge held by the people who have lived on and cared for it the longest. The AbCF is not your typical carbon market company – and they wouldn't call themselves that either; they see their work as community development. They connect Aboriginal communities, supplying carbon credits with councils and businesses seeking to offset their carbon emissions.

The foundation isn't just another player in the carbon market; they reject the 'eco-colonial models' that dominate much of the sector. Instead, the AbCF has developed the world's first Indigenous-owned and operated carbon trading platform, focusing on more than just carbon. Here, carbon credits aren't the prize – they're the bonus. Or, as Foley puts it, 'the icing on the cake'. 'The core benefits – environmental, social and cultural – are what truly matter,' he says.

At the heart of the AbCF's work is cultural burning – a practice woven into the wisdom and land management traditions of Aboriginal peoples for tens of thousands of years. Using small, controlled fires – known as 'cool burns' – early in the season, cultural fires reduce fuel loads, prevent catastrophic wildfires and encourage biodiversity. These fires release only a fraction of the carbon dioxide released by massive bushfires, striking a delicate balance between fire and renewal and making them an ideal method for generating carbon credits that are truly rooted in Country. Foley explains the practice of cultural burning with characteristic bluntness: 'Cool fires are the ones you can stand beside. Hot fires are the ones you're running from.' The result? Less carbon dioxide, less smoke and healthier ecosystems that can breathe again.

In Australia, cultural burning has been formally incorporated into the country's efforts to reduce greenhouse gas emissions through the 'savanna fire management method'. This approach, adopted as part of Australia's carbon market, is built on Indigenous knowledge to manage fires in a way that prevents catastrophic wildfires and reduces emissions. Despite this, the majority of savanna burning projects listed by the Australian government are carried out by non-Indigenous landowners who often follow only the technical guidelines, whereas Indigenous peoples are driven by deeper cultural practices that make these fires so effective and meaningful.[32]

'Fire is often seen as a threat,' says Foley. But this, according to him, reflects a 'European mindset'. 'For Traditional Owners, fire is a friend – a vital tool for caring for Country.'

'Cultural burning isn't just fire management,' he says. 'It's a holistic way of looking after Country.'

Not just that, it works.

In Cape York, for example, there's been a 50 per cent reduction in bushfires in areas managed with the savanna fire management method since 2012.[33]

Crucially, the money from projects driven by the AbCF flows directly to Traditional Owners. 'Our priority is, and always will be, Mob,' says Foley.

The savanna burning projects that Foley and his team support flip the typical approach to the carbon market on its head. It's about building an economy that suits Traditional Owners. 'Instead of forcing Traditional Owners to move to where the jobs are – in mining or tourism – the jobs are coming to them, where they can stay on Country, using traditional knowledge and skills to look after it. And people pay them to do that.'

The AbCF operates with a peer-to-peer, strengths-based model: 'blackfella to blackfella,' says Foley. This rejects the colonial notion of top-down 'help' for Indigenous communities. 'When you use a peer-to-peer strength-based approach, you're decolonising the whole process,' he says. This is about 'high-integrity' carbon credits, rooted in transparency, free, prior and informed consent and verified social, cultural and environmental outcomes. 'Good stewardship of the land must come first,' Foley insists. 'If you get that right, the carbon credits will follow.'

For Foley and his team, partnerships with Traditional Owners are essential to achieving environmental and social outcomes and must be authentic rather than token representation. He critiques the long shadow of settler-colonialism and the persistent 'problems narrative' that continues to justify external control over Indigenous lands and decisions.

'Indigenous Australians have legal rights and ownership of 58 per cent of Australia's land mass, where much of the biodiversity thrives,' he points out.[34] 'If you want to achieve climate or biodiversity outcomes, you need to work with and be led by Traditional Owners.'

Yet Foley knows that not everyone in the carbon market shares his commitment to integrity. 'Now, most people will do the right things with a carbon credit,' he says. 'Some companies will misuse a carbon credit,' but, 'that's on them.'

'If I go to Traditional Owners in Cape York and say fellas, you've got to stop burning Country because we're generating carbon credits and some people in the cities are misusing them. So, stop burning Country, stop doing traditional fire practices,' he says, the response would be 'Rowan, you've spent too much time in the sun.' They would say, 'We're here looking after Country, Rowan. This Country is beautiful. We're going to keep on looking after

Country, regardless of whether a carbon credit is involved or not.'

'Our values are true, our practices are right, and the benefits are local. That's high integrity,' he says.

When it comes to the commercialised, colonial and performative aspects of the carbon industry, Foley offers a vivid analogy. 'Sharks are good for the environment, they're a sign of a healthy eco-system. However, you should be careful not to be recycled into their habitat,' he says.

'My mother never allowed us to swim in the water, but we did bathe at low tide and catch fish on the coming in tide. You can live quite well with sharks, just be careful and respectful of their nature,' he says.

For Foley and his team, the AbCF's approach isn't revolutionary – it's a return to enduring, proven practices that are guided by 65,000 years of knowledge of Country. As Indigenous peoples work to heal the land, they're offering a blueprint for a more honest, culturally rich and ecologically sound path forward.

'We're shining a light on the amazing work of our Mob all over the country,' Foley says. 'It's about utilising our existing knowledge and Lore to connect and heal Country while developing new ways to build strong, healthy communities.'

'If you want to walk with us, walk with us,' he says simply.

Around the globe, quirky and inventive approaches are emerging to drive finance towards nature, climate solutions and communities. In South Africa, 'Rhino Bonds' link investors' returns to the survival and growth of black rhino populations. Kenya's 'Lion Carbon' credits channel funds into lion habitat preservation while offsetting carbon emissions. Ecuador's forests are safeguarded through a debt-for-nature swap, trading national debt relief for conservation commitments. While in Hawaii, coral reef insurance triggers rapid payouts after severe storms, providing funds to restore damaged reefs.

Despite these efforts, climate and nature remain critically underfunded – with funding towards biodiversity and nature trailing far,

far behind the funds that are allocated to climate. And very often, communities – particularly Indigenous communities – only see a fraction of that funding. To add insult to injury, most enthusiasm is reserved for financing activities that destroy nature.

For example, around $200 billion flowed into nature-based solutions in 2022.[35] Yet, $7 trillion is invested globally each year in activities that have a direct negative impact on nature.[36] This includes more than $2.6 trillion of government subsidies that provide direct support – often in the form of tax breaks – for deforestation, water pollution and fossil fuels.[37] The noble (and sometimes ignoble) attempts to channel investment into environmental activities – whether it be through carbon markets or rhino bonds – are exceedingly difficult to achieve when governments and financiers are still supporting, permitting and financing activities that destroy nature.

The logic underpinning the dominant approach to addressing climate and nature crises is often baffling. Indeed, investors could be forgiven for scratching their heads when the same government promising returns on koala protection credits is also handing out tax breaks to those who turn koala habitat into woodchips.

The most obvious way to protect nature and make environmental incentives work would, of course, be to have some actual rules. Environmental laws and regulations could set clear boundaries and outline a unified path to restoring nature, making it easier for investors and communities to trust that environmental investments aren't undermined by conflicting policies. Rules act like a compass, pointing everyone in the same direction. They ensure that environmental activities and investments actually work to protect the environment rather than just redecorate it. Regrettably, at this moment, that compass is directed very far away from nature.

Polly Hemming, director of the Australia Institute's Climate and Energy program, captures the dominant approach to nature protection aptly through the wisdom of her seven-year-old son. She says, 'If I asked my son how to save trees his response would not be, "Mum, what about creating a complex market-based mechanism based on an unprovable counterfactual." His response would be, "Mum, if you want to save trees ... don't let people cut down trees."'[38]

Perhaps a seven-year-old is best placed to point out the things we'd rather not admit: sometimes, our solutions aren't solving things. And often, they're less about tackling the root causes of our environmental destruction and more about clinging to the comforting thought that someone, somewhere is cleaning up the mess we're still enthusiastically creating. Sometimes, you really can't offset infidelity.

Chapter 9

Invisible Creatures: The Curse of Ugly

> What may be a distressingly hideous feature
> to humans often belies evolutionary genius.
>
> Sami Bayly[1]

The vulture is not a great beauty. Nor is the naked mole rat. Nor the wrinkle-faced bat. The proboscis monkey is, to the casual observer, growing a penis on its face. The wolffish looks like it would eat its young. The silver-headed antechinus has the sort of face you forget while you're still looking at it. And the purple frog has the visual appeal of an open wound. The blobfish looks like it was pulled from a clogged drain and, according to evolutionary psychologist Geoffrey Miller, 'looks like if you handled it … at the very least you'd get some kind of rash'.[2]

Of course, beauty is in the eye of the beholder. Or is it?

There is a well-known bias that exists among the world's species: the blessing and curse of the charismatic megafauna. These animals, also known as 'flagship species', are the large, headline-grabbing species that are used as symbols of conservation. In the grand theatre of conservation, charismatic megafauna like elephants, pandas and lions take centre stage. They are, in a sense, conservation's celebrities, blessed with traits that appeal emotionally and aesthetically to the general public: the soulful eyes, the plush fur, the air of majesty.

In short, these species make people feel more feelings. And when people feel more feelings, they give more – both in money and attention. Species that appeal aesthetically to humans – particularly colourful, large-bodied animals – receive more conservation attention and funding.[3] For example, a charity appeal featuring flagship species or charismatic megafauna is more likely to receive higher donations compared to those featuring non-flagship species.[4] This bias has also translated to conservation research, where large, charismatic species with high public appeal consistently receive the most attention.[5]

The panda is a delight. With a fluffy round belly and playful nature, it's charmingly clumsy and endearing. By contrast, the jewel wasp injects venom into the brain of a cockroach, partially eats it, then makes a home and has its babies inside the dead body. Then the babies hatch and eat each other's organs. The donor list for the jewel wasp is, unsurprisingly, short. The panda, on the other hand, has a full-time team of staff, graces magazine covers and draws substantial support from a wide base of donors. Securing financial and political support to save a panda is infinitely easier than rallying funds to protect a species that makes homes out of dead cockroach bodies, despite the latter's essential role in regulating populations of other insects.

Human perception is a powerful filter, shaping our relationship with the natural world and dictating which species receive funding and attention and which do not. Nature, it turns out, is bound by the same stringent beauty standards that plague humanity, and this impacts how conservation and protection efforts are prioritised. The result is that conservation of species can become a hierarchy driven by beauty (or charm) which excludes species that don't fit humanity's ideal standards of beauty.

One study reviewed 17,502 research articles published in four top conservation-focused journals. It found that mammals comprised seven of the ten most studied species between 2010 and 2019, and five of the seven mammals were charismatic megafauna: lions, panthers, tigers, wolves and leopards.[6]

Even the world's premier extinction barometer reflects this bias. The International Union for Conservation of Nature (IUCN) Red List of Threatened Species is a global information platform that assesses the risk of extinction for animals, fungi and plants. It is a key indicator of the health of the world's biodiversity. Not only that, the Red List guides conservation decision-making and is a crucial tool that governments and conservationists use to understand the biodiversity crisis and target conservation efforts towards threatened species. But, as the IUCN acknowledges, its Red List is heavily biased. Land species and forest ecosystems have been the most comprehensively assessed for extinction risk by the IUCN, and overall, there is a 'strong bias towards animals, rather than plants or fungi' and 'major gaps in understanding and assessment of invertebrates and freshwater and marine species'.[7]

This bias towards charismatic species has real-world implications. For example, freshwater species – heavily under-researched and under-funded in general – have suffered a dramatic 85 per cent decline in populations over the past fifty years.[8]

The world's invertebrates – from spiders to jellyfish to crabs – make up 97 per cent of all animal life on Earth. They are the quiet majority and unseen caretakers of the Earth: they pollinate the flowers that become our food, recycle the nutrients that enrich our soils and sustain countless other creatures in the delicate web of life. Without them, the world as we know it would simply unravel. Yet they only form 31 per cent of all animals that have been assessed on the IUCN Red List.[9] This immense data gap means that thousands, perhaps millions, of invertebrates could be fading towards extinction unnoticed.

The bias that humanity has towards species that seem 'cute' or 'noble' may be tied to our anthropocentric view of the world, where we subconsciously elevate creatures that we deem relatable or valuable to us personally. Anthropomorphism is our tendency to attribute human traits, emotions or intentions to non-human animals. We gravitate towards animals who mirror our own ideals of beauty, strength or virtue, often at the expense of creatures who are ecologically important but may seem specifically designed to activate our gag reflex. In this way, we prioritise the needs and experiences of creatures most like ourselves – animals with faces we can empathise with, eyes we can connect to and behaviours that are easy to relate to. While this tendency may stem from a genuine desire to understand and empathise with others, it also highlights our struggle to accept that animals may experience the world in ways fundamentally different from us.

Our relationship with nature and, indeed, our conservation priorities are deeply influenced by our evolutionary history, as well as cultural and social narratives, which have shaped which animals we value and, by extension, which animals we protect. Many ancient societies revered powerful animals as emblems of strength, divinity or wisdom, embedding these associations in their religions, myths, superstitions and folklore.* The eagle symbolised the might of the Roman Empire and was a 'point of singular focus and devotion for the men who carved out Rome's empire'.[10] Across Mesopotamia, lions were celebrated as symbols of courage, strength and nobility, tied to the power of kings and goddesses. Meanwhile, creatures like hyenas and vultures often skulked in the shadows of myth, cast as omens of death, decay or impurity.[11]

These cultural depictions were rarely uniform – different societies and time periods assigned varying, and sometimes even dual or contradictory, meanings to the same species. Nevertheless, these historical narratives played a powerful role in creating a hierarchy of species, shaping perceptions based on how they fit with human

---

* There is no single, static or straightforward interpretation of species across cultures and time periods. Animal symbolism is often multilayered and contradictory, evolving over time, and subject to reinterpretation.

ideals of beauty, esteem and power. This mindset has persisted throughout the ages, reinforced by art, religion, literature and even modern media, creating a lasting legacy of bias that shapes the way we view the natural world today.

The bat has become one victim of this bias. Over 125 years ago, Bram Stoker introduced *Dracula* to the world and unknowingly doomed the bat to stereotypes that would forever plague efforts to protect and conserve it. Ecologist Kelly Sheldrick says bats have long been tied to myth and the supernatural – creatures like vampires – and the COVID-19 outbreak only deepened that stigma.[12] As she says, 'People associate bats with drinking blood.'[13] It turns out that invoking fear or disgust does not endear donors or policymakers to the bat. This bias has had far-reaching consequences for the bat, and ecosystems at large. Bats play an essential role in pest control, pollinating plants and dispersing seeds. In fact, 'bats eat enough pests to save more than $1 billion per year in crop damage and pesticide costs in the United States corn industry alone'.[14]

In 2006, an amateur caver in upstate New York stumbled upon a haunting sight deep within the damp, shadowy walls of a cave. A cluster of hibernating bats hung in their usual winter slumber, but something was wrong. One of the tiny creatures had a strange white powdery substance clinging to its muzzle. It was an odd detail – small, almost easy to overlook – but it marked the beginning of something far more sinister. Soon, reports started trickling in from people who were noticing bats behaving strangely, waking from hibernation early in a state of serious starvation, weak and disoriented. Then, more grim discoveries were made. Reports came in of small brown lumps littering the ground of various caves across the USA: thousands and thousands of dead bats, all with a white substance on their nose. Dubbed 'white-nose syndrome', this deadly disease caused by fungus was killing bats in droves. Within five years, 6.7 million bats had died in North America due to white-nose syndrome.[15] The disease has killed 90 to 100 per cent of the bat population at some affected sites in North America.[16]

This was, according to experts, the gravest threat in memory to bats in the USA. 'We are basically monitoring and watching one of the greatest population declines through disease that's ever been recorded for a mammal species,' said Jeremy Coleman, the national white-nose syndrome coordinator for the US Fish and Wildlife Service, in 2012.[17]

Efforts to save the bat may have been hampered by charisma bias – the unfortunate arrangement where pandas get documentaries and bats get fungal diseases. According to Jordan Hadlock and Brent Z Kaup, researchers in environment and sustainability at William and Mary university, 'the economic and public appeal of charismatic species significantly influences their classification under endangered species frameworks', and can often 'influence what gets classified as endangered and how much funding a species receives for protection'.[18] They argue that the US *Endangered Species Act* (ESA) is designed to primarily focus on individual species protection rather than broader ecosystem health. This narrow focus complicates efforts to understand and manage complex wildlife diseases like white-nose syndrome – a disease that doesn't just attack bats, but weaves itself into the larger ecological web. Added to that, the ESA tends to prioritise species with economic or public appeal, meaning that non-charismatic species, like the much-maligned bat, may receive less funding and attention despite their critical ecological roles.

What does this mean? In many cases, species survival may be contingent on humanity's whims, tastes and ideas of beauty at the expense of 'uncharismatic' – yet ecologically vital – creatures whose existence is crucial to a functioning planet.

It's not just bats that have a PR problem. Any animal that incites fear, revulsion, apathy or boredom is susceptible to being overlooked. Perhaps no species has been more maligned than one that has existed for millions of years, predating even the dinosaurs.

'They are darlings,' says conservationist Valerie Taylor. 'They're wonderful to dive with, a big smile full of teeth.' She is, of course, talking about sharks. The grey nurse shark, according to Valerie, is 'gentle, polite, and not at all dangerous'. You need to be careful around the great white, though. 'The great white investigates the unusual. And they don't have hands, so they use their teeth,' she says matter-of-factly.

Valerie Taylor is a protector of sharks, a marine pioneer, a filmmaker and an icon of the ocean. She has swum into the belly of a whale carcass with a school of oceanic whitetip sharks during a feeding frenzy. She has been a pack member to great whites. She is a fearless campaigner for ocean protection. She's had museum exhibitions in her honour, marine parks established in her name and documentaries that came to life through her and her late husband, Ron's, work. They worked with Steven Spielberg and their filmmaking both inspired and featured in the 1975 blockbuster movie *Jaws*. Valerie is the reason the grey nurse shark is a protected species. She has been bitten and bumped and nudged and caressed by sharks. And she has loved every minute of it. 'I have had a great life,' she says.

Valerie dived and documented the ocean and its prehistoric creatures for more than sixty-five years, most of them with Ron, even as arthritis set in – 'thousands and thousands of dives,' she says. She began her underwater career as a spearfisher, hunting in the same seas she would later protect – often the only woman on the boat. Alongside Ron, Valerie evolved from hunter to defender, devoting her life to understanding, documenting and conserving marine life.

In the 1950s, Valerie and Ron began capturing life beneath the waves, fascinated especially by sharks. Their remarkable underwater footage soon attracted Hollywood's attention. Steven Spielberg, looking for authentic footage of great white sharks for the production of *Jaws*, hired Valerie and Ron to film in South Australia. Spielberg was so captivated by their footage that he had the script altered to include more of it, ultimately creating one of *Jaws*' most suspenseful scenes. Valerie and Ron, to their delight, found themselves working on what was to become the highest grossing movie of its time. Following the film's release, the couple went on a tour funded by Universal Studios to educate the public about sharks, aiming to dismantle harmful myths that were circulating in the movie's wake.

'We told the public not to be afraid,' Valerie said. 'Jaws was a mechanical shark and a fictitious story. People didn't have to stay away from the beaches. People were already fearful of sharks. The media had a lot to do with it. The saying at the time was, "The only good shark is a dead one." Ron used to say, "You don't go to New York and expect to see King Kong on the Empire State Building. Neither should you go into the water expecting to see Jaws."'

Fear of sharks has fuelled policies and practices that are devastating to shark populations – and, by extension, to the marine ecosystems essential to life on Earth. Despite the fact that sharks are critical to maintaining ocean health, the global response to sharks has been largely punitive. Fear-fuelled and reactionary initiatives like shark culls, beach netting and so-called 'shark control' programs have led to the indiscriminate killing of these prehistoric apex predators, driven largely by deeply ingrained biases and misconceptions and a tendency to project human emotions onto creatures that don't fit our idea of 'friendly'.

Each year, around the world, sharks are responsible for an average of ten deaths. Notably, these incidents typically involve just a handful of specific shark species (out of more than 500), and most are the result of curiosity or a case of mistaken identity, not malice. Statistically, your lifetime risk of dying in a shark attack is astronomically low compared to just about anything else.[19] In fact, you're far more likely to die by choking on a hot dog[20] or from taking a selfie (259 people died taking selfies between October 2011 and November 2017).[21] For a little perspective, more people die each year from being hit by a flying champagne cork (twenty-four deaths) than from shark attacks.[22] As for the animal kingdom's true villains, sharks don't even make the list of the top ten species most dangerous to humans.[23] Mosquitoes cause one million deaths per year. Dogs cause 30,000 deaths per year. Freshwater snails cause 20,000 deaths per year. Even cows kill more people than sharks.[24, 25] But, as Valerie says, 'people don't like the thought of being eaten alive'.

When it comes to sharks, Valerie is emphatic: their behaviour is not personal. 'They aren't here to hurt you!' she says. 'They're bottom feeders. Their mouth is under their head! They'd have to turn upside down to eat you!' To her, the loathing towards sharks is unthinkable and unfounded. 'A shark doesn't come up on the beach and chase you across the road. The decision is yours. You make the choice to go into their domain. You caused it,' she says. 'Swim in a swimming pool or a lake or river! Don't blame the shark.'

The decline in shark populations is not a small or isolated issue; it's a systemic crisis. Sharks, which have survived all five mass extinctions, are now spiralling towards complete collapse under humanity's assault due to a combination of overfishing, bycatch and targeted hunting. For 450 million years, these primordial predators roamed the seas, existing even before the first trees. Yet in a mere half-century, humanity slashed their populations by a staggering 71 per cent.[26]

In certain regions, species like hammerheads and great whites have seen populations decline by up to 90 per cent over the past few decades.[27]

As Valerie says, sharks play a specific and crucial ecological role: as apex predators, they maintain the health of marine environments by preying on the weak. 'The great white shark eats large things. Its regular prey is the old, the sick, and the unwary. It keeps the species strong. It's the job nature put it there to do and they do it very well,' she says.

Sharks aren't just the villains of horror movies; they're intelligent, curious creatures, with memories and instincts as finely tuned as any other animal – if not more so. 'They're not just a piece of swimming meat,' Valerie says, pointedly. 'They have a brain, and they have a memory, and they know what's bad and they know what's good. They're just not given their due.'

She reflects with some urgency on the state of the oceans. 'Everybody should be interested in the marine world. The marine world dies, we die. There is no two ways about it. All life on the planet comes from the ocean originally. That's a scientific fact, and we treat this mother of all life with great disrespect,' says Valerie. 'And unless we change our destructive ways we are going to pay the ultimate price.'

But, as Valerie knows, this fear and bias towards sharks isn't insurmountable. In the 1970s, she says, the grey nurse shark was inaccurately branded as a killer. 'For some reason, media, who loves to be dramatic, said grey nurse sharks were attacking people.' This was completely false, she says. In 1984, Valerie made history when she campaigned successfully for the grey nurse shark to become the world's first protected shark species. Today, she says, 'they're big money makers; tourists come from all over the world to see our beautiful grey nurse', she says. Not to mention, they're 'gentle darlings'.

Valerie had her final dive in November 2023. She says her arthritis has gotten too bad. But her conservation efforts have left a lasting legacy, which she continues to nurture today. The area where she and Ron filmed their *Jaws* sequences is now a marine park named in their honour. The park is an internationally significant site for the protection of great white sharks.[28]

She has had the most incredible life, she says. 'It has been the greatest adventure.'

'By the way,' she adds. 'If you ever got to get bitten by a shark, make sure you are working for Hollywood, okay? They have the best plastic surgeons.' And, as she reminds people, a shark bite is never personal – they're just curious. And they really would have to turn upside down to eat you.*

---

* This comment refers to the position of a shark's mouth, which in most species is located on the underside of the head. To bite something above them – like a swimmer at the surface – sharks may need to tilt or roll their bodies, particularly bottom-associated species such as wobbegongs or nurse sharks.

Public perceptions of species can both help and hinder conservation efforts. When it comes to the shark and the bat, conservation organisations are fighting an exhausting battle to grant these species the most basic protections.

Sadly, conservation organisations are bound by public perception of species and, in some cases reluctantly and sometimes inadvertently, perpetuate them. It is an endless cycle that these conservation organisations are well aware of. Faced with limited funds and an increasingly worrying extinction crisis, conservation campaigns are often forced to strategically select charismatic species that align with public preferences in order to drive donations. And it works.

The public adores the koala and inadvertently perceives it to be more worthy of protection than, for example, the vulture, which urinates on itself to keep cool before feasting on a rotting carcass. By drawing on the emotional appeal of the koala, conservation organisations can raise much-needed funds to carry on their work. There is a striking difference between the funds a photo of a koala can raise and the funds a photo of a vulture can raise, despite both species facing extinction.

The media also reinforces this cycle of bias. After all, a story featuring a polar bear or orangutan is much more likely to be read and shared than a story of a naked mole rat.

Yet as Helen Buckland, CEO of the Sumatran Orangutan Society, says, flagship or charismatic species can also be considered 'umbrella' species – 'by protecting their habitat, a huge range of other species are also protected', she says.[29] Flagship species can help to shine a spotlight on often distant wild landscapes and the rich biodiversity that they support. She argues that:

> Very few, if any, conservation organisations are working to save one individual species to the exclusion of all others.[30]

Despite this, the evidence for 'trickle-down conservation' is mixed. It isn't that charismatic species are getting *too much* funding. It's that right across the board, species and ecosystems are underfunded. The most iconic and charismatic species in the world are not immune to extinction. As researchers note, the conservation status of the ten most charismatic species in the world is grave.[31] Iconic species like tigers, leopards, giraffes and polar bears remain highly vulnerable to extinction, despite their profiles. Added to this, conservation organisations wrestle with the never-ending challenges of securing funds, vying for donor attention and attracting and retaining talent in a landscape where donors often baulk at the idea of nonprofit staff even earning a competitive salary. Having to navigate the public's fixation on animal charisma and beauty presumably only adds to the already staggering weight on their shoulders.

Fortunately, the blobfish catapulted to stardom with its gelatinous sag to disrupt society's rigid beauty standards, and with it, became a veritable ambassador for all things ugly. Any beauty the blobfish might have has, regrettably, been lost in translation. A deep-sea denizen, the blobfish favours function over form. It has no muscles to actively hunt, but rather, it passively feeds by floating and swallowing whatever passes it by.[32] Commentators have compared it to an 'anthropomorphised piece of colon'[33] and a 'frowning pile of goop'.[34] Yet this deep-sea fish from Oceania has become an icon for the uncharismatic species of the world, captivating the world with its absence of charm.

In 2013, the Ugly Animal Preservation Society embarked on a noble mission: to find and amplify the most underwhelming members of the animal kingdom. The society launched a global contest to identify the world's ugliest endangered animal. It was like a beauty pageant in reverse, where each creature's grotesquerie was celebrated rather than concealed. Thousands of people from around the world took to the polls to vote on their preferred ugly animal. In what was a nail-biting contest with highly competitive election campaign videos featuring the finalists, the winner was finally announced on 12 September. The blobfish had been voted

the world's ugliest animal, defeating the likes of the 'scrotum frog'*
and pubic lice. Simon Watt, the society's president, said, 'We've
needed an ugly face for endangered animals for a long time and
I've been amazed by the public's reaction. For too long the cute
and fluffy animals have taken the limelight but now the blobfish
will be a voice for the mingers who always get forgotten.'[35]

Some might question whether a blobfish or scrotum frog has the
same right to be saved as a tiger, but the Ugly Animal Preservation
Society would argue that it's not really up to us to determine
which species get to stick around. As Watt and his unique society
are teaching us, the world is richer – and a bit stranger – for the
survival of its so-called ugliest inhabitants.

While the 'curse of ugly' places aesthetic and emotional judgements on animals, impacting how society values and prioritises their conservation, so too does the 'curse of cute'. As it turns out, being beautiful has a dark side.

In 2009, a video appeared online that would go on to shape the fate of an entire species. It featured a slow loris – a small, wide-eyed primate found across Southeast Asia – raising its arms as a human gently 'tickled' its sides. The creature's enormous eyes, small hands and hesitant movements made it an instant hit; viewers around the world cooed over the loris, declaring it 'adorable' and 'cuddly'. In a matter of weeks, the video went viral, racking up millions of views on YouTube and other platforms. To the untrained eye, it looked like the loris was enjoying the attention. But for conservationists, alarm bells were ringing.

Anna Nekaris, a professor of primate conservation and one of the world's leading loris experts, saw the video as a flashing red warning sign. What seemed like 'cute' behaviour was, in fact, a sign of distress.[36] The slow loris was not 'enjoying' being tickled; it was terrified. Its raised arms were not an invitation to play but a defensive

---

* Also known as the Titicaca water frog.

posture, a natural reflex when the animal feels threatened. For this small, nocturnal creature, the bright lights and human contact were torture. But the damage was done. The viral video sparked an explosive demand for slow lorises in the exotic pet trade. Social media was flooded with comments from people eager to own one as a pet, while influencers and celebrities shared the clip, further glamourising the idea of having a 'pet loris'. What followed was a surge in poaching across Southeast Asia, where hunters targeted the wide-eyed primates to sell them to unwitting buyers across the globe.

This rise in demand was catastrophic for lorises. Lorises are nocturnal and solitary animals adapted to the dense forests of Southeast Asia. They rely on darkness to hunt and avoid predators. But as the illegal pet trade grew, poachers found the animals easy to capture. Unlike other primates that can flee into the trees, the slow loris moves, well ... slowly. Often, hunters would simply pluck them from branches and cage them for transport. Once captured, lorises are subjected to brutal treatment to prepare them for life as pets. Their sharp teeth – one of their only defences – are clipped or pulled out to prevent biting, a procedure that often leads to infection, immense pain and even death. In captivity, they are fed improper diets, exposed to harsh light and deprived of the dense forest canopy they need to thrive. The stress of this unnatural environment leads to high mortality rates; few lorises survive long as pets.

The original video's unintended consequence is still felt today. Demand remains high for these 'adorable' creatures, especially on social media platforms like Instagram and TikTok, where videos of lorises 'smiling' or 'dancing' continue to attract millions of views. Conservationists have pleaded with these platforms to remove videos showing lorises being kept as pets, but responses have been slow, absent or largely ineffective.[37] 'Almost always when [lorises] are pets outside of their range countries, it's because they've been illegally smuggled,' says Nekaris.[38] It's not cute – it's exploitation masquerading as entertainment.

Now the slow loris is on the edge of extinction. All species of loris are classified as vulnerable, endangered or critically endangered. What began as a viral video turned into a nightmare for these gentle creatures, revealing the disturbing power of social media to influence demand for exotic pets. Conservationists hope that some day, the loris can be seen not as a cute 'pet' but as the remarkable and complex wild animal it truly is – best admired from afar, in its natural habitat, and far from the harsh lights of a viral video.

This 'curse of cute' isn't a new phenomenon, but social media has taken it to new, troubling levels. And it doesn't stop with lorises. Tigers, monkeys, hedgehogs and even otters are routinely stripped of their wildness and treated as props or pets, paraded in viral videos and exotic 'pet cafes' where people pay to pet and feed them. Many of these species suffer in captivity, their complex needs disregarded in favour of spectacle. Experts warn this social media trend is driving demand for exotic pets and normalising the exploitation of wild animals as pets, props and entertainers. Too often, it leads to harm and abuse – all for followers, clicks and profit.

The panda may be the quintessential example of both the blessing and curse of being a 'charismatic megafauna' species. The panda is, quite literally, the global face of conservation. It is an icon, a global brand, a political tool. Pandas are an anthropomorphic goldmine. They are clumsy, perpetually confused and spend most of their time eating and falling off things – behaviours that, to the human mind, register as endearing rather than deeply impractical for an animal on the brink of extinction. If the bias of beauty determines which species get noticed, then the panda is the ultimate beneficiary. But being noticed comes at a cost, and for a species so endlessly adored, the line between protection and commodification has become increasingly difficult to distinguish.[39] After all, it's one thing to be beautiful. It's another to survive it.

The Chinese panda conservation program is renowned as a global symbol of wildlife conservation leadership. Pandas are carefully managed 'ambassadors', loaned to foreign zoos in exchange for financial support and, often, political goodwill. American zoos reportedly pay up to $1.1 million annually to rent a pair of pandas from China, and cubs generate massive visitor numbers, merchandise sales and media coverage. Meanwhile, China collects substantial rental fees and, according to reports, offers bonuses to its breeders for every cub born. The panda attracts disproportionately large amounts of funding compared to other species. This funding ostensibly supports habitat preservation and species conservation but, as a recent investigation by *The New York Times* reveals, all is not what it seems.

China's panda conservation program was launched in the 1990s, the intent being for future generations of pandas to be released into nature. But thirty years into its conservation program, more pandas have been removed from the wild than released. Instead, the program has focused on captive breeding, with more than 700 pandas now living in enclosures. The exact number of wild pandas remains elusive. [40]

The commodification of pandas goes beyond mere zoo exhibits. Given the high public interest in panda cubs and pandas' reluctance to mate in captivity, zoos often resort to artificial insemination to boost birth rates, a practice that has raised serious ethical concerns. *The New York Times* revealed that procedures on the pandas sometimes involve electrified probes, ketamine-based sedation and repeated inseminations within a single breeding cycle. 'We choose how they breed. If they don't want to breed, we make them breed,' critiqued Dr Heather Bacon, a veterinarian and director of the Bear Care Group. 'All we're doing,' she said, 'is producing more pandas to live in captivity and have those same experiences over and over again'.[41] In some cases, anaesthesia was used recklessly, leaving pandas partly awake and in pain. 'They used dangerously high voltages and too many stimulations on male Ping Ping after we left,' wrote JoGayle Howard, a scientist at the US National Zoo, in her 1999 research notes.[42] One panda, sedated with ketamine, even came close to waking mid-procedure.

For pandas, charisma is a double-edged sword. They are beloved but commodified, turned into symbols of conservation that are rarely actually conserved in the wild. While captive breeding plays a huge role in expanding scientific understanding and often provides an insurance population for vulnerable species, the reality for this program is the creation of a generation of pandas who know little of their natural behaviours or habitats, and likely never will. Zoos, the investigation noted, continue to clamour for pandas, benefiting from a steady influx of visitors, while pandas are paraded for profit in air-conditioned domes in Qatar and theme parks in China. As the article reported, 'What the program does best is make more cubs for zoos.'[43]

Anthropomorphism, this human tendency to project our own emotions and social values onto animals, shapes far more than just *which* species we protect – it also determines how we interact with them, for better or worse. While the panda showed us the power and peril of human projection, it was the chimpanzee that forced us to examine what that projection really means.

A chimpanzee never forgets the face of an enemy. And chimpanzees don't just get mad – they get even. It's an eye for an eye, and then some. In chimp society, no grudge goes unanswered, and no slight is too small to provoke retaliation.

Wildlife vet Dr Chloe Buiting counts chimps among her favourite species to work with. They are 'incredibly violent and vicious,' she says affectionately, 'scary as hell' and 'alarmingly intelligent'. Chimps are our closest living relatives and are susceptible to human disease. Which means that sometimes, Buiting must dart a chimp to undertake disease investigation. It's a delicate operation that requires cunning, speed and nerves of steel. The first problem is that chimps don't like to be darted. The second problem is that a vet must outsmart the chimp, and Buiting isn't quite convinced that is possible. 'They are a whole different kettle of fish,' she says.

The first rule of darting a chimp is simple: don't let them see you. The second is: don't let them see you. For a vet working with chimps, disguises aren't optional – they're survival gear. If a chimp gets a glimpse of your face, you can forget working with them again – they'll hold a grudge for life. 'When you're going to dart a chimp, you have to wear a disguise,' she explains earnestly. 'They will remember you and will target you for the rest of their lives,' she says.

Buiting says you also have to be quick to dart a chimp and stand as far away from them as possible. Unfortunately, many a vet has underestimated a chimp and lived to regret it. You see, if you dart a chimp while in close proximity, the chimp will yank the dart from its body and hurl it back at you with alarming accuracy. 'They have shockingly good aim,' Buiting says, laughing. 'I haven't been darted yet by a chimp, but it is coming. For sure.'

Buiting and her team now employ a variety of tactics when handling and caring for chimps. Aside from their lust for revenge, chimps are often violent and will rip each other to shreds in one of their many brawls. When this happens, her team performs surgery to stitch up the combatants, which means placing a chimp under anaesthetic. But post-op care comes with its own challenges: chimps wake up angry. Very angry. And they love nothing more than undoing a vet's hard work. 'They'll rip at their stitches as soon as they wake up,' Buiting explains. Thankfully, she and her team found a solution to this: manicures. 'We paint their nails. A manicure for them really. They often wake up from surgery with pink sparkly nails,' she says. 'And they sit there, so fascinated by their nails that they leave their incision site alone for at least a week,' she says. Chloe says other vets make fake incision sites to distract from the real incision, but she personally prefers the manicure method.

With an unwavering enthusiasm for violence and signature pink sparkly nails, chimps are the loveable troublemakers of the conservation world – and they were also the unwitting stars of a philosophical debate that raged through the 1960s and 1970s. And it all started when a woman named Jane Goodall did the unthinkable: she named a chimp.

Jane Goodall is the world's most influential chimp expert and advocate. And in the 1960s, she up-ended everything we thought we knew about non-human animals. You see, Goodall took an unorthodox approach to her research: rather than assigning chimps a number, as was scientific convention, she named them. Then, she started attributing human personality traits to them; they were playful, aggressive, affectionate. At the time, the prevailing view was that attributing emotions, thoughts or individuality to animals was a sentimental trap – a distortion that serious researchers should avoid at all costs. Animals were to be studied dispassionately, catalogued in numbers and behaviours, not in narratives or relationships. Goodall was, as she says herself, 'ascribing human characteristics to nonhuman animals and was thus guilty of that worst of ethological sins – anthropomorphism'.[44]

But on 4 November 1960, Goodall witnessed something extraordinary. A chimp, whom she had named David Greybeard, was stripping leaves from a twig and then inserting it into a termite mound to extract the insects.[45] This was the first time someone had observed a non-human species using a tool. This discovery sent shockwaves throughout the world. Goodall's findings challenged centuries of thought about what it meant to be human and up-ended foundational beliefs about the distinction between humans and other species. In her unconventional and anthropomorphic approach to chimp research, Goodall set off a seismic shift in anthropology, primatology and understandings of animal intelligence. Not just that, her work spurred the publication of hundreds of scientific papers about chimpanzees, laid the groundwork for enhanced conservation efforts, bolstered animal welfare advocacy and influenced ethical standards in scientific research on primates.

Jane Goodall challenged the assumption that attributing emotions to animals was inherently unscientific. By viewing animals through her own human lens, she opened a gateway to better understanding of, and empathy towards, other species. Importantly, she also pushed the world to move beyond the simplistic notion that anthropomorphism is either entirely right or entirely wrong. And in the process, she revealed unexpected insights into human evolution, history and culture.

Anthropomorphism itself isn't inherently good or bad – it shapes conservation and our relationships with other species in complex ways, sometimes in ways we intend and sometimes in ways we don't. Indeed, our human instincts and projections can be powerful tools for strengthening our relationship with nature. Yet they carry with them a web of biases – biases that shape how we value species and, in turn, how we act towards them. For some, this bias draws us into action, leading to greater protection efforts, like those who rally the world around the chimpanzee or the blobfish. For others, it leaves a veil over creatures slipping into obscurity – such as the dwindling populations of insects – or drives destructive practices, like the installation of shark nets or the mass exploitation of species like the loris.

Viewing animals through a human lens is both an instinctive impulse and a response shaped by social and environmental forces. It stirs compassion and allows us to extend our empathy across the boundary of human experience. But recognising our inherent biases can add even more depth to this empathy. By shifting our understanding of what is worth protecting, we can honour life forms not just for their appeal, but also for their ecological significance and intrinsic right to exist. And in doing this, we can embrace a broader, more even-handed approach to nature that sees values beyond charisma, aesthetics and human attributes.

For some of us, this might start with choosing to advocate for a species that gives birth inside the dead body of a cockroach.

Chapter 10

The Rights of Nature: The River That Is a Person

> E rere kau mai te Āwanui, Mai i te Kāhui maunga ki
> Tangaroa. Kō au te Āwa, kō te Āwa kō au.
> (The great river flows from the mountains to the sea.
> I am the river, the river is me.)
> Whakataukī (Māori proverb)[1]

In 1545, a colony of beetles was summoned to appear before court, where they were accused and indicted for their crimes. The beetles – or, more specifically, the weevils – were accused by wine growers of ravaging the vineyards of Saint-Julien-Beychevelle, a hamlet in the Bordeaux region of France. Counsel was appointed for both the plaintiffs and the defendants, and both sides presented their cases with great solemnity. The judge, having heard the case, decided, in this instance, to show mercy to the insects. He proclaimed it would be 'unbecoming to proceed with rashness' and recommended the plaintiff appeal to 'the mercy of heaven and to implore pardon for our sins.'[2] Without delay, he instructed the public to pray, whereby they should 'turn to the Lord with pure and undivided hearts'. And with that edict, the public indeed offered prayers for the insects. They gathered on 20, 21 and 22 May for high mass, which involved a 'solemn procession with songs and supplications round the vineyards'.[3]

According to the official record, the decree was carried out in good faith and the insects, undoubtedly sensing the gravity of the situation, promptly vanished from the area.

But divine intervention, it seems, was fleeting. Thirty years later, the weevils returned to Saint-Julien and 'resumed their depredations', causing 'incalculable injury'.[4] Winemakers were compelled to take legal action against the weevils once again. This time, the case proceeded in earnest and went to trial, with strict adherence to legal procedure. Regrettably, the original defence lawyers had died since the previous hearing, compelling the presiding court to appoint new counsel on behalf of the insects, who would defend them with great dignity.[5]

Pierre Rembaud, the weevils' newly appointed defence counsel, made a motion to dismiss the case, arguing that God had created the insects before humans and that they were merely 'exercising a legitimate right conferred upon them at the time of their creation'.[6] Therefore, he said, his client had a prior right to the vineyards. The plaintiff was unsatisfied with this defence and insisted that the court visit the site of the weevil invasion, observe the damage and then proceed with a verdict befitting the severity of the crime: excommunication.

Months passed, visitations to the site were undertaken and the legal battle intensified until, after seemingly endless months of deliberations, it was decided that a meeting of the townspeople and affected parties should be held to expedite the process and settle the matter once and for all. On 29 June 1587, a public meeting was called at noon immediately after mass on the great square of Saint-Julien to which all townspeople – as well as the plaintiffs and defendants – were summoned by the ringing of the church bell. It was decided that a section of land would be set aside for the defendants (the weevils), which would be sufficiently suitable to provide them with their required sustenance without devouring and destroying the town's precious vineyards.

The selected site was named 'La Grand Feisse' and was chosen based upon a 'topographical survey, not only as to its location and dimensions, but also as to the character of its foliage and herbage'.[7] The plaintiffs and their counsel were permitted to inspect the site, and it was hoped they would 'express themselves satisfied with such an arrangement'.[8] To the weary townspeople, it seemed the prolonged feud between man and insect might be coming to an end.[9] Alas, the defence counsel and the weevils rejected this generous offer, claiming La Grand Feisse was insufficiently fertile. They returned to court, refusing the 'sterile' sanctuary and demanding dismissal of the case.[10] The plaintiff's attorney insisted that the land was, in fact, suitable and insisted upon adjudication in favour of the complainants. The judge decided to reserve his decision and appointed experts to examine the site and submit a written report on its suitability.

After nine months of meticulous proceedings, the final decision of the case was regrettably lost to time 'by the unfortunate circumstance that the last page of the records has been destroyed by rats or bugs of some sort'.[11]

Criminal prosecution of animals was a common occurrence in the medieval courts of Europe. All manner of species were put on trial: snails, flies, worms, beetles, caterpillars, dolphins, horses, birds, wolves, chickens, goats, rats, dogs, pigs and cows. The crimes they were accused and sometimes convicted of ranged in severity: chickens disturbing the peace, insects charged for property damage, rats charged with robbery, pigs accused of manslaughter, cows charged with aggravated assault. Often, they would be summoned to appear in court at a specified time to answer for their conduct, as was the case for an army of caterpillars in northern Italy in 1659 who were 'ordered to appear in court on the twenty eighth of June, at a specified hour, where they would be assigned legal representation'.[12] The caterpillars, according to official records, were accused of 'trespassing upon the fields, gardens and orchards and doing great damage therein'.[13] A summons had been duly issued and copies notifying the caterpillars of this summons were posted on trees in the affected towns.

The trials and sentencing of animals were conducted with full pomp and ceremony, rigidly adhering to legal precedent and established procedure.[14] Animals would have the opportunity to appear in court and plead their case. They would be appointed legal counsel, who would defend their clients with great honour. Evidence was heard on both sides, witnesses were called, and the judge ensured that cases were reviewed with careful consideration, 'befitting the dignity and solemnity of the occasion'.[15] Such prosecutions often resulted in the accused being excommunicated and sometimes sent to the gallows. But occasionally, an animal would be acquitted of charges on good behaviour.

In his 1906 book *The Criminal Prosecution and Capital Punishment of Animals*, Edward Payson Evans documented more than 191 prosecutions and excommunications of animals between the 9th and 20th centuries. This list referenced the excommunication of a glacier for the crime of advancing on the homes of inhabitants of Mont Blanc in 1875.[16]

Such medieval trials and tribulations reveal the centuries-old struggle to define humanity's relationship, both legally and morally, with the natural world and the many creatures that inhabit it. And importantly, they may illustrate an early – if strange – interpretation of legal personhood, a concept that is now challenging our modern understanding of legal rights of non-humans. Today, a complex dance between animals, nature and the law is weaving its way through courtrooms, legislation and communities and prompting the question: should nature have rights?

To understand the evolution of the relationship between nature and the law, it is important to first be aware that you, your grandmother and McDonald's all have something in common: you are all considered legal persons. Legal personhood is the recognition by (Western) law of an entity – human or non-human – as having rights and responsibilities, including the ability to hold property, enter into contracts and sue or be sued. Legal personhood is a Western construct that traces back to the Romans, who decided to distinguish 'persons' from 'things' – a helpful distinction unless you're a thing. In the Middle Ages, canon law expanded this idea by allowing religious groups like monasteries to act as 'persons' in legal matters. Christian beliefs nudged this idea further, shaping the view that (certain) individuals naturally hold rights. In the 19th century, corporations were granted legal personhood, enabling them to wield many of the same rights as the select group of human beings who were afforded full legal rights at the time.

Today, this framework born of ancient and ecclesiastical traditions underpins the Western legal system – the most influential globally. Yet how legal personhood is applied and interpreted varies widely across jurisdictions – and very often, it reveals and reflects the biases of those in power. As legal scholar Jens Kersten says, legal personhood is 'often driven by the interests of the ruling class'.[17] Historically, these 'interests' translated to a highly selective view of personhood. Until relatively recently, the full benefits of legal personhood were reserved primarily for white property-owning men. Of course, if you were an enslaved individual, a woman, a working class man, a person with a disability, an Indigenous person, a Black person, a brown person, an LGBTQIA+ person or a child, you would likely not have been considered a full legal person at one point or another, in which case you would broadly (or sometimes strictly) be considered property.[18]

Over time, as legal rights expanded to include a broader swathe of humanity, people began questioning whether the boundaries of legal personhood should be pushed even further. This line of thinking gave rise to a series of provocative questions, chief among them a now famous 1972 article by law professor Christopher Stone titled, 'Should Trees Have Standing? Toward Legal Rights for Natural Objects'.[19] Stone argued that natural objects such as trees should be bestowed with legal rights. He noted that whenever the circle of personhood expands to any new prospective legal persons, there's a sense that it's ridiculous and 'unthinkable'. He drew parallels to the United States Supreme Court, which could 'straight-facedly tell us' that an enslaved Black man could be denied the rights of citizenship due to being deemed 'a subordinate and inferior class of [being], who had been subjugated by the dominant race',[20] and that the highest court in California 'explained that Chinese [people] had not the right to testify against white men in criminal matters because they were "a race of people whom nature has marked as inferior, and who are incapable of progress or intellectual development beyond a certain point."'[21] Given this, he said, 'I am quite seriously proposing that we give legal rights to forests, oceans, rivers and other so-called "natural objects" in the environment – indeed, to the natural environment as a whole'.[22] He went on to propose a guardianship model, wherein natural features have rights of their own, but humans or 'guardians' can speak and act on their behalf.

Shortly after Stone's article was published, the US Supreme Court rejected a lawsuit by the Sierra Club that aimed to stop Disney from building a resort in the Sierra Nevada mountains. The case prompted a famous dissent by Justice William O Douglas, who cited Professor Stone's work, noting that 'both corporations and ships had long been parties in litigation, despite being artificial and inanimate. So it should be as respects valleys, alpine meadows, rivers, lakes, estuaries, beaches, ridges, groves of trees, swampland, or even air that feels the destructive pressures of modern technology and modern life.'[23]

This thinking formed part of the foundation for a growing global movement that began calling for the legal rights of nature. The movement proposes that Earth's natural elements, from forests to mountains to rivers, are not simply resources for human use – they are entities with inherent rights to exist and thrive. They call for the law to recognise that nature has value beyond its utility to humanity.

Yet this idea is hardly a new one. Around the world, for millennia, many Indigenous peoples have viewed rivers, mountains, forests and other natural entities as relatives, ancestors or deities, with whom they have reciprocal relationships and obligations. These relationships and views of nature far precede any Western frameworks and recognise natural entities as living beings with intrinsic rights.

As Michelle Bender of the Earth Law Center says, 'To be clear, the Rights of Nature movement is learning from an Indigenous worldview rather than the other way around. The use of rights of nature can help to reorient the law around Indigenous relationships and responsibilities to nature.'[24]

Dr Terri Janke, a Wuthathi, Yadhaigana, and Meriam woman and lawyer specialising in Indigenous cultural and intellectual property in Australia, says that this relationship is deeply rooted in kinship and caring for Country:

> First Nations people have cared for Country since the beginning of time – time immemorial … It's a fundamental belief of First Nations people that Country is kin. So, you treat the earth, the waters, the plants, the animals, each other with that reverence – that you are part of an ecosystem, that we as humans are part of that circle of life and that we are not at the top of the pyramid that wield our power over all things.

She says that 65,000 years of Indigenous cultural practice around caring for Country offers a blueprint for restoring the damage done to the environment and communities over the past two centuries. Central to this is recognising the link between ancient knowledge and care for Country. Indigenous peoples' culture and heritage, she says, stem from 'knowledge of Country, story, songlines, art – all of that is part of our kinship obligation'. Indeed, Indigenous knowledge systems, deeply attuned to the rhythms of lands and waters, recognise the interdependence between humanity and nature and hold answers for how we can better care for nature. This way of thinking is being increasingly reflected in global legal movements to grant rights to nature.

And that is how a river came to be a person.

THE RIGHTS OF NATURE: THE RIVER THAT IS A PERSON

E rere kau mai te Āwanui, *Mai i te Kāhui maunga ki Tangaroa. Kō au te* Āwa, *kō te* Āwa *kō au*: 'The great river flows from the mountains to the sea. I am the river, the river is me.' This is the proverb that defines the Māori iwi (tribe) of the Whanganui River and region. Along its length, this river has sustained and shaped the lives of Te Ātī Haunui-a-Paparangi people for over forty generations. It is the longest navigable river in Aotearoa New Zealand.

In March 2017, after more than 150 years of legal battle, the Whanganui River became the first river in the world to be given the same legal rights as a human being. Dr Rāwiri Tinirau, advisor on Māori and Indigenous human rights, said, 'The river is entrenched in our history. It has ancestral names. It is viewed as an ancestor. We follow the river to get home.'[25]

Tinirau is the deputy chair of a trust that is responsible for the health and wellbeing of the Whanganui River and its people. He says, 'We see the river as a connector, as a giver of life, as a food source, as a place where our people pray, as a playground for our children. We cross the river to get to work every day. We swim in the river. We take food from the river.'[26]

But the river, and the people bound to it, came under threat with the colonisation of Aotearoa in the mid-1800s.

In 1841, the first settlers from England, Scotland and Ireland arrived in Whanganui. They sought to alter the river to accommodate their needs, clearing rapids for steamers and extracting gravel, which destroyed places of cultural and environmental significance. These disruptions culminated in a petition to the New Zealand Parliament in 1883, where Whanganui Iwi – the Maori custodians of the Whanganui River – asked for the restoration of their river's cultural, social and environmental integrity. After almost eight decades of litigation and resistance, a 1962 Court of Appeal decision ruled that Māori customary ownership of the riverbed had been extinguished: a painful setback. Undeterred, Whanganui Iwi pursued a new avenue in the 1980s through the Waitangi Tribunal, a commission established to address grievances relating to breaches of New Zealand's founding document, the Treaty of Waitangi. In 1999, the tribunal released its report: the Whanganui people were indeed custodians of the Whanganui River 'in its entirety'.[27]

From there, Whanganui Iwi worked with the New Zealand government to write up a settlement. And so it was on 20 March 2017 that New Zealand Parliament passed a historic bill and the Whanganui River became the first river in the world to be recognised as a legal person.[28] Two guardians were appointed to act on behalf of the Whanganui River, one representing the Crown and one representing Whanganui Iwi. For the first time in history, a river owned itself, which meant, 'Any discussions regarding the river must be discussed with the river itself, via its legal representatives.'[29]

If there is any kind of harm or damage inflicted upon the river, the river can sue. In an event such as this, the river's appointed guardians are empowered to protect the river and have the authority to take legal action on its behalf to seek redress or enforce its rights.

Naturally, immediately after the river became a person, questions were raised such as: can the river vote? (it cannot); can it buy a beer? (not likely); and, importantly, could it be charged with murder if a swimmer drowned in it?[30] The last question is actually important. While the river has been granted legal personhood, this status does not mean that the river is responsible for accidents or can be held liable for them in the way a human person might be. The concept of legal personhood in this case allows the river to hold rights and have its interests represented through appointed guardians, but it doesn't assign the river liability for harm to individuals who interact with it.

The river that became a person represents, according to environmental lawyer and former UN special rapporteur David R Boyd, 'a legal and cultural revolution in the way that we relate to the natural world'.[31] In the case of the river, the law was applied in a way that could be used as a tool to protect nature and offer reconciliation, re-establishing Māori governance and management over the river that they had held for centuries before European colonisation.[32]

THE RIGHTS OF NATURE: THE RIVER THAT IS A PERSON

Five days after the Whanganui, the Ganges and Yamuna rivers and their tributaries in India were granted personhood rights.[33]

In fact, all over the world, nature is being granted rights. It was Ecuador that made history by becoming the first country in the world to enshrine the rights of nature in its constitution in 2008.[34] This legal process is designed quite differently from that of the Whanganui River. Rather than granting legal personhood, Ecuador took it one step further. They codified the rights of nature, recognising nature as its own legal entity, existing in its own right. By doing this, Ecuador effectively granted *every* citizen the right to speak on behalf of nature and, importantly, take legal action on its behalf.

This foundational step set a precedent for other nations to follow. And follow they did.[35] In 2010, Bolivia passed the 'Law of the Rights of Mother Earth'.[36] In 2016, Colombia recognised the rights of the Atrato River within its constitution.[37] In 2019, Uganda became the first nation in Africa to recognise the rights of nature in national legislation.[38] In 2022, Panama signed the rights of nature into national law.[39] These shifts weren't limited to national governments. Localised initiatives brought rights to rivers, forests and even pollinators, offering nature a voice in the courtroom. In Cartagena, Colombia, the court ordered the protection of bees.[40] In Australia, the Birrarung (Yarra River) was recognised as a living entity.[41] The state of Guerrero in Mexico recognised the rights of nature in its constitution.[42] In Curridabat, Costa Rica, pollinators, trees and native plants were afforded citizenship,[43] recognising their essential roles in sustaining ecosystems. The recognition of rights for nature is sweeping the globe. Nepal, Brazil, the Philippines, Panama, Peru, Pakistan, Spain and Guatemala are just some of the other countries that are recognising the rights of nature in one form or another.[44]

And then, in 2024, the movement reached another historic milestone: whales and dolphins – ancient guardians of the seas – became legal persons. In a treaty formed by Pacific Indigenous leaders from the Cook Islands, French Polynesia, Aotearoa New

Zealand and Tonga, whales and dolphins were granted legal personhood. This treaty, He Whakaputanga Moana, or, 'Declaration for the Ocean', grants whales and dolphins 'the right to freedom of movement, a healthy environment, and the ability to thrive alongside humanity'.[45]

'The songs of our ancestor, the tohorā (whale), who have navigated these very waters for generations, grow fainter,' said Kīngi Tūheitia, Māori King. 'He Whakaputanga Moana is a Hinemoana Halo – a cloak of protection for our taonga, our ancestor – the whales.'[46]

There is a common thread woven through almost all of these legal campaigns: Indigenous peoples and local communities overwhelmingly lead them.

Rikki Dank, known by her Country as Lludibpina Noralima, is a Gudanji and Wakaya person and a proud Gudanji Traditional Owner. As the director of Gudanji for Country, an Australian First Nations organisation dedicated to safeguarding Country from overgrazing, mining and fracking, her work is both deeply personal and profoundly communal. It is a calling steeped in history, responsibility and an enduring connection to Country.

For Dank, Country is 'a place where my family have lived for 65,000 years'. It is a place of love, warmth and joy, but also of sadness; bittersweet sadness. A sadness that carries the weight of Australia's history and ongoing legacy – of stolen lives, stolen lands and the deep scars of colonisation.

Dank says, 'We forget that, without Country, we cease to exist. We cannot continue to separate ourselves from nature. Without our land, who are we anymore? I can't call myself Gudanji if I haven't fought to protect my Country like my great Aunts and Uncles. Who are we if our birthplace has been destroyed?'

'This is why Country is so important,' she says. 'It keeps these memories, it keeps these feelings, these places and times for us as Gudanji People. This is where we rest, whether it's our bones or our souls. We rest here.'

This deep, spiritual connection to lands, waters and skies – shared by many Indigenous peoples – is woven throughout the Rights of Nature movement.

The idea of granting rivers, mountains and ecosystems legal personhood – allowing them to have a voice in court – has captured imaginations worldwide. Yet, as inspiring as it sounds, does this concept truly translate into meaningful protection of nature?

Dr Rónán Kennedy, an expert in environmental law, explains that granting legal rights to nature isn't a 'quick fix' for environmental issues. He says it is certainly not a 'substitute for strong institutions and meaningful enforcement'.[47] But it's beginning to yield meaningful results and see some success. Indeed, the law can be slow to penetrate.

For example, women's right to vote, an early human rights victory, marked a milestone in legal breakthroughs but didn't immediately dismantle the broader social and institutional barriers that continued to restrict women's full participation in society.

A similar pattern is emerging in environmental law. In Ecuador, for example, where nature's rights were embedded in the constitution, the Constitutional Court revoked mining permits in Los Cedros, a richly biodiverse cloud forest, in 2021. The court asserted that these permits 'not only violated the rights of local residents (such as the right to clean water and a healthy environment) but also the rights of the forest itself'.[48] While progress has been slow, Ecuador's legal system has gradually begun to interpret and apply these rights, particularly as courts and legal professionals grow more familiar with the concept. In fact, since then, Ecuadorian courts have ruled on at least fifty-five cases involving the rights of nature, half of which occurred between 2019 and 2022.[49]

But if Ecuador offers a cautious case of progress, other attempts have faced more immediate resistance. In the US, for example, the Lake Erie Bill of Rights sought to recognise the lake's right to 'exist and flourish', granting residents legal standing to protect it from pollution. Yet almost as soon as the bill passed, farmers challenged it, fearing they'd be held accountable for the inevitable fertiliser that would enter into waterways. The court deemed the bill 'unconstitutionally vague', and the law was overturned. The problem, however, wasn't in the desire to protect Lake Erie; it was in the ambiguous, sprawling language that couldn't withstand legal scrutiny.

Still, the Rights of Nature approach is not without its critics. Some argue that forcing nature into the framework of Western law risks reducing it to another commodity, stripping it of its inherent worth and making it vulnerable to the very human systems that have long driven its exploitation.[50] By forcing nature into a system that prizes ownership and liability, we may be reducing it to something we can control, rather than honouring its intrinsic worth.

This tension between nature and Western legal frameworks has also been criticised by some Indigenous scholars. Dr Virginia Marshall, a Wiradjuri Nyemba woman and Research Fellow at the Australian National University's School of Regulation and Global Governance, argues that the rights-of-nature approach often references Indigenous worldviews as a source of legitimacy, but it can, in fact, undermine Indigenous sovereignty and cultural obligations. She says this could create 'a new tool of colonisation' that is 'antithetical to the Indigenous rule of law and their cultural obligations'. She says 'the answer to inadequate or poorly performing government policies and law is not to retreat to legal personhood' but to 'meaningfully consult and engage with Indigenous communities and give those communities a significant and central role in the management of land, waters and resources'.[51]

The legal system is indeed attempting to redefine how it addresses the complex web of relationships in nature. The Rights of Nature movement may seem like yet another human attempt to bend nature into a legal structure, but for the first time, a river might hold the same legal status as a corporation or a ship. It's a slow, imperfect evolution, riddled with questions we may not yet have the tools to answer. Still, as lawyers and judges learn to interpret these laws, we may see a shift – a gradual turning that allows nature, along with communities, to wield the law in a way that leads to a more harmonious relationship between humanity and nature.

But difficult questions remain: how do we know what natural entities want or need? And who has the authority to speak for them? This is where we must broaden our sources of knowledge. By centring Indigenous knowledge and scientific insights and taking in diverse forms of wisdom, we may be able to approach the natural world through a more holistic and respectful lens – even if it is within the confines of a Western framework. After all, in the race to protect nature, everything is on the table.

Assigning rights to nature could, however, mark a powerful new frontier for environmental activism.

As it stands, poisoning a river or flattening a forest is typically treated not as a crime, but as an acceptable byproduct of economic progress. Meanwhile, peacefully protesting environmental harm will trigger a swift punishment. For example, in Australia, fossil fuel extraction, such as coal mining and offshore drilling, is legal and heavily subsidised, despite the environmental damage it causes. Yet over the past two decades in Australia, forty-nine laws have been passed that restrict, reduce or remove the freedom to protest in public.[52] Individuals who protest environmental harm or abuse often face life-disrupting fines, imprisonment and criminal records.

Corporate entities, you see, are endowed with personhood under the law and enjoy protections that frequently outweigh the preservation of ecosystems and the rights of those who protest against their actions. While a human activist can be restrained, jailed or silenced, a corporation has no physical form to be punished or detained.* As 18th-century British parliamentarian Lord Edward

---

* In general, CEOs and executives of these companies are not personally liable for the actions of their corporations – unless they are found to have directly engaged in illegal activity, gross negligence or breached specific duties (like fiduciary duty). This principle is part of what's called the 'corporate veil' – the legal separation between a corporation and the people who run it.

Thurlow famously said, corporations have 'no soul to be damned, and no body to be kicked'.[53] This legal advantage allows corporations to avoid many of the repercussions that individual protesters face – one that often tilts the scales in favour of corporate interests over environmental protection.

The idea of the rights of nature could potentially up-end this imbalance. By granting rivers, mountains and ecosystems legal personhood, activists could be given a defence rooted in the duty to protect a vulnerable entity. In Ecuador, where the rights of nature have been enshrined in the constitution, any citizen can stand as a guardian for nature in court. As a result, lawsuits have been filed – sometimes successfully – to stop mining, oil drilling and deforestation projects.[54]

This may mean that legal recognition of the rights of nature could provide avenues to strengthen human rights, empower communities and grassroots movements and safeguard democratic principles. In the eyes of the law, defenders of nature could be viewed as protectors and guardians rather than criminals.

Humanity's tangled relationship with nature – when viewed through the lenses of culture, ethics, philosophy and law – reshapes our understanding of environmental and social justice, and challenges us to rethink our connections with the other beings who share this planet.

For some, the rise of the Rights of Nature movement has raised the question: if nature can have rights, and if nature includes animals, could individual animals be granted rights?[55] Could a whale sue for grievous bodily harm? Could a goat take a farmer to court?

The Rights of Nature movement, in its ideal form, creates a bridge between human (Western-centric) legal frameworks and those that recognise the intrinsic value of nature and all its inhabitants. It is a more expansive and inclusive approach to rights – one that not only serves human interests but also respects the dignity and interconnectedness of all life on Earth. In fact, the evolution of these questions has sparked interesting overlaps between animal rights law and environmental (or nature) law.

The Rights of Nature movement generally takes a holistic approach to the environment. Rather than focusing on individual animals and their wellbeing, the rights of nature generally apply to entire ecosystems, natural entities like rivers and mountains, or wild species. Animal rights law, by contrast, zeroes in on individual animals, with the goal of protecting them from cruelty, abuse and exploitation. It advocates for their welfare, freedom from suffering and, in some cases, legal personhood. These two fields of law have an uneasy relationship, one that may be marked by mutual irritation. At a basic level, they are linked by the fundamental observation that animals – human and non-human – exist in the environment. And they're both grappling with the unenviable task of figuring out how to protect the non-human world within a human-centred legal system. But there are a few tensions.

Animal rights law has traditionally aimed to secure rights or protections for animals based on their sentience, intelligence, emotional life and capacity to feel pain or suffer. This focus on the individual often leads to clashes with environmentalists, who prioritise the health of ecosystems and entire species in the wild over the wellbeing of any one creature. To illustrate this point of tension most saliently, it might be helpful to indulge in a hypothetical scenario laced with some reasonably offensive stereotyping.

If an animal rights advocate and an environmental advocate were driving through the Australian countryside together and happened upon a free-roaming (feral) cat, they would likely respond in divergent ways. The environmentalist would immediately see a threat: cats in Australia are considered an invasive species, responsible for killing two billion wild animals each year, wreaking havoc on ecosystems and contributing to the extinction of over thirty native Australian species.[56] The animal rights advocate, on the other hand, may see a sentient creature that is capable of suffering, just like us. Both would likely agree that humanity created this problem when it introduced and then mismanaged cats. Nevertheless, they would respond differently. The environmentalist would promptly eject themselves from the car and clock the cat over the head with a nine iron. This would, conceivably, upset the animal rights advocate.

Of course, that is a radical hypothetical scenario that doesn't consider the fact that the majority of these two groups likely sit somewhere in the middle – and likely don't carry golf clubs in their cars. But it illustrates that environmentalists may be accused of prioritising the whole over the individual. And animal rights advocates may be accused of focusing too narrowly on the suffering or rights of individual animals at the expense of broader ecological considerations.

Another point of tension arises over the topic of sentience. Animal rights law frequently argues that certain animals deserve moral and legal protection because they can feel pain, experience emotions and demonstrate intelligence – in short, because they are 'sentient'. But to some environmentalists, this emphasis risks creating a moral hierarchy based on how closely a species resembles humans. It suggests that animals capable of experiencing suffering or pleasure (like mammals and birds) are entitled to more protection than animals considered less sentient or non-sentient (like invertebrates or plants). Rights of Nature laws do not typically rely on sentience as a criterion. Instead, they view ecosystems, rivers, mountains and forests as worthy of legal protection based on their intrinsic ecological value, regardless of whether these entities are sentient.

In short, an animal rights advocate might advocate for the release of a chimpanzee from captivity based on its 'human-like' intelligence. Environmentalists, on the other hand, may bristle at this anthropomorphic bias, before launching into a lengthy sermon about the ecological importance of the sea sponge.

Despite these tensions, the momentum of the Rights of Nature movement might offer areas for both fields to achieve mutual goals. In fact, according to Harvard Law professor Kristen Stilt,

THE RIGHTS OF NATURE: THE RIVER THAT IS A PERSON

environmental law could be 'instructive, intellectually and practically, to the cause of animal protection and animal rights',[57] affording more robust protections to certain animals.

Both environmentalists and animal rights advocates would likely agree with the notion that humanity positioning itself as superior over other beings has led to harmful outcomes for the planet and all of its inhabitants.

Generally speaking, individual animals are not yet fully included in the framework of the rights of nature. However, as Professor Stilt says, 'provisions recognizing the Rights of Nature still implicitly acknowledge that a nonhuman can have rights'.[58] And this may just provide an avenue for non-human animals to be granted rights as individuals.

In the eyes of the law in most jurisdictions, animals are property, whether they're dogs or goldfish or kangaroos or turtles. For a long time, 'the only way under the law to protect an animal from wanton abuse was to characterise the abuse as harm to property, essentially no different from the sort of law that prevents you from smashing your neighbor's wheelbarrow'.[59] However, in some cases, they may be afforded varying levels of protection under animal welfare law, though this varies across countries, states and territories.

A patchwork of efforts has been made to recognise legal personhood for animals in some unique cases. Argentina has been at the forefront of granting legal personhood to great apes and other sentient animals. Sandra, a captive orangutan in a Buenos Aires zoo, was declared a 'non-human person' by an Argentine court in 2014, followed by Cecilia the chimpanzee in 2016.[60] And in India and Pakistan, animals have been granted a limited form of legal personhood in some regions.[61]

In the US, legal personhood for animals has been elusive, but that is not for lack of trying. The Nonhuman Rights Project (NhRP) is an organisation dedicated solely to securing rights for non-human animals. They demand recognition of the legal personhood and the fundamental right to bodily liberty of animals. Their client list includes Happy the elephant and Hercules, Leo and Kiko the chimpanzees, among others. They focus primarily on elephants, dolphins and whales living in captivity. These animals, they argue, are 'members of species for whom there is ample, robust scientific evidence of self-awareness and autonomy – qualities the common law already purports to value where humans are concerned'.[62]

People for the Ethical Treatment of Animals (PETA) have arguably employed more theatrical strategies to secure rights for animals. In 2011, a crested macaque in Indonesia snapped a now iconic photo of himself using a wildlife photographer's camera. The image went viral, but so too did the legal battle that followed. PETA filed a lawsuit on behalf of the monkey, arguing that the monkey should own the copyright to its selfie, not the photographer. While the case was eventually dismissed, it raised questions about the legal standing of animals and their rights within human constructs of law.[63]

That same year, PETA filed a lawsuit on behalf of five orcas at SeaWorld, Florida. Orcas, often described as intelligent, emotionally complex creatures, are known to suffer extraordinary psychological stress in captivity. PETA accused the theme park of violating the Thirteenth Amendment (the constitutional amendment that abolished slavery) by keeping orcas in captivity.[64] The suit was dismissed.

Advocates of animal legal personhood have achieved very limited victories, whereas the Rights of Nature movement – demanding that nature or some parts of nature be given rights or legal personhood – has been quite successful in many parts of the world.

But a parrot might be about to show that these two approaches can work hand in hand.

Verdinho is a blue-fronted Amazon parrot from Brazil who could well become the poster bird for fusing animal rights law and environmental law (or, more specifically, Rights of Nature law). For over two decades, Verdinho lived in captivity with his 'owner', Maria. But in 2018, what became known as the 'Wild Parrot Case' arose

THE RIGHTS OF NATURE: THE RIVER THAT IS A PERSON

after allegations surfaced that Maria had been keeping Verdinho in inadequate living conditions.⁶⁵ The court was tasked with making a crucial decision regarding the parrot's future: whether to allow Verdinho to remain with Maria, provided she improved his living conditions, or to transfer him to a wildlife sanctuary.

How the court came to its decision is curious – and subtle. It marked a pivotal moment in merging animal rights with the evolving concept of nature's rights.

Traditionally, the law in this case would view an animal in relation to human interests; its protection is only valid when it is of benefit to a human. But in this case, the court did something unconventional. It drew on constitutional protections for animals in Germany and Switzerland, and also the rights of nature frameworks from Ecuador and Bolivia. Then, it blended them into a declaration that said the parrot's welfare was important not because it is of benefit to humans, but simply because the parrot has inherent dignity and intrinsic value. What the court said, in essence, was, 'The parrot deserves its own rights, care and dignity because it's a parrot.' This subtle shift in language is significant. It redefines the purpose of environmental and animal protection laws, demonstrating that the law can be used to protect individual animals *as well as* entire ecosystems and natural entities.

Verdinho was permitted to continue with Maria, who agreed to improve the parrot's living standards.

~

In 2004, Katya was sentenced to life in prison after being found guilty of causing bodily harm to two people. She was sent to an all-male penal colony in Kazakhstan where she was imprisoned alongside 730 other inmates for fifteen years. She was the only female. She was also the only bear.⁶⁶

In 2016, the 'testimony' of an African grey parrot helped solve a murder case. Martin Duran was shot five times in his home in Michigan, US, in 2015. His wife, Glenda, suffered a head wound.

Police originally thought the couple were the victims of a violent break-in. That was until the couple's pet parrot was overheard repeating the last words of his owner: 'Don't f*cking shoot.' This turned the trial on its head, leading the prosecutor to consider whether the parrot's squawking could be used as evidence in the murder trial. This was later dismissed, but Glenda was ultimately found guilty of killing her husband.

In 2008, a dog named Scooby became the first animal to appear as a witness in a murder trial in a French court.[67] In 2009, police arrested a goat on suspicion of attempted armed robbery in Nigeria.[68] In 2022, a sheep was sentenced to three years in jail in South Sudan after being found guilty of headbutting and killing a woman.[69]

In trying animals and even glaciers in court, medieval societies – and indeed, modern societies – have acknowledged, perhaps unwittingly, that nature and non-human animals are participants within a shared moral and legal universe.

While it may seem far-fetched to consider granting rights to nature or even individual animals, it's worth remembering that, for much of history, the idea of recognising the rights of women, Black people, Indigenous peoples, people with a disability and many others was also seen as implausible. And if a ship or a corporation can legally be considered a person, why not a whale – or even a tree?

The movement to afford nature rights is slow, as all paradigm shifts are. But it is underway. It took centuries for the law to fully recognise all humans as legal persons, and even now, that recognition remains tenuous or unfulfilled for many.

Granting rights to nature is not a panacea; it is not the sole solution to the nature or extinction crises. But it is forcing humanity to confront its relationship with the world and, in some cases, is reframing humanity's relationship with nature through legal frameworks.

Around the world, the rights of nature are slowly redefining the boundaries of environmentalism, challenging longstanding assumptions about the natural world's place in human society. In Ecuador, Bolivia, New Zealand, Colombia, the US and beyond, people are beginning to see that protecting nature may mean something more profound than just conservation. It may mean listening – really listening – to the rivers, mountains and creatures that sustain life on Earth, and finding a way to let them speak.

After all, I am the river and the river is me.

PART 4

Battlegrounds: Politics, Power and Nature

Chapter 11

Dumping Grounds: Let Them Sink

> Too small for people to see, too far for people to reach, and a number of 52,634 people too little for people to care. Our islands are not just barely-there dots on the maps for many to turn a blind-eye to; they are our home. And our home is [the link between] our ancestors and us today.
>
> Selina Neirok Leem[1]

We have statistics and we have stories, but none of them work. Stories like the turtle that can fit in the palm of a hand that dragged itself ashore with sunken eyes and laboured breath before dying painfully with 104 pieces of plastic lodged in its intestinal tract. Stories like the baby bird that wore a balloon string around its neck, which tightened like a noose and choked the sky from its wings. Stories of children with poisoned lungs from smoke piles who wade through rivers of rubbish to make a penny. Stories of oceans and rivers and waters buried by waste, of communities staggering under its weight. Stories of bodies that are swollen and strangled and suffocated and bloodied. Stories of torment and pain and life flickering out. We have statistics, like the ones that tell us that 10 million tonnes of plastic enter the ocean each year, and that soon it will outweigh fish.[2] That plastic is in our blood, our brains, our breast milk, and it is killing almost one million people every year.[3]

That one in six people die from pollution – that pollution kills more people than all the wars and violence in the world.[4] And that nations are sinking under the weight of a contaminated Earth.

But the stories and the numbers blur. And nothing changes. It's getting worse, in fact. We sit atop an empire of waste, and we bury the world's most vulnerable. And though we may be last to be buried, buried we will be.

~

It was 2017 and China was at its wits' end. For almost four decades, the country had served as the dumping ground for the world's waste, taking on 56 per cent of global waste plastics and paper for recycling.[5] But no more. In an announcement to the World Trade Organization, they informed the world that they would no longer be accepting 'foreign garbage'.[6]

To underscore their discontent, China dubbed the new legislation 'Operation National Sword'. For years, the arrangement had seemed straightforward: wealthy nations shipped their trash to China, and China recycled it. In theory, it was a win-win – China's booming manufacturing sector gained cheap raw materials, and wealthy countries got a cheap way to offload their rubbish. But in practice, the deal fell apart. Contaminated plastics and paper flooded in, much of it unrecyclable, leaving China to bear the brunt of the environmental, social and health damages associated with such large quantities of rubbish.

Operation National Sword up-ended the global waste trade, leaving wealthy countries positively beside themselves at the horrifying revelation that they might have to dispose of their own trash. Among the hardest hit were major exporters of rubbish like the US, Japan, Germany, Canada, the UK and Australia. Australia alone had sent 1.25 million tonnes of recycled materials to China in 2016 and 2017 – roughly 52 kilograms per person.[7]

Indeed, China's ban on other countries' rubbish peeled back the veneer on a global waste industry built on questionable economics and politics – and it wasn't pretty. The world produces about 2 billion tonnes of waste each year.⁸ Of this massive mountain of garbage, roughly one-tenth takes a world tour, entering the global trash trade.⁹ This waste that countries can't (or won't) recycle within their own borders is known as surplus waste.

Following the ban, the UK, Europe and the USA started burning plastic, driving an increase in pollution, while Australia's recycling system was in crisis and at risk of collapse.¹⁰ The result was that wealthy countries who were reliant on outsourcing their rubbish ended up with more plastics in landfills, incinerators, and being stockpiled as countries grappled with growing mounds of trash that would have previously been sent to China.

Then, inspiration struck. The most logical solution, these countries concluded, was to find poor nations. Yes, they agreed. Poor people could take the trash! It would be a gift to them – a treasure trove of opportunity and income. And that is how these countries graciously gifted 2 million tonnes of waste to Southeast Asia.¹¹

Today, an estimated 20 million individuals worldwide earn their livelihoods through collecting and recycling waste in the informal waste sector. These waste workers are paid through the recyclables they collect and sell.¹² It can provide crucial income for people and households while providing reusable materials to other enterprises. Yet, as the United Nations Development Programme notes, they work 'under hazardous and deplorable conditions, making a living out of things that are discarded as waste by others'.¹³ In addition to economic hardship, they face discrimination, violence, harassment and limited labour rights in a largely unregulated sector. Without basic safety and protective gear, injuries, respiratory illnesses and allergies are common. Exposure to harmful chemicals places workers at heightened risk of contracting severe, long-term diseases such as cancer, diabetes or reproductive disorders. The vast majority of waste pickers are in the Global South.

When wealthy nations export overwhelming amounts of contaminated or non-recyclable plastic under the guise of recycling, it's the importing countries that bear the brunt: choked ecosystems, polluted landscapes and a social and environmental cost to communities that arguably outweighs any financial benefit. Writer and sustainability consultant Aja Barber captures the essence of this dynamic in her book *Consumed: The Need for Collective Change: Colonialism, Climate Change, and Consumerism*.[14] She says, 'They literally are taking our resources from our lands, selling it back to us and burying garbage next to us; it's colonialism at its finest.'[15]

In this modern-day trade, waste is transferred primarily from wealthy, industrialised nations in the Global North to less developed, low-income nations in the Global South, a practice that plastic pollution expert Sedat Gündoğdu says 'constitutes a form of 21st century colonialism'.[16] The term 'waste colonialism' emerged in 1989 during a UN Basel Convention meeting in which African nations protested against wealthy countries, particularly in Europe and North America, for dumping hazardous waste in poorer regions like Africa, the Caribbean and Asia. This practice exploited low-income countries as cheap and unregulated waste disposal sites. 'Some of the top waste producers in Europe, like the UK, France and Germany have to find ways to deal with this issue. And the way they've found is exporting to poorer countries without effective waste management systems or environmental legislation and regulations. This is waste colonialism,' says Gündoğdu.[17]

By 2019, following China's ban on 'foreign trash', the global waste trade had descended into what can only be described as chaos. Southeast Asia, which was cast as the next convenient receptacle for the waste of wealthy nations, quickly reached its limit, drowning in an influx of contaminated plastic. Soon, these countries decided they'd had enough too.[18]

'Malaysia will not be a dumping ground to the world … we will fight back. Even though we are a small country, we can't be bullied by developed countries,' said Malaysia's environment minister, Yeo Bee Yin.[19] Meanwhile, in the Philippines, then president Rodrigo Duterte threatened to forcibly ship back dozens of containers of garbage to Canada, which he said was turning his nation into a 'dump site'. 'For Canada's garbage, I want a boat prepared,' he said. 'They better pull that thing out or I will set sail to Canada and dump their garbage there.' He added, 'Let's fight Canada. I will declare war against them.'[20]

Soon, avenues for wealthy countries to dump their waste upon poorer nations started to dry up. As the UN said, wealthy countries 'will, at last, have to face up to the true cost of their plastic addiction' instead of shipping the problem abroad.[21] This, they said, could offer the opportunity to drive 'much-needed investment in domestic recycling facilities as well as innovation in plastic manufacturing to make products more suited to repurposing. It could also invigorate the vociferous public campaign to change our throwaway culture.'[22]

For some countries, this prompted action: setting new rules on waste exports, taking steps towards a circular economy – recycling, reusing and rethinking waste – and even converting waste into energy where possible.[23] Others felt that the best way to tackle the problem was to do nothing, but with greater conviction.*

China's National Sword policy and the subsequent shifts in waste trade shone a fresh light on some of the unequal dynamics between the Global North and the Global South. It is within this dynamic that the links between nature harm and human harm are revealed. After all, harm to nature is harm to people – but it's usually the poor who bleed first.

The outsourcing of waste to poor nations is just one facet of a much broader and more insidious system – one in which wealthy nations and businesses insulate themselves from the consequences of their extraction, production and consumption by offloading

---

* For example, the European Union introduced new rules on waste exports, Japan ramped up its domestic recycling systems and Sweden continued expanding its waste-to-energy infrastructure, while the USA and Australia spent a very long time discussing the issue at length.

environmental harm to poor nations and communities. Central to this system are 'sacrifice zones'. Sacrifice zones are, in short, the parts of the world society has decided are okay to ruin the most.* Coined during the Cold War to describe areas destroyed by nuclear testing, the term has now come to refer to dumping grounds for pollution, waste and ecological carnage. These zones don't just accidentally emerge; they are carefully selected, deliberately placed in low-income, vulnerable and/or marginalised communities that lack the political, social and economic clout to resist. These are decisions made by institutions, industries and governments – often on behalf of those who will never have to live with the consequences.

In these zones, one can establish toxic facilities, landfills, industrial sites, mining projects and so on. They are the engines of 'progress', just with a hefty price tag paid by poor people whom you'll likely never dine with nor see at a decision-making table. And that is the intent, of course. In this way, the unpleasantness of human misery and environmental decay can be hidden away, nestled between an open-pit mine and a toxic waste dump.

In 2022, the UN special rapporteur on human rights and the environment, Dr David R Boyd, presented a report to the Human Rights Council and General Assembly on sacrifice zones, outlining their devastating implications for human rights and the environment.[24] Sacrifice zones, according to the report, 'can be understood to be a place where residents suffer devastating physical and mental health consequences and human rights violations as a result of living in pollution hotspots and heavily contaminated areas'.[25] The most heavily polluting and hazardous facilities included 'open-pit mines, smelters, petroleum refineries, chemical plants, coal-fired power stations, oil and gas fields, steel plants, garbage dumps and hazardous waste incinerators, as well as clusters of these facilities',

---

* In this context, 'society' refers to certain governments, industries and institutions.

which, the report noted, 'tend to be located in close proximity to poor and marginalized communities'.[26]

Boyd's findings reflect global patterns in pollution-related mortality: '92% of pollution-related deaths occur in low-income and middle-income countries'.[27]

Hidden in the world's poorest parts of the world, sacrifice zones stretch across nearly every country, disproportionately affecting the world's most vulnerable people. These zones both mirror and deepen racial, gender and economic inequalities. Toxic pollution, degraded ecosystems, cancer, heart disease, respiratory illness, strokes, reproductive health problems – this is life in a sacrifice zone. To harm nature is to harm humanity, but we've managed to concentrate that harm within the most disadvantaged communities.

The ledger of sacrifice zones, as detailed by Boyd and other human rights organisations, is as long as it is grim – a catalogue of places where the cost of progress is measured in poisoned land and broken lives.

Agbogbloshie is a sprawling electronic waste scrapyard in Accra, Ghana, known as one of the most polluted places on Earth. Here, thousands of workers, many of them migrants from northern Ghana or members of marginalised ethnic groups, collect, dismantle and burn discarded electronics to salvage precious metals like gold, copper and silver. Toxic chemicals – including lead, mercury, cadmium and chromium – seep into the air, soil and water, turning the region into a toxic wasteland.[28]

In Grassy Narrows, Canada, the Dryden Chemical Company dumped over 9000 kilograms of mercury into the river system of an Indigenous community in the 1960s and 1970s. Nine out of ten members of the community experience symptoms of mercury poisoning, including neurological disease and facial paralysis.[29]

In Wittenoom, Australia, asbestos mines left behind three million tonnes of waste rock laced with deadly asbestos, turning 46,000 hectares of Banjima Country – the ancestral lands of the Banjima people in the Pilbara region of Western Australia – into a toxic waste dump.[30] For nearly two decades, Aboriginal men worked in the mines, bagging blue asbestos by hand. Almost all of them have died from mesothelioma – an aggressive cancer caused by asbestos exposure. Their mortality rate from asbestos exposure is among the highest recorded anywhere in the world. In 1966, the operation shut down – not for safety, but for financial reasons. In 2007, the government erased Wittenoom from the map. Despite decades of inquiries urging remediation and clean-up, toxic minerals continue to spread, and Banjima Traditional Owners remain cut off from large areas of their ancestral lands.[31]

In Anosy, Madagascar, one of the world's most biodiverse regions, Rio Tinto partnered with the Malagasy government to establish an ilmenite mine (a mineral used in sunscreens, fabric, inks, food and cosmetics). Farmers, pastoralists and fishers were displaced, forests were destroyed, and water was contaminated. Uranium levels are now 350 times higher than the local average and lead levels are ten times higher. These heavy metals are known to cause organ damage and developmental delays in children. The loss of forests has affected residents' ability to hunt and grow manioc, their staple food source.[32]

In a section of the Mississippi River, US, known as 'Cancer Alley', predominantly Black communities exist side by side with some 200 fossil fuel and petrochemical operations.[33] Toxic air pollution has resulted in one of the highest cancer risks in the nation, as well as documented risks of maternal, reproductive and newborn health harms and respiratory ailments.

Nauru, once known as 'Pleasant Island', has suffered devastating environmental and health consequences from a century of phosphate mining and 'has been transformed by mining into a scarred wasteland'.[34] Almost all mined phosphate goes into crop fertilisers. This has created an extreme health crisis for its inhabitants, including cadmium poisoning due to the contamination of groundwater and high rates of diabetes due to the eradication of traditional foods and agricultural land. In 1989, Nauru took legal action against Australia in the International Court of Justice (ICJ) for failing to address the environmental destruction caused during its administration of the island.

Elevated arsenic levels for one in six residents in Tsumeb, Namibia, because of copper mines.[35] Birth defects in Sudan from the Thar Jath oilfield. Children with elevated cancer risk in the Democratic Republic of the Congo from cobalt mines. Miscarriages in Ethiopia among the Indigenous Guji Oromo people from the Lega Dembi goldmine. Cardiovascular dysfunction in Togo from phosphate mining. Breathing difficulties, skin rashes, nosebleeds and headaches at Karachaganak oilfield in Kazakhstan. Five thousand children with breathing difficulties, nausea, vomiting and dizziness from hazardous waste dumping in Kim Kim River in Malaysia. Lead poisoning in Kanchanaburi Province, Thailand. Kids with leukaemia from the Vaca Muerta oil and gas project in Argentina. Kidney failure in people of the Atoyac and Santiago river basins in Mexico from industrial pollution.

This is the modern global arrangement, where poorer nations and communities – and, to a disproportionate degree, Indigenous peoples – contend with the fallout of an environmental crisis they didn't create and that they are ill-equipped to deal with.

Climate change is carving out new sacrifice zones, particularly among small island developing nations and countries with economies that are heavily reliant on nature-sensitive sectors, like agriculture and fisheries. These regions are, unsurprisingly, largely in the Global South. Indeed, seventeen of the world's twenty most climate-vulnerable and least climate change-prepared nations are low-income countries. Since the 1970s, these nations have shouldered 69 per cent of global climate-related deaths.[36] Rising sea levels threaten to submerge entire island nations, while droughts, floods and fires wreak havoc on food systems, infrastructure and livelihoods.

Tuvalu is a small Pacific island nation. And soon, it will be gone. This will be the first country in the world to be uninhabitable due to climate change, with 95 per cent of the country expected to be underwater by 2100. If global emissions continue on their current trajectory, the people born today in Tuvalu will be the last remaining generation to live there. In 2021, Tuvalu's minister of foreign affairs, Simon Kofe, stood knee-deep in water and delivered a speech to the UN. 'We are sinking,' he said. 'Climate change and sea level rise are deadly and existential threats to Tuvalu and low-lying atoll countries ... No matter if we feel the effects today like in Tuvalu, or in a hundred years, we will all still feel the dire effects of this global crisis one day.'[37]

But we won't feel it equally.

Faced with the knowledge that their country will no longer exist seventy-five years from now, Tuvalu is building a digital twin to remember it by. In 2022, they announced that Tuvalu would 'digitally recreate its land, archive its rich history and culture, and move all governmental functions into a digital space.'[38] Because soon their country will not physically exist. The country has amended its constitution to reflect a new definition of statehood, an effort to ensure that Tuvalu can 'retain its identity and continue to function as a state, even after its physical land is gone'.[39]

Similarly, in Fiji, a major operation is underway to relocate forty villages from the shoreline as climate change and nature degradation drive rising sea levels, flooding and more frequent and severe cyclones.[40] For the past six years, a special government task force in Fiji has been trying to work out how to move these villages. The plan is called the 'Standard Operating Procedures for Planned Relocation'.[41]

The impacts of the climate crisis are also felt unequally within communities. After all, environmental catastrophes like climate change are 'threat multipliers' – they amplify pre-existing threats and inequalities. Women, especially Indigenous women and women of colour, face disproportionate risks in the face of climate change and nature degradation. In times of environmental crisis, they are exposed to increased risks of displacement, violence, sexual and domestic abuse, trafficking and conflict. In fact, when an extreme climate disaster strikes, women and children are fourteen times more likely to die than men.[42]

Harmful gender norms often dictate women's survival. For example, in many parts of the world, women are less likely to have access to bank accounts, healthcare, education and mobility, limiting their ability to respond to disasters. The unequal toll of disasters is stark: in the 2004 Indian Ocean tsunami, four times as many women as men were killed in Indonesia, Sri Lanka and India.[43] The reason? Men were taught to swim and climb trees at a young age, while women were not. In Vanuatu, there was a 300 per cent increase in new domestic violence cases after two tropical cyclones hit Tafea Province in 2011.[44]

Despite ample research indicating that peace and environmental agreements brokered by women have a substantially higher degree of success and longevity,[45] women's voices, particularly those from marginalised communities, are often excluded from key environmental decision-making structures.* Currently, only twenty-five countries have a woman serving as head of state or government,[46] just 17 per cent of climate-related ministerial positions are held by women,[47] and a mere twenty-eight of the world's top 500 companies are led by women.[48]

This exclusion does more than perpetuate inequality – it fundamentally weakens our collective capacity to respond to and mitigate environmental crises. In an era when climate change and nature degradation threaten the fabric of societies, embedding diverse voices and perspectives into decision-making is not just a matter of justice but a necessity for resilience. If we are serious about tackling these colossal challenges, we cannot afford to keep ignoring the voices of those who have firsthand experience of these crises. Without them, our ability to confront the urgent challenges of our time is critically impaired.

---

* According to the UN, peace agreements brokered with women's participation are 35 per cent more likely to last at least fifteen years.

Very often, the destruction of nature goes hand in hand with the exploitation and marginalisation of vulnerable communities, forming a reinforcing cycle that amplifies inequality. That is, violence to Earth mirrors violence to communities.

At the heart of this dynamic is a tangled web of historical processes and enduring legacies: colonialism's systems of exploitation, the Industrial Revolution and the expansion of consumption-driven capitalist economies, and the deeply held belief that humanity has mastery over nature – and, by extension, those communities that are closely tied to nature. Together, these forces have shaped, justified and extended the exploitation of nature and the communities intimately tied to it – entrenching systemic inequality, creating enduring cycles of dependency and leaving a legacy of environmental destruction that continues to define humanity's relationship with nature.

In today's global economy, natural resources – that is, the stuff we take from nature to wear, eat, warm our homes and build things – are predominantly dug up from poor nations to meet the consumption demands of wealthy nations, who reap most of the benefits from this arrangement. In this way, poorer nations are expected to serve as the 'world's factory', extracting and producing goods at minimal cost to satisfy the ever-growing consumption demands of wealthy economies.[49]

High-income countries consume six times more resources and generate ten times the climate pollution of low-income nations.[50] In 2015 alone, the Global North consumed US$10.8 trillion worth of resources – materials, land, energy and labour – extracted from the Global South, a sum large enough to end extreme poverty seventy times over.[51] Over the twenty-five years from 1990 to 2015, the Global North appropriated $242 trillion worth of resources and labour from the Global South. For every dollar in aid given to the Global South, the Global North takes $30 worth of resources through trade.[52]

This inequity is particularly pronounced in Africa. Despite

being a continent of immense natural resources, Africa has the highest rates of extreme poverty in the world, with two-thirds of the global population living in extreme poverty residing in Sub-Saharan Africa.[53] Yet much of this wealth is controlled by foreign multinational corporations that export profits to their home countries. Most mining companies in Africa are internationally operated and companies headquartered in the US, Canada and Australia dominate ownership of mines in Africa.[54]

Today, Nigeria is one of the largest economies in Africa and its top crude oil exporter.[55] More than 93 per cent of the crude oil produced in Nigeria is exported.[56] Despite fuelling much of the world through their vast natural resources and oil and gas reserves, at least 40 per cent of the Nigerian population lack access to electricity[57] and 38.9 per cent of the population lives in poverty.[58]

Crude oil exploration and production in Nigeria began under British colonial rule, when the country's resources, including oil, were exploited for the benefit of the empire.[59] In 1938, the colonial government granted the Royal Dutch Shell Group (which would later become Shell-BP and then Shell) an exclusive exploration licence to search for oil throughout Nigeria.[60] This gave Shell a 'virtual monopoly over oil exploration in the country and Shell has remained the dominant oil company in Nigeria'.[61]

The oil-rich Niger Delta region has been the world's epicentre of oil spills. An average of 240,000 barrels of crude oil are spilt in the region every year – enough to fill more than fifteen Olympic swimming pools.[62] Between 1976 and 1992, more than two million barrels of oil were spilled across Ogoniland in the Niger Delta, in 2976 separate incidents.[63] These spills have devastated the region, contaminating soil and water to the extent that traditional Ogoni fishing and farming practices are no longer viable. Rivers in the Niger Delta can no longer sustain marine life, and crops grown in the soil are overwhelmingly contaminated with carcinogens and toxins. While the Niger Delta is one of the world's ten most important wetland and coastal marine ecosystems, it is also the world's most severely petroleum-impacted ecosystem.⁶⁴ Endangered species in the region, like the Niger Delta red colobus monkey, Sclater's monkey and West African manatee, are threatened with extinction from habitat destruction, pollution and ecosystem disruption.⁶⁵

The Indigenous Ogoni people have been engaged in a decades-long battle for justice against Shell and other oil companies operating in the region.⁶⁶ The peaceful protest of the Ogoni people – speaking against environmental and human rights abuses – has been met with brutal repression and violence. According to Amnesty International, 'Shell repeatedly encouraged the Nigerian military to deal with community protests, even when it knew the horrors this would lead to – unlawful killings, rape, torture, the burning of villages.'⁶⁷ In 1995, nine Ogoni activists were hanged by the Nigerian government. In 2017, Amnesty International accused Shell of 'complicity in the unlawful arrest, detention and execution' of these men.⁶⁸

In recent years, Shell has been selling off its remaining onshore oilfields in the region, citing 'community unrest', 'sabotage' and a 'company-wide refocus on promoting green energy' as the reasons.⁶⁹ Instead, the company will focus on 'deepwater and integrated gas businesses' in Nigeria, which means it will drill into the ocean floor to extract oil and gas, away from the scrutiny of local communities.⁷⁰ Locals and lawyers see this move as Shell dodging its responsibility to clean up after itself, leaving the burden of social and environmental damage for locals to deal with.

Nigerian environmental activist Celestine Akbopari says, 'Shell has to restore our environment and lost livelihoods before selling anything. Our environment should be restored to the level Shell met it ... Our people enjoy their fishing and farming business but can't do that anymore.'⁷¹ Osai Ojigho, director of Amnesty International Nigeria, says, 'Had this level of contamination and pollution occurred in Europe or North America, it is hard to imagine that there would not have been swift and severe consequences and legal redress. Shell should clean up the pollution the oil has caused in these communities and compensate those whose livelihoods have been devastated and whose health has been harmed.'[72]

In 2022, the Court of Appeal in the Hague held Shell Nigeria liable for the spills and ordered the company to pay US$15.9 million[*] in compensation to affected communities. 'The settlement is on a no admission of liability basis, and settles all claims and ends all pending litigation related to the spills,' Shell said.[73] That same year, Shell reported a net income of US$42 billion. The total annual remuneration of Shell's CEO (in salary, benefits and bonuses) in 2022 was US$12.4 million.[**][74]

By 2023, more than 13,500 residents from communities in the Niger Delta had filed claims against Shell demanding that the company clean up oil spills, which they say have wrecked their livelihoods, poisoned their wells and polluted their land and water, meaning they can no longer farm or fish.[75]

Over the span of seven decades, 7.95 billion litres of oil have been spilt across the Niger Delta.[76]

Today, the world's poorest nations and communities bear the brunt of the impacts of environmental degradation, biodiversity destruction, pollution and climate change, despite contributing the least to these crises and being least equipped to endure the consequences. They aren't just expected to produce and pollute to meet the demands of wealthy countries, businesses, and communities – they are also left to clean up the mess.

The story of environmental inequity is one of shifting burdens. Wealthier nations, while claiming environmental progress or advances in waste management, have largely achieved these successes by exporting harm to the poorest communities and countries. It's a system that allows consumption patterns in the richest countries to thrive unchecked while condemning millions to live amidst pollution, waste and the effects of a rapidly warming planet.

---

*   Converted from €15 million.
**   Converted from £9.7 million.

As David R Boyd says, 'The current economic and business paradigms are based on exploiting people and nature. Among the fundamental flaws of these paradigms are a belief in limitless growth, short-term thinking, a narrow focus on maximizing profits for shareholders, and the externalization of social, health and environmental costs onto society.'[77]

At its core, this is a moral and economic failure: a world where the wealth of one nation is built upon the ruin of another, where the luxury of a few depends on the sacrifice of many and where waste and pollution become weapons of inequality.

Yet out of the shadow of environmental colonialism, towering voices have emerged: leaders and movements that are turning the tides of exploitation and inequity into rallying cries for justice. Among them stands Mia Amor Mottley, the prime minister of Barbados. Described as 'bold, fearless' and an 'embodiment of [Barbados's] conscience', she is the first female prime minister of Barbados and is, quite simply, a force to be reckoned with.[78] When Mottley takes to the stage, she is formidable, relentless and unforgettable. Not one to mince words, she was once asked by a reporter why vulnerable countries were still so debt-stricken and riddled with corruption. 'Do you really want me to answer you?' Mottley replied. At the reporter's insistence, Mottley delivered a blistering response:

> Why is it that every time we talk about countries from the South, the first allegation is corruption? Last time I checked, in the USA and UK, and Europe, they're riddled with corruption but nobody says that they're not capable of achieving their objectives because of corruption.
>
> Why is it that we are not talking about the fact that these countries became independent having allowed those countries that colonized them to extract significant portions of their wealth? Such that we had no proper housing, no proper education, no proper health care systems, no proper legal systems ...

Now when our blood, sweat, and tears finances the industrial revolution, and the industrial revolution then causes a climate crisis, and then I have to pay for the consequences of the climate crisis because of the industrial revolution, financed by our blood, sweat, and tears. Then I think that they have no moral authority to tell me anything about the financing of the climate, or about why we don't have enough.[79]

As Mottley says, 'We in the islands are the canaries.'[80] The first to suffer, the first to die – victims of a climate crisis they did little to create.

Mottley has risen as an icon of environmental justice. Representing a country of fewer than 300,000 people, she takes to the world stage and electrifies audiences. From condemning wealthy nations for failing to meet their environmental obligations to challenging global leaders with the piercing question, 'When will leaders lead?' Mottley has become a powerful voice for small island nations – and perhaps for the entire world.[81]

Indeed, few people have changed the rules of the game more than Mia Amor Mottley. Through her powerful advocacy, Mottley is leading a push to rewire the global financial system, ensuring poorer nations can access the funds they desperately need to deal with the consequences of climate change and nature destruction. In 2022, she unveiled the Bridgetown Initiative, a groundbreaking plan to reform how wealthier nations finance climate responses in the Global South. Named after Barbados's capital, the initiative calls for emergency liquidity to address crippling debt crises and proposes expanding multilateral lending by $1 trillion – offering a lifeline to nations on the frontlines of environmental crisis.[82]

And it's not just Mottley. Leaders from small island nations, on the frontlines of environmental collapse, are demonstrating to the world a brand of bold, compassionate and fearless leadership that is reverberating around the world.

As Samoan climate activist Brianna Fruean, says, 'We are not drowning – we are fighting.'[83]

National leaders are stepping up, from the foreign minister of Tuvalu, who stood knee-deep in water to highlight the plight of his nation;[84] to the prime minister of the Cook Islands, who declared to world leaders, 'Our survival is being held to ransom at the cost of profit and an unwillingness to act despite the ability to do so';[85] to the prime minister of Saint Kitts and Nevis, who said, 'We are small, yes, but our voices carry the weight of rising seas, thundering storms and livelihoods teetering on the edge of erasure';[86] and the prime minister of the Bahamas, who said that while states 'can somehow quickly find eye-watering sums of money for bullets and bombs', they only 'rattle small change in their pockets' to finance the costs of environmental repair and recovery.[87] Together, small island nations are fighting.

Indeed, it was small island nations that fought the hardest for the UN global climate treaty, playing a pivotal role in promoting the inclusion of the 1.5 degrees Celsius temperature limit in the Paris Agreement. They laid the foundation for the Fossil Fuel Non-Proliferation Treaty Initiative, calling for a global phase-out of fossil fuels. They initiated the largest legal case in history, seeking a groundbreaking judgement from the ICJ on nations' obligations to protect their citizens from climate harm. Small island nations, united under the Alliance of Small Island States, are amplifying their voices to demand ambitious climate action. They co-founded the High Ambition Coalition for Nature and People to pressure major polluters to raise their climate commitments. At the forefront of negotiations, these countries championed the establishment of a historic Loss and Damage Fund, demanding that wealthy nations help fund the cost of the devastating impacts of climate change on the world's most vulnerable communities.[88]

Small island nations, though small in landmass and population size, have become giants in the fight for environmental justice.

Defiant against the forces of environmental colonialism, they refuse to be the world's sacrifice zones. Instead, they are charting a path not just for their survival, but for the survival of us all. And, in the process, they are showing the world what true leadership looks like.

Chapter 12

Conspicuous Consumption: The Emperor's Jewels

> Clothes do not merely make the man, the clothes
> are the man; that without them he is a cipher, a
> vacancy, a nobody, a nothing.[1]
>
> Mark Twain

Caligula was a pervert and a tyrant. If the Roman emperor had any redeeming qualities, they weren't immediately obvious. Unencumbered by things like morality and ethics, his charm lay largely in his fondness for excess, incest and violence.

Born as Gaius Caesar Augustus Germanicus, he is better known as Caligula, a nickname that has come to be synonymous with extravagance and sexual deviance.*

He enjoyed sex with his sisters and had a great fondness for watching lions and panthers eat humans.[2] He declared himself a god, erected statues in his own honour and waged a war against Neptune, the god of the sea.[3]

But his power and esteem were best demonstrated through his wealth and decadence. He rolled around in piles of money, drank

---

* Much of the information about Caligula's actions comes from ancient sources that were written after his death, some of which may have been exaggerated.

CONSPICUOUS CONSUMPTION: THE EMPEROR'S JEWELS

wine infused with dissolved pearls, feasted upon meals sheathed in gold and erected a marble palace for his horse – equipped with a retinue of slaves and luxurious furniture.[4] He once ordered the construction of a two-mile floating bridge which had the sole purpose of allowing him to theatrically gallop back and forth on his horse in front of awed onlookers.[5] Another time, he built a floating pleasure palace in which he could host grand orgies.[6] Before he was stabbed to death by his own guards, Caligula had emptied Rome's treasury and sent the empire into economic crisis.

But Caligula knew one thing: silk, jewels and riches didn't just decorate a person – they defined them. He wore his riches as his crown, sword and shield, knowing that in the eyes of the empire, worth was defined by what glittered around him. Extravagance wasn't just vanity; it was strategy. Power was a spectacle: the more it dazzled, the more it dominated.

The Roman Empire didn't just shape Western culture – it built the scaffolding for much of it. Modern legal systems, government institutions, religious traditions, engineering feats, philosophies and even languages bear Rome's imprint. But just as important as its ideas was its material world – coins, statues, textiles and luxury goods. These weren't just accessories to Roman life; they were tools that shaped its society. In Rome, material culture was a language of power, a way to communicate status, authority and imperial reach.

Jeremy Tanner and Andrew Gardner, editors of *Materialising the Roman Empire*, argue that material culture played a central role in both the rise and fall of the Roman Empire. 'Material culture was integral to the process of imperialism,' they said, 'both as the Empire grew, and as it fragmented.'[7] Gold, silks, spices and art were not mere indulgences in the Roman Empire – they were tools of empire: markers of social status and instruments of superiority and power. Trade routes pulsed with treasures that were proof of the empire's might, while banquet tables groaned under the weight of decadent feasts, the meat of rare and exotic animals paraded for a few fleeting bites. Public spectacles – lavish, violent and utterly mesmerising – were statecraft masquerading as entertainment. Gladiatorial combat, triumphal parades and grand festivals reinforced Rome's social order, turning blood and theatre into instruments of governance.

Roman material culture was embedded into the very fabric of daily life as an expression of power, embedding the empire's dominance into daily existence and reinforcing a hierarchical society where material wealth equated to authority.

From sprawling infrastructure and monumental architecture to statues and monuments, to coins, artworks, clothing, furniture and glittering adornments – these materials were a physical manifestation of the empire. They were symbols, that were as heavy with ideology as they were with stone – a reminder of Rome's omnipresence.

Roads weren't just for travel; they were arteries of empire, stone-carved reminders of Rome's reach. In fact, poorly maintained roads were 'symbolic of an ageing empire in decline'.[8] Aqueducts didn't just deliver water; they demonstrated how even water would bow to Roman engineering. Coins, stamped with the emperor's face, weren't just currency; they were 'monuments in their own right', reflecting political messages and imperial achievements and showcasing the divinity and prestige of the empire.[9]

Through its material culture, Rome projected its ideology, structured its society and legitimised its authority. The empire's presence was made tangible through stone, metal, cloth and spectacle. The currency of power became the accumulation of material goods, and the empire spent freely.

The empire's reliance on status-driven consumption entrenched its hierarchies. Luxuries weren't just indulgences; they were symbols of rank, tools to separate the elite from the masses. Wealth had to be performed. Banquets dripping with opulence, homes adorned with paintings and silverware, and slaves paraded for their aesthetic appeal – all these were public declarations of social superiority. To fail in this theatre of excess wasn't just embarrassing; it was disqualifying.

While extravagant displays fuelled trade and artistic innovation, they also sowed the seeds of decay. Wealth legitimised power, but it also eroded the foundations of stability as greed tore apart the natural world and consumed the intellectual and civic pursuits that had once defined Roman greatness. The line between material culture – the objects that shaped social, economic, and political life – and materialism – the pursuit of luxury for spectacle – began to blur.

Roman naturalist and philosopher Pliny the Elder saw this danger. He condemned the empire's unrelenting appetite for luxury, warning that greed was both wasteful and morally corrosive. To him, material excess hadn't saved the empire, it had poisoned it. 'The wide expanse of the [Roman] world and the fullness of things has become a curse,' he lamented.[10] In developing its society, Rome was not only depleting the Earth's resources – it was, he feared, digging its own grave.

'To satisfy the requirements of luxury,' he said, 'we trace out all the veins of the earth ... We penetrate into [the Earth's] entrails, and seek for treasures.' This, he warned, will 'urge us to our ruin, that send us to the very depths of hell'.[11] The material culture of the Roman Empire was the backbone of its identity and development. But therein lay the paradox of the empire's wealth: the more Rome conquered and extracted from the environment, the more it consumed itself.

There are troubling parallels between Rome's trajectory and our own world. Modern society thrives on the same principles of material signalling that underpinned Roman excess. Today, luxury goods aren't just physical materials – they're aspirational narratives and tools of power. Designer brands, high-end gadgets and influencer-driven lifestyles serve the same function as Caligula's gilded banquets: they transform consumption into identity. The global consumer market doesn't cater to necessity; it caters to status. Just as Roman subjects expressed loyalty, power or resistance through material choices, modern consumers use purchases to signal cultural affiliations or personal values. These material goods carve hierarchies and define power. And in excess, they erode societies.

Wealth isn't just accumulated, it's broadcast. Contemporary advertising and social media amplify the ideology of materialism, creating a feedback loop where possessions define value, and value drives possession. The result is a society in which happiness, success and even self-worth are measured by what we can display.

Just as Rome built its empire on unsustainable consumption, so too does our world stagger under the weight of its own appetite. Mass production, over-reliance on nature and its finite resources and an endless thirst for more – our gilded age risks ending in the same slow collapse. Rome's lessons aren't in its glory; they're in its ruins. And modern society is no less captivated by spectacle than the Romans were.

For $25,000, you can turn your feet into tombstones by encasing them in a curated selection of endangered and exotic animal species. The Air Jordan 1 'Brooklyn Zoo' is a custom-made, limited-edition 21st-century sneaker that allows wealthy people to strap a small wildlife reserve to their limbs. It is accented with gold stitching and is painstakingly constructed from the skins of nine dead animals: elephant, ostrich, python, anaconda, crocodile, alligator, lizard, stingray and cow. Launched in 2012, only ten pairs of these shoes were ever made, making them as exclusive and coveted as they come – the very height of extinction couture for your feet. They boldly declare, 'I have so much money that my morally flexible toes are cradled by elephants.'

Caligula's extravagant legacy never truly ended; it just took on new forms in modern society.

In New York, the Golden Opulence Sundae is swaddled in edible gold leaf and sells for US$1000. It's a more modest version of the

CONSPICUOUS CONSUMPTION: THE EMPEROR'S JEWELS

Frrrozen Haute Chocolate ice-cream sundae from the same restaurant, which was made of the rarest cocoa sat atop a diamond bracelet, costing US$25,000.

But such culinary frivolities barely scratch the surface of today's theatre of excess.

The Diamond Himalaya Birkin handbag by Hermès is constructed out of niloticus crocodile skin dyed in a white-and-grey gradient intended to 'evoke the white-capped peaks of the Himalaya Mountains' and is encrusted with over 200 diamonds. One of the bags sold at auction in 2022 for over US$450,000.[12] Meanwhile, the Stuart Hughes iPhone 5 Black Diamond edition – a gilded gadget encrusted with over 600 white diamonds and adorned with a 26-carat flawless black diamond as its home button – sold for over US$15 million.[13] It has no additional functionality from a regular iPhone.

Then there's Fendi's €1 million fur coat, made from the fur of about forty sables – small furry mammals from Russia. The fur was tipped with pure silver, 'giving a unique and contemporary luminous metallic effect'.[14]

The Graff Diamonds Hallucination watch is a US$55 million-dollar multi-coloured timepiece set with over 110 carats of 'extraordinarily rare' coloured diamonds. It is the world's most expensive watch and may require its own security detail.[15]

Fragrance enthusiasts can opt for the Oud perfume collection, distilled from endangered agarwood trees and priced at around $4500 per bottle.[16] Its allure is so potent that illegal trafficking of agarwood is now on the rise.[17] Spritz it on your wrist and you can smell like ecological collapse and irreparable loss.

The Sultan of Brunei has a collection of 7000 luxury cars, one of which is a US$14 million gold-plated Rolls-Royce.[18] This is in addition to his gold-plated private jet, known as the 'flying palace', and his own private zoo.[19]

Amazon founder Jeff Bezos has a US$500 million custom-designed superyacht that is almost certainly compensating for something. So large is the yacht that the city of Rotterdam announced it would have to dismantle a historic bridge to allow it to fit through.* [20]

Billionaire businessman Mukesh Ambani built himself a twenty-seven-storey skyscraper mansion in Mumbai valued at around US$2 billion and equipped with helipads, 168 parking spaces, a salon, an ice-cream parlour, a private movie theatre to accommodate fifty people and a room that produces artificial snow.[21] The property soars above the nearby slums and is reportedly built atop illegally sold land that belonged to an orphanage.[22]

These aren't just homes or vehicles. They're monuments to excess and status, built on the back of a planet stretched to its limits and its most vulnerable people.

Global spending on luxury goods and experiences reached €1.48 trillion in 2024, with the market expected to grow 4 to 6 per cent annually until 2030.[23] This growth is fuelled by a culture (and economy) that glorifies excess and consumption, and equates material wealth with personal worth and success. The higher the price, the more status you are buying. These aren't merely luxuries – they're mirrors, reflecting a world that knows the price of everything and the value of nothing, one that is exacting a heavy toll on the environment in the process.

There is a term to describe the peculiar human compulsion to flaunt wealth: 'conspicuous consumption'. It was coined by American sociologist and economist Thorstein Veblen, and it describes the way individuals display their wealth to gain social standing. He said, 'It is not sufficient merely to possess wealth or power. The wealth or power must be put in evidence, for esteem is awarded only on evidence.'[24] In other words, if your Rolex isn't seen by someone who knows what a Rolex is, it might as well be a sundial. It's a form of 'wealth activism', a decidedly unsubtle performance of our consumption habits.

---

* After significant public backlash and media attention, the Rotterdam government confirmed that the bridge would not be dismantled.

CONSPICUOUS CONSUMPTION: THE EMPEROR'S JEWELS

Many of us, whether consciously or unconsciously, find ourselves complicit in a culture of unrestrained consumption – a hallmark of a capitalist economy that measures progress through the production, consumption and accumulation of resources. We want more: oversized homes, annual mobile phone upgrades, platinum frequent flyer memberships, wardrobes full of clothes with their tags still on. The flaunting of wealth and consumerism has become so deeply ingrained in our societal fabric that it often escapes notice. And resisting this culture proves to be a formidable challenge.

This behaviour stems from a complex interplay of psychological drives, societal expectations and economic systems. French philosopher Jean Baudrillard observed that in the 20th century, capitalism underwent a clever rebranding, shifting its focus from not just producing goods but also creating demand for them. It turned its focus to shaping our desires, persuading us to see products not just as useful but as mirrors of our status and identity. He called this 'sign value' – the idea that what we buy is not just about practical use but what the item says about us, like wealth, style or prestige.[25] Building on Veblen's idea of conspicuous consumption, he suggested that modern life is built around this display of goods. Just as words have meaning based on their context, the value of what we own is tied to its place in a system of status and prestige. The luxury watch isn't just a timepiece; it's a declaration of your place in a social hierarchy. Capitalism doesn't just encourage conspicuous consumption but requires it, turning desires into products, products into status and status into profit. In short, we consume not just to live, but to define and advertise who we are.

Evolutionary psychologists have helpfully waded into this topic to explain that conspicuous consumption may not just be a capitalist indulgence but an ancient mating ritual.[26] To illustrate their point, they referred to none other than Charles Darwin and his theory of sexual selection. As we know, it was Darwin's frustration with the peacock's cumbersome but beautiful tail that led him to the theory of sexual selection, which explains the evolution of physical and behavioural traits that improve reproductive success, even if those traits may be a hindrance to an individual's survival. Elaborate courtship displays are intended to irresistibly draw potential paramours to engage in lovemaking. Much like the kākāpō who waddles up a hill, digs a hole and bellows into it to attract a mate, billionaires may be engaging in their own form of bellowing and feather-waving with admirable enthusiasm: They are hoping that somewhere, someone is appreciating just how terribly desirable they are.

This led these evolutionary psychologists to wonder: are billionaires buying superyachts for the same reasons peacocks grow tails? Is the Sultan of Brunei, with his fleet of 7000 luxury cars, engaged in his own avian-esque display? Are we all, in some way, performing our version of a wildlife mating call?

In 2011, ScienceDaily posed the question, 'Does driving a Porsche make a man more desirable to women?' The article was based on research that found 'men's conspicuous spending is driven by the desire to have uncommitted romantic flings'.[27] If you've ever scrolled through dating apps like Tinder, Hinge or Grindr, you have likely seen this phenomenon illustrated with alarming precision. It is here that you will indubitably find a potential match casually reclined in a first-class plane seat, sprawled across a luxury car or artfully displaying a wrist embellished with a Rolex watch. The lead author of the paper, Jill Sundie, explained that, 'This research suggests that conspicuous products, such as Porsches, can serve the same function for some men that large and brilliant feathers serve for peacocks.'[28]

Evolutionary psychologists, whose job is to explain why humans are so peculiar, believe that conspicuous consumption serves two primary evolutionary goals for men and women alike: to attract mates and gain social status. According to them, we are driven by an evolutionary need to signal our desirable traits – like wealth, abundance, status or beauty – which enhance our chances of securing mates, forming alliances and navigating social hierarchies. These signals allow us to demonstrate fitness and ensure that we stand out in competitive environments. In essence, consumption becomes a form of communication, a way to broadcast how good we are at surviving. Or, more accurately, communicate who we are and what we bring to the table, both biologically and socially. In other words, we may be hardwired to show off.

After all, a Hermès handbag or a diamond-encased iPhone are not functional or necessary for survival – they are ornamental. Just like the peacock's tail, they are signals – beautiful, seemingly unnecessary for survival and entirely dependent on others noticing them.

However, evolutionary instincts are only part of the story. While evolution provides a compelling framework for understanding certain aspects of consumption behaviour, it is far from a complete explanation and is often oversimplified. It fails to fully account for the cultural, historical, economic and social forces that shape what we consume, who participates in conspicuous consumption and how these signals are interpreted.

As we know, the theory of evolution has often been misapplied to justify inequalities and stereotypes. Much like Charles Darwin's theories being distorted and weaponised to justify Social Darwinism – a pseudoscientific rationale for eugenics and racial hierarchies that was ultimately brandished by Hitler. Similarly, gender and social biases underpinned the formulation of much of evolutionary theory, which is only recently being challenged by scientists via bonobos and their bulging clitorises.

In the same way, evolutionary assumptions about conspicuous consumption – such as men's worth being tied to their financial status and women's worth being tied to their beauty – are social constructs underpinned by historical biases that are often mistakenly framed as biological truths. These are cultural and social constructs, not immutable truths.

While evolution may partially explain our subtle desires to signal wealth or attractiveness, it is not encoded in our genetic makeup to buy designer handbags or superyachts. These behaviours are mediated by cultural norms, economic systems and personal choices. This means that the Sultan of Brunei didn't purchase 7000 cars because his DNA or ancient forebears told him to. He did it because his immense wealth and the economic, social and cultural context made it possible.

According to evolutionary psychologist Geoffrey Miller, the modern capitalist system exploits these evolutionary instincts, creating endless cycles of consumption that prioritise signalling over substance.[29] As Miller says, income inequality, luxury brand marketing and the aspirational narratives of social media further fuel this phenomenon, making material excess a marker of success and worth. While Veblen's original critique focused on the wealthy elites of the Gilded Age in 19th century America, conspicuous consumption is now more widely distributed. Social media platforms have turned everyone into potential participants in this performance. Indeed, social media (and even traditional media) has transformed conspicuous consumption into a twenty-four seven global stage.

A holiday or a diamond ring isn't just experienced; it's documented, shared and liked – feeding an algorithmic loop of validation. This creates what some sociologists call 'aspirational inequality', in which individuals feel compelled to spend (and consume) beyond their means to keep up with influencers, celebrities or even friends. In this way, conspicuous consumption goes well beyond the realm of individual vanity: it profoundly shapes collective societal behaviours and financial decisions. Brands understand this and market accordingly, crafting products as lifestyle props and extensions of one's identity.

At its core, conspicuous consumption is humanity's most glittering paradox – a collision of evolutionary instincts, economic systems and societal constructs dressed up in designer couture and shared on Instagram. While it may be partially born of an ancient drive to display fitness, it has been refracted through capitalism and algorithmic culture, and now manifests as curated excess, divorced from utility or survival. A designer handbag doesn't just carry our wallet; it carries the unspoken narratives of our worth, our aspirations and our place in the social cosmos.

Evolutionary social motives certainly influence much of modern behaviour. But culture, society and economic systems dictate the specifics of how we choose to demonstrate our perceived worth, value, status and desirability: often through luxury cars, designer handbags and opulent mansions.

Ultimately, the true cost of consumption is borne by the planet and its most vulnerable inhabitants. Every act of consumption – whether eating, shopping, driving or flying – leaves a mark on ecosystems and vulnerable communities, shaping the world for future generations. These impacts can be measured in many ways, but two common methods are the ecological footprint and the carbon footprint.

A carbon footprint represents the total amount of greenhouse gas pollution that we emit through our actions (and lifestyle). Greenhouse gases are a natural part of the Earth's system and help keep the planet at a stable temperature to support life. But an excess of these gases disrupts the Earth's energy balance, with dire consequences. And humans are creating a lot of excess gas.

Think of it like this: greenhouse gases form a layer around the Earth, similar to a cosy blanket. Normally, this blanket traps just the right amount of heat to keep the planet at a liveable temperature. But as we burn fossil fuels like coal, oil and natural gas, we're making the blanket thicker and thicker. Soon, the blanket stops being cosy and becomes suffocating. It traps too much heat, altering the Earth's entire climate system. This creates a kind of 'climate chaos' where different regions experience different effects: the trapped heat fuels extreme weather events like stronger and more frequent storms and fires, prolonged droughts and intense heatwaves. It also accelerates the melting of glaciers and ice caps, raises sea levels and alters ecosystems, making it harder for plants, animals and people to adapt.

Globally, the average person emits 6.7 tonnes of greenhouse gases each year. However, this varies greatly between countries. In Qatar, the average person emits over 68.5 tonnes of greenhouse gas pollution. In Australia, the average is 22 tonnes per person. In the USA, it is 17.2 tonnes per person, while China and Singapore both sit at 9.8 tonnes per person. The lowest emitters include countries like Burundi, Rwanda and Yemen, which emit less than 1 tonne of greenhouse gases per person.[30]

An ecological footprint, on the other hand, measures how much land and water are used to support our lifestyles, from the forests cleared to grow food or make clothes to the minerals used for our technology. Globally, the average person uses 2.77 global hectares of land (and seas) through their current consumption habits in a single year. That is roughly 5.2 football fields of land. Yet this, too, varies widely by country. In the US, it's 7.8 hectares per person (14.6 football fields), while in Australia, it's 6.1 hectares (11.4 football fields). In China, it's 3.5 hectares, while in Kenya, Nepal, Yemen, Afghanistan, Ethiopia, Burundi and Rwanda, it's less than 1 hectare per person.[31]

Globally, we are consuming natural resources (like forests, water and arable land) and generating waste (like greenhouse gas pollution) at a rate that is 1.75 times faster than the planet's ecosystems can replenish. This means that to sustain current average global consumption levels, humanity is already using resources at a rate that would require 1.75 Earths. If everybody on the planet consumed like the average person in the US, we would need 5.1 planet Earths to sustain our current lifestyle. If everybody consumed like the average Australian, we would need 4.5 planets to sustain our current lifestyle.[32]

Within this data lies a clear story about the connection between wealth, consumption and environmental harm. Unsurprisingly, wealth tends to correlate with higher levels of consumption, which in turn creates a much larger environmental impact. In short, the more wealth you have, generally, the bigger your footprint is.

In fact, just fifty of the world's richest people generate more carbon dioxide pollution from their consumption habits (including their private jets, superyachts and investments) than the combined consumption emissions of the poorest 155 million people.[33] These numbers don't include ecological footprint, nor do they include other greenhouse gases like methane – they are only measuring carbon dioxide.

The environmental impact of these fifty people is so immense that the carbon dioxide pollution resulting from their investments alone – over just a decade (2018 to 2028) – will cause an estimated $250 billion in economic damage by 2050, due to climate change wreaking havoc on crop production and health.[34]

For example, Russian oligarch Roman Abramovich has a $600 million superyacht that belches 22,000 tonnes of carbon dioxide each year.[35] To put this in perspective, it would take the average person over 3300 years to emit this much carbon pollution.* And it would take about 1500 years for someone in the bottom 99 per cent of global income earners to produce as much carbon dioxide pollution as the richest billionaires do in a single year.[36]

Though it isn't just ultra-wealthy. While the wealthiest 0.1 per cent of the world's population have an enormous carbon footprint – with the average billionaire emitting over 200 tonnes of carbon dioxide each year[37] – it is actually the wealthiest 10 per cent of the global population that accounts for half of all global carbon dioxide emissions.[38] This is problematic. You see, the Earth has a 'carbon budget', which is essentially a limit on how much pollution we can put into

---

* This estimate uses global average per capita carbon dioxide emissions of approximately 6.6 tonnes per year (Oxfam, 2024). Over an average global lifespan of 73 years (World Health Organization, 2024), this amounts to roughly 482 tonnes of CO_2 per person in a lifetime. Roman Abramovich's superyacht emits an estimated 22,000 tonnes of CO_2 annually – equivalent to the lifetime emissions of about forty-six average people, or what one person would emit over more than 3300 years.

the atmosphere before causing increasingly dangerous changes to our climate. If the top 10 per cent of pollution emitters maintain their current consumption and pollution levels, they alone will use up the entire carbon budget by the year 2046.[39]

Making matters worse, conspicuous consumption is no longer limited to the top earners. Lower- and middle-class consumers are increasing spending, while also entering the luxury goods market at a faster pace than ever before. Today, they account for nearly half of the luxury market globally.[40] In the fashion industry – which accounts for 10 per cent of all greenhouse gas emissions – luxury brands are expanding their reach to access new consumers by marketing their more affordable luxuries to middle-income consumers, while fast-fashion brands are mass-producing clothing that mimics luxury goods, driving increased over-consumption across middle-income earners.[41]

The result is a world that is living well beyond its ecological means. Nature provides resources like food, water and energy, but if we use them faster than they can be replenished, we're effectively overdrafting on nature while also trapping greenhouse gases in the atmosphere to such a degree that the planet becomes virtually uninhabitable for millions of people (and species) around the world.

Of course, the consequences are felt differently across the world. In Bordeaux and Burgundy in France, climate change is threatening wine production. Rising temperatures and unpredictable weather are disrupting grape-growing seasons, reducing the quality and yield of wine. Climate change is not only threatening the livelihoods of winemakers but also centuries of intergenerational knowledge and cultural heritage tied to these world-famous regions. Even olive oil is affected. As Spain, Greece and Italy battle against severe droughts and heatwaves, olive yields dwindle, while the price of olive oil climbs steadily upward.

Extreme heat and climate-driven disasters have killed and displaced millions across the globe. Floods in Pakistan in 2022 impacted more than 33 million people, while Australia's Black Summer bushfires killed or displaced an estimated three billion animals, destroyed nearly 3000 homes and left scars on entire ecosystems. In recent years, wildfires have raged across Canada, hurricanes have battered southern USA and drought in the Horn of Africa has pushed millions into acute hunger, while relentless heatwaves have plagued Europe and overwhelmed hospitals. Right now, in fact, you are likely reading about a fresh disaster that is 'unprecedented'.

And even if you don't believe in climate change – despite consensus among scientists – your insurers certainly do. Just ask Californians. As wildfires grow fiercer and more frequent, thanks to prolonged droughts and rising temperatures, billions of dollars in damages have prompted insurers like State Farm and Allstate to exit high-risk markets or hike premiums.[42] In 2023, both companies stopped issuing new home insurance policies in California, citing wildfire costs as unsustainable. This isn't just a California problem – a disaster anywhere affects insurance everywhere. As Günther Thallinger of Allianz SE says, if global temperatures rise by 3°C – which is where we're currently headed – the insurance industry will collapse. 'The financial sector as we know it ceases to function. And with it, capitalism as we know it ceases to be viable,' he says.[43] Globally, insurers are reworking their models to keep pace with increasingly catastrophic hurricanes, floods and fires. According to the global ratings agency Moody's, global insured losses from natural disasters have averaged about US$100 billion over the past five years.[44] Insurers don't deal in opinion – they deal in data. And the data is clear: climate change and nature decline aren't up for debate; they're a reality that you are witnessing. Whether or not you buy the science, the financial consequences are impossible to ignore. Your premiums have already noticed.

This isn't normal. These events are becoming more unpredictable, more intense and more deadly. Climate change and the destruction of nature are combining to create the perfect storm, fuelling disasters while stripping away our capacity to endure them.

The Amazon rainforest, considered the lungs of the Earth, is so damaged that in some areas it is now releasing more pollution than it absorbs.[45] The ocean, which produces at least half of the oxygen we need, is under threat: warming and acidifying sea water is causing detrimental changes to life underwater and on land, reducing the ocean's ability to absorb carbon dioxide, provide safe habitats for species and safeguard life on the planet.

Together, the result is large-scale disruption that is destabilising ecosystems, threatening livelihoods and undermining the planet's capacity to sustain life. And behind this environmental chaos is a complex and entrenched web of institutions, industries, economic structures and societal narratives that convince us that our value, worth and identity is measured by the glitter and gold of our environmentally harmful consumption.

This system, refined and perfected over millennia, finally reached its absurd zenith with the birth of a golden toilet.

At precisely 4.50am on 14 September 2019, under the cover of darkness, a small band of thieves snuck into Blenheim Palace in Oxfordshire, England – the birthplace of Winston Churchill – and executed a daring heist of a golden toilet that would confound authorities for years to come.

This marked the beginning of the great golden toilet heist mystery.

The toilet in question, a fully functioning 18-carat gold commode titled 'America', was the work of Italian artist Maurizio Cattelan, valued at around US$6 million. Famous for his provocative, tongue-in-cheek creations, Cattelan is rumoured to have designed the toilet as both a satirical critique of wealth and a mirror (quite literally, in this case) to modern consumer excess.

Two days before the theft, the golden toilet opened to the public and was available for guests to experience in all of its glory. Visitors were invited to form an orderly queue for the privilege of defiling a fully functional golden toilet in the name of art. Edward Spencer-Churchill, a resident of Blenheim Palace, was of the very strong opinion that the toilet would absolutely never be stolen. Shortly before the heist, he said, 'It's not going to be the easiest thing to [steal]. Firstly, it's plumbed in and secondly a potential thief will have no idea who last used the toilet or what they ate. So no, I don't plan on guarding it.'[46] This was, of course, a regrettable oversight.

With meticulous timing, the thieves breached the palace's defences and made their way to the exhibit housing this unusual treasure. They severed it from the palace's water system before fleeing the scene with the 100-kilogram golden toilet in tow.

The heist took mere minutes, but its impact would ripple across headlines worldwide for years to come. Investigations were launched, arrests were made and interrogations were conducted with the utmost seriousness. But investigators were left scratching their heads. The golden toilet heist simply could not be solved.

Almost five years later, in February 2025, two men were found guilty of stealing the toilet.[47] To this very day, the golden toilet remains missing.

The gold toilet has become a cultural icon. The Guggenheim Museum, who had loaned the toilet to Blenheim Palace, stated that the piece was:

> making available to the public an extravagant luxury product seemingly intended for the 1 percent. Its participatory nature, in which viewers are invited to make use of the fixture individually and privately, allows for an experience of unprecedented intimacy with a work of art. Cattelan's toilet offers a wink to the excesses of the art market but also evokes the American dream of opportunity for all – its utility ultimately reminding us of the inescapable physical realities of our shared humanity.[48]

In 2017, prior to the heist, the White House under the Trump administration requested to borrow a Vincent van Gogh painting from the Guggenheim. Nancy Spector, the museum's chief curator, declined the request. Instead, she offered to loan the president the gold toilet. She noted it was 'extremely valuable and somewhat fragile' but assured the president that the museum would provide all necessary instructions for its installation and care. The White House did not respond to inquiries about this offer.[49]

In the annals of art and audacity, few heists rival the bizarre drama of the golden toilet. For all its satirical charm, the 18-carat gold toilet is a striking symbol of social inequity and the environmental cost of excess. And behind the golden intrigue is a story of resource depletion, pollution and ethical dilemmas tied to the materials and processes that made it possible. The journey from raw material to a gleaming golden toilet reveals the hidden impact of extravagance.

To create such a statement piece, approximately 100 kilograms of gold had to be mined – no small feat, and certainly no clean one. Goldmining is an environmental calamity, driving deforestation, habitat destruction and soil erosion. Hundreds of tonnes of earth must be disembowelled to yield even a fraction of the roughly 100 kilograms of gold needed for the toilet. The process relies heavily on chemicals like cyanide to separate the gold from ore, generating toxic waste that often leaches into rivers and groundwater, polluting drinking supplies and harming human health as well as the wildlife that depends on these ecosystems.[50]

The act of mining in and of itself wreaks havoc on nature. Forests are razed to make way for mines, while species are displaced, soil is left barren and vital habitats are lost. Added to that, the heavy machinery used to dig deep in the earth burns through huge amounts of fuel, releasing greenhouse gases into the atmosphere to helpfully contribute to climate change. Goldmining can also carry a grim human cost, especially in developing regions. Workers, including children, often labour in perilous conditions with inadequate safety protections. Reports have documented instances of violence, exploitation and forced labour within the industry.[51]

Once the gold was out of the ground, it had to be refined and shaped into the toilet. Refining gold means heating it to extremely high temperatures to remove impurities, a process that uses even more energy, often from fossil fuels. This creates more pollution and releases yet more greenhouse gases. After refining, the gold was melted down, poured into moulds and polished to a mirror-like shine. This final polishing step uses chemicals and abrasives that generate chemical-filled wastewater that can flow into rivers and lakes, poisoning aquatic life and disrupting ecosystems.

Transporting the finished toilet added another layer of environmental impact. It was not only enormously heavy but also valuable, requiring secure transport – likely by high-emission vehicles or even planes. Every stage of its journey added to its carbon footprint.

The golden toilet wasn't just a symbol of the extravagance of luxury; it was a reflection of the hidden environmental toll behind the arbitrary things we consider precious.

Indeed, everything comes with a cost.

To produce a single cotton T-shirt, up to 7000 litres of water is used and around 2.6 kilograms of greenhouse gases are emitted.[52] Often, chemical dyes leach into waterways, harming ecosystems and marine life.

To produce 1 kilogram of beef, about 15,400 litres of water and 326 square metres of land are used, and around 60 kilograms of greenhouse gases are emitted.[53]

And rarely are goods made to last. The modern economy thrives on a little thing called 'planned obsolescence', a perfectly irrational scheme in which goods are effectively designed to develop a slow, wheezing death rattle just as you're starting to get attached to them, forcing you to replace them out of frustration rather than necessity. Companies like Apple and Samsung release new phone models every year, often with minor upgrades, while software updates can slow down older devices, nudging users to buy the latest version, made from parts dug from the earth. Fast-fashion items are often designed to endure only a few runs through a washing machine, while everyday items like razors are designed to wear down quickly – or, at the very least, razor blades are engineered with the ingenious flaw of being entirely incompatible with the old cartridges.

Luxury goods are not immune to planned obsolescence, but they also come with their own unique range of challenges. They typically have larger carbon footprints due to their use of rare materials and energy-intensive production processes, contributing disproportionately to environmental degradation.[54] This, combined with over-consumption and conspicuous consumption, translates to a spectacular burden placed on the environment.

The answer, of course, isn't to stop consuming and live entirely off a diet of existential dread. The real trick is recognising that most of what we buy isn't meant to improve our lives but rather to impress someone we probably don't even like all that much. Being aware of the impact and drivers of our consumption behaviours invites us to rethink the culture and messaging around material goods – and how we may use these items to assign value and worth to ourselves and one another.

Everything we consume has an impact. But not all impact is proportionate. And at what point do we ask, does the shine justify the stain?

While consumptive behaviour, of course, has catastrophic implications for the environment and millions of people around the world, it can also foster a deep sense of insecurity and selfishness in us. It preoccupies us with accumulating 'stuff' for our own personal gain, and it obscures the needs of the world around us. It prevents us from valuing and being grateful for the things we have. The other lives that we share the world with become distant objects. The world isn't our responsibility. We owe it nothing.

Wealth contributes significantly to happiness when it ensures access to basic needs such as food, shelter, healthcare and education. People living in poverty often experience lower wellbeing due to the stress, insecurity and lack of safety that come with financial instability. Yet there's a tipping point – somewhere between US$75,000 and US$100,000 a year for most Western societies

– where the link between income and happiness starts to fray.*⁵⁵ Beyond this, the accumulation of wealth becomes less about joy and more about optics. Despite knowing that excessive amounts of wealth and material consumption don't deliver happiness, we continue to define ourselves, our societies and our economies by this metric.

Imagine, though, if desirability, worth, value and progress were measured not by the size of our yachts or wardrobes but by the health of our ecosystems, the vitality of our communities or through measures of wellbeing, connection, health and purpose.

In some corners of the world, this isn't just idealistic daydreaming – it's happening. A growing number of countries are boldly reimagining how we measure societal worth, moving beyond gross domestic product (GDP) and consumer-driven metrics to frameworks that champion wellbeing, sustainability and equity. It's a quiet rebellion but one that hints at an intriguing possibility: perhaps our true legacy won't be the things we accumulate but the world we leave behind – and the meaning, purpose and connection we cultivate along the way.

In 1972, Bhutan's king declared to the world that GDP was not a meaningful measurement of wellbeing and said the country should instead look at gross national happiness (GNH). In 1998, the country's prime minister took this idea to the UN as an alternative measurement for progress and development, sparking a global conversation on how we define and measure humanity's progress, value and worth. In 2008, Bhutan formally adopted the GNH framework into its constitution. This framework emphasises psychological wellbeing, health, balanced time use, education, cultural diversity and resilience, good governance, community vitality, ecological diversity and resilience and living standards. This process has led to notable achievements and lessons for the rest of the world. Bhutan has maintained over 70 per cent of its

---

* This study is from 2010 and has not been adjusted for inflation.

forest cover, is the only country in the world that absorbs more greenhouse gases than it emits and has seen a reduction in poverty rates and, importantly, an increase in happiness among citizens.[56] While there have been challenges for Bhutan, it is their pioneering spirit, imagination and willingness to challenge traditional measures of 'progress' that offer a blueprint for the world.

In 2013, the UN General Assembly adopted a resolution entitled Happiness: Towards a Holistic Approach to Development. This recognised that the pursuit of happiness was a 'fundamental human goal, and recognised that the gross domestic product (GDP) indicator was not designed to and did not adequately reflect the happiness and wellbeing of people'.[57]

Countries around the world are starting to rethink the way they measure progress, recognising that GDP is insufficient as a primary measure of progress and often drives many of the behaviours that are depleting the planet's resources.[58] From New Zealand's groundbreaking 'Wellbeing Budget' to Costa Rica's consistent happiness achieved with minimal ecological impact, to the EU's recognition of the need to shift to an 'Economy of Wellbeing', countries are redefining what success means by putting human wellbeing, environmental health and social equity at the centre of progress. Cities like Amsterdam, Copenhagen, Brussels and Berlin are embracing 'doughnut economics', ensuring human needs are met without overburdening the planet, and even the UAE has created a Ministry of State for Happiness to hardwire wellbeing into policy. The movement is spreading: Canada and Australia are incorporating wellbeing into national frameworks, while countries such as Finland, Iceland, Scotland and Wales aim to transform their economies by 2040 to prioritise sustainability, equity and happiness. In 2018, leaders and changemakers from these nations – many of them women – formed the Wellbeing Economy Alliance partnership to champion this new vision.[59] The message is clear: the movement of valuing people and the planet over perpetual growth is gaining momentum.

CONSPICUOUS CONSUMPTION: THE EMPEROR'S JEWELS

Billionaires have become focal points in discussions of inequality and environmental degradation, veritable emblems of the excesses within a consumption-driven society. This is not without reason: their lifestyles leave an environmental footprint that dwarfs those of lower-income individuals, and even countries. Yet their resources also hold unparalleled potential to drive transformative change.

US billionaire Michael Bloomberg is 'perhaps the world's single largest funder of climate activism,' according to *The New York Times*.[60] His philanthropic work with the Sierra Club's Beyond Coal campaign has helped retire over 70 per cent of coal plants in the USA.[61] He has contributed billions of dollars to fund environmental initiatives, healthcare, education, innovation and the arts. Similarly, Patagonia's billionaire founder, Yvon Chouinard, donated 98 per cent of his company's stock to fund environmental protection. 'Earth is now our only shareholder,' Patagonia announced in 2022.[62] Chouinard has been an avid environmentalist and rock climber all his life. He wears old clothes, drives a rundown Subaru and doesn't own a computer or mobile phone.

With their vast resources, the ultra-wealthy can drive the adoption of sustainable technologies, fund renewable energy and nature protection measures, and support policies that promote equity. Of course, the news reports of tech billionaires buying up luxury bunkers to escape planetary collapse while the rest of the world burns likely doesn't endear people to their cause.[63] Nonetheless, the potential for this wealth to catalyse meaningful change remains undeniable.

In recent years, the idea of taxing billionaires as well as excessive corporate profits has gained traction, largely in response to rising inequality, environmental crises and concentrated wealth. Wealth inequality has reached unprecedented levels, surpassing even the excesses of Caligula's reign.[64] Advocates for a billionaire tax and more innovative corporate taxation argue that those holding extreme wealth should shoulder a larger share of the tax burden to address systemic inequities, redistribute resources, fund essential public goods and mitigate environmental damage. Economist

Gabriel Zucman estimates that a coordinated global wealth tax of even just 2 per cent on ultra-high-net-worth individuals – approximately 3000 people with assets exceeding US$1 billion – could generate $200 to $250 billion annually.[65] This proposal received a landmark endorsement at the 2024 G20 summit, where member nations signalled agreement on implementing a minimum tax rate for billionaires.[66]

Efforts to reimagine wealth redistribution extend beyond taxing individual fortunes. One striking model of leveraging wealth for societal benefit is Norway's sovereign wealth fund. Valued at over 2.78 trillion, it is funded by revenues from the oil and gas industry and is designed to channel wealth from resource extraction into investments that serve long-term public interests.[67]

Measures such as taxing luxury emissions from private jets and superyachts, curbing corporate tax avoidance, instituting wellbeing economies and limiting windfall profits aim to reframe wealth as a tool for collective environmental and social health, rather than individual opulence. They are increasingly being considered and, in some cases, implemented.

Changing consumption patterns and deeply ingrained beliefs around wealth, consumerism and status requires major economic and policy shifts. But more than that, it demands alternative models that challenge the status quo and demonstrate their viability on the ground.

The circular economy is one model that offers a radical departure from our current 'take, make, dispose' approach to production and consumption. A circular economy is about fundamentally rethinking how we use resources, aiming to minimise waste and reduce the need for constant production by focusing on keeping materials in use – through repairing, reusing, sharing, leasing and recycling – while ensuring products are designed to last. In short, it envisions a world where yesterday's trash isn't just reduced at the source but also becomes tomorrow's treasure. A worn garment becomes new fabric, food scraps regenerate the soil and outdated

CONSPICUOUS CONSUMPTION: THE EMPEROR'S JEWELS

electronics are restored rather than discarded. Central to the circular economy model is the embrace of bold and innovative thinking.

This is precisely how the poo bus came to make its debut in Bristol, UK, in 2015. Powered entirely by human faeces and food waste, the Bio-Bus – or the 'poo bus', as it is affectionately known – is a transit system that not only moves people but also demonstrates, with alarming clarity, that we are perfectly capable of turning yesterday's cuisine into today's commute.

Indeed, it is in the face of only the gravest challenges that humanity rises to the occasion with unparalleled ingenuity. While the humble poo bus is ensuring the movements of people, other inventive solutions are cropping up worldwide, spearheaded by startups, communities, corporations and political leaders.

Over 3000 repair cafes have sprung up across the globe, transforming communities with their simple yet powerful mission: to breathe new life into broken items.[68] These volunteer-driven hubs are places of creativity and connection, where curious minds and skilled (or even unskilled) hands join forces to repair electronics, clothing and furniture, saving countless items from making their way to landfill.

In Germany, the startup Qmilk is spinning spoiled milk into silky fabrics, while Italian startup Vegea is collaborating with wineries to turn discarded wine grape skins into synthetic leather for shoes and handbags, now sold by H&M. In Australia, charities like SecondBite are rerouting quality food destined for landfill to ensure it reaches people facing hunger and food insecurity. Fishing nets are being repurposed into luxe carpets, cigarette butts are being transformed into park benches, agricultural waste is being repurposed into construction materials and furniture, and used tennis balls are being repurposed into tennis courts.

You don't even have to be alive to participate in the circular economy either – composting oneself is now also an option. Coffins made out of mushrooms allow you to find your final resting place on a bed of soft moss inside a 'living cocoon'. The nature-based coffin turns corpses into compost that enrich the soil, meaning you can keep helping the planet long after you die.[69]

The world over, innovation is not just reimagining waste, it's rewriting the story of consumption and sparking an extraordinary social and environmental movement redefining the boundaries of human ingenuity.

New laws and policies are bolstering these efforts. Many countries are adopting 'polluter pays' policies, which means that if a company causes pollution, they must pay to clean it up or fix the damage. Taking this one step further, the Extended Producer Responsibility policy is being implemented across the EU and some states in the USA, as well as Korea, China and Japan.[70] This means that manufacturers and companies are responsible for the environmental impact of their products throughout their lifespan, including what happens after the products are used. For example, if a Mars Bar wrapper ends up in the ocean, Mars Incorporated may be held accountable for that. The goal is to make polluters think twice before developing products that harm the environment, while encouraging them to adopt more sustainable practices. Across the world, countries are trialling policies and experimenting with ways to curb our current cycle of over-consumption. In 2016, France became the first country to implement a law that requires supermarkets to donate unsold food to charities and food banks, while Singapore is revolutionising construction by treating buildings like giant Lego sets. Entire rooms, complete with walls and interiors, are built in factories and then slotted together on-site. This clever approach makes construction faster, safer and much less messy, while generating less waste.[71]

In an age of ecological collapse and growing inequality, conspicuous consumption has become a distorted mirror, showcasing excess at the expense of nature and equity. The golden Rolls-Royces and diamond-encrusted iPhones are less symbols of success than they are monuments to misplaced priorities. While humans, like peacocks and kākāpōs, may be wired to signal our fitness for reproduction to others, how we signal this fitness can be changed.

CONSPICUOUS CONSUMPTION: THE EMPEROR'S JEWELS

One of the most potent arenas for this change may lie in our relationship with social media.

It was a casual Saturday in September 2020 when a couple in California, aglow with the news of their pending child, ventured into the forest and set off a smoke bomb to reveal the gender of their baby, which they intended to share with their social media followers. The smoke bomb sparked the El Dorado Fire. This fire burned for seventy-two days, destroyed more than 9000 hectares of land, destroyed twenty buildings and killed one firefighter. The couple was charged with manslaughter.[72]

This tragic episode wasn't the first of its kind. Yet it serves as a stark reminder of how the relentless pursuit of social media validation can spiral into catastrophic consequences. Social media doesn't just amplify conspicuous consumption through material excess; it incentivises and rewards performative acts, no matter the cost.

From celebrities showcasing elaborate birthday parties adorned with thousands of balloons to influencers posing with – or owning – exotic wildlife, fuelling the illegal wildlife trade, to the flaunting of mega-mansions and luxury car fleets – while the intent may not be to cause harm, that is ultimately the outcome.

Social media platforms like Instagram and TikTok democratise the ability to display wealth and curate aspirational lifestyles, turning society into a spectacle of consumption. This economy – the 'attention economy', in which likes and followers translate into social capital – rewards conspicuous displays. Here, visibility equates to value and consumption to identity.

But this culture of endless comparison and consumption has a corrosive twist. Aside from the hefty environmental costs of consumption, research consistently shows that social media not only makes people spend and consume more, it makes them miserable. A 2024 study published in the *International Journal of Marketing Studies* found that frequent social media use is directly tied to increased feelings of envy and decreased life satisfaction.[73] The reason is simple enough: constantly comparing yourself to the carefully curated, picture-perfect lives of others tends to leave you feeling unworthy. Similarly, studies have repeatedly demonstrated links between social media and depression, eating disorders, conspicuous consumption and addiction.[74] These platforms are designed to maximise engagement and foster a cycle of desire and dissatisfaction. And the result is a collective disintegration of self-esteem and happiness.

Governments and platform owners must, of course, be held accountable and play a role in creating environments that foster healthier relationships with consumption. Regulations requiring transparency in advertising and algorithm design could counteract the current bias towards hyper-consumerist content. Additionally, public campaigns promoting media literacy can empower users to recognise and resist manipulative narratives.

But individuals have an enormous opportunity to help shift this dynamic.

And the first step may just be awareness. Recognising that much of what we see online is meticulously curated – often to sell products, ideas or lifestyles – helps us critically assess what we consume. Social media platforms do not simply reflect culture; they shape it, often in ways that reinforce consumerism, amplify inequality, harm the environment and drive deep individual dissatisfaction. And the truth is, the pursuit of this idealised material lifestyle is unattainable, designed to leave people always wanting more.

Ultimately, rethinking our relationship with social media is not about rejecting it outright, nor is it about vilifying influencers who promote certain lifestyles. It is about reclaiming individual agency and, for some, using it to champion a new set of societal values. If we can shift the signals of status from what we possess to what we contribute to the world – creatively, intellectually, socially or environmentally – then we have the potential to transform conspicuous consumption from a force of waste into a better reflection of our shared values. It is about recognising that what we see online is often a polished illusion, and about disentangling our self-worth from this constructed reality.

This is also an opportunity for us to rethink what we find aspirational. Being conscious of who and what we choose to elevate in society is a small but powerful step towards ensuring that society reflects the needs, priorities and wellbeing of everybody, including the planet. This may be as simple as reviewing the social media accounts you follow and asking yourself: does that person reflect the society that you want to live in? Imagine if charities like the Red Cross, WWF and Oxfam had more followers than any one of the Kardashians. Or if social media was filled with stories of hope, resilience, community, acts of kindness, people driving positive change around the world, natural wonders or wildlife and environments being protected.

This is no criticism of any individual celebrity or influencer; in fact, they have thrived in a landscape that we, as a society, have created. But it does give us an opportunity to reflect on who and what we choose to elevate and admire – and consider the sort of society we want to work towards building.

Humanity's appetite for extravagance is as old as civilisation itself. From the opulence of Roman emperors who gilded their lives in gold and pearls, to modern billionaires who invest in yachts larger than naval destroyers, the pursuit of conspicuous consumption has been a defining feature of power and identity. But history has a habit of offering uncomfortable truths: societies that hinge their worth on material excess often find themselves buried beneath it. Rome's collapse was driven by a host of factors, among which were societal excess, decadence and over-exploitation of nature – a pattern that is reflected and repeated in modern society today.

And while we have a shared global responsibility for addressing social inequity and environmental destruction, the truth is, wealthier countries and wealthier people are linked to higher levels of materialism and wastefulness, which is unsustainable and harmful to the planet and its most vulnerable people. Wealthier countries have, in large part, been the architects of the environmental crisis. In order to address this issue, we could rethink the relationship between wealth, materialism and wastefulness and recognise we have a moral duty to the planet and its people – not just to ourselves.

Prosperity and economic growth can still be realised – and, in fact, improved – if we dedicate more resources, more brainpower and more attention towards, for example, addressing the connected crises of environmental degradation, global hunger and poverty, and inequality.

This is not a manifesto for austerity. It is not a rejection of comfort, beauty or indulgence. Instead, it is an exploration of how we might shift from consuming for status to consuming with intention. The solutions are within reach – some are already in motion – but they demand creativity, courage and a willingness to question the narratives we've inherited about what it means to live well.

The environment – and humanity – requires a profound cultural reimagining of what we prioritise as a society. And it might just start with updating your Tinder profile picture. After all, travelling first class wearing a Rolex shouldn't be sexy. What should be sexy is connection to community, acts of kindness and service, innovation to solve world problems, care for causes and nature and learning, and wearing the same outfit more than once.

Chapter 13

Tales of Deceit and Cunning: The Propaganda Playbook

> Plot idea: 97% of the world's scientists contrive an environmental crisis, but are exposed by a plucky band of billionaires and oil companies.
> Scott Westerfeld[1]

'The average man thinks another man who wears a wristwatch is decidedly a "sissy boy" and belongs in the same class as females,' read an article in the *Evening Independent* in 1917.[2]

Prior to World War I, wristwatches were taboo for men. Considered delicate, feminine accessories, they were deemed wholly unfit for the manly wrist. And for a man, there is and was no greater insult than being likened to a woman. The cultural disapproval that takes place when a man uses a product that is associated with women has a name. It's called 'gender contamination'.

The idea of gender contamination traces back to ancient cultural taboos that banished menstruating women to huts for fear they'd infect the menfolk with their menstrual energy.

Put simply, culture dictates that men risk denigrating themselves by shopping or consuming like a woman. While gender contamination can apply to men and women, it's usually more prominent among men. Women are happy enough to aspire to the so-called higher status and power associated with men – from wearing suits to driving SUVs. But men would often baulk at the thought of carrying a handbag or using a shampoo infused with the faint but unmistakable scent of lilac.

Gender contamination presents a very real concern for advertisers and marketers. You see, they want to ensure products are sold to the largest number of consumers possible – regardless of their gender. This presented an existential problem for watch manufacturers in the early 20th century. To increase sales of their watches, they had to convince men that wristwatches weren't effeminate frippery. How could they persuade these wood-chopping, fence-mending paragons of masculinity to strap a dainty timepiece to their wrists without putting their manhood in peril?

There was only one man who could do this.

Edward Bernays: the father of public relations and perhaps the most notorious marketing guru of all time. He was also, relevantly, the nephew of Sigmund Freud, the founder of psychoanalysis.

In 1918, a watch manufacturer approached Bernays and tasked him with the job of making wristwatches more manly. And so, he set out on a noble quest to make wristwatches desirable to men. He knew that he wasn't just selling a product. He had to sell a story.

To crack this particular PR problem, Bernays dove headfirst into the epicentre of masculine identity: the military. Soldiers, after all, were considered the epitome of masculinity. If Bernays could tie wristwatches to soldiers' valiant image, he could dismantle the stigma of femininity that was associated with them. But he needed more than a loose association – he needed a compelling reason why soldiers *had* to use wristwatches.

He began researching the behaviours of the American soldier. Soon, he discovered a promising detail: soldiers carried pocket watches on the battlefield. And that was his opening. Through carefully crafted stories, Bernays began to circulate the idea that pocket watches, long the trusted timepiece of men, were a liability for soldiers in war. Articles began to appear in newspapers and magazines, describing how soldiers found themselves fumbling for their pocket watches at critical moments. These tales were gripping, vivid and invoked stories of soldiers in the trenches – and may or may not have been entirely invented.

Soon enough, Bernays had convinced the US Army that wristwatches could save soldiers' lives. And with that, they became standard issue in the army: wristwatches were no longer seen as delicate or feminine. Instead, they were imbued with the ruggedness, heroism, masculinity and reliability of men on the battlefield. For men who hadn't served, the wristwatch became aspirational – a way to embody the discipline and bravery of the soldiers they admired. To wear a wristwatch was to align oneself with a new ideal of masculinity: modern, strong, efficient and capable.[3] By the 1920s, wristwatches had transitioned from taboo to essential for men, and sales soared.[4]

Edward Bernays had an illustrious career in manipulating public perception. He convinced women to start smoking by marketing cigarettes as feminist 'Torches of Freedom',[5] and persuaded the American public to use disposable cups by scaring people into thinking they were the only sanitary option, often using subtle sexual messaging. This later linked reusable glasses with venereal disease.[6]

Bernays synthesised ideas from psychology, sociology and mass communication to create a systematic approach to influence and manipulate public opinion. His method was structured around several key steps:

Step 1: Shape unconscious desire and leverage fear and anxiety. Appeal to emotion, fear and desire rather than rational thought or sound logic.

Step 2: Create social proof. Associate ideas and products with influential people or trends to give the product credibility and desirability.

Step 3: Sow the seeds of doubt. Doubt doesn't have to be definitive – it only needs to disrupt certainty and create hesitation over an oppositional product or idea.

Step 4: Link products to aspirational lifestyles. Connect goods to values like prestige, success, status and desirability.
Step 5: Shape culture. Associate the product with social and cultural trends.
Step 6: Control and dominate the information ecosystem. Flood the media with a consistent message.
Step 7: Create or exploit a crisis to sell solutions. Highlight a problem – whether real or exaggerated – and present a product or policy as the solution.
Step 8: Undermine opposition. Shape public opinion by influencing trusted authority figures or discrediting those who oppose the interests of the product.

At the heart of Bernays's strategies were the concepts of propaganda and 'engineering consent'. He described the mechanics of propaganda in his 1928 book on the topic, and published an essay called 'The Engineering of Consent' in 1947 and a later book by the same name. These techniques involve guiding public opinion in subtle, often invisible ways, so people believe they are acting on their own free will while actually following a carefully orchestrated path.

In *Propaganda*, he said:

> The conscious and intelligent manipulation of the organized habits and opinions of the masses is an important element in democratic society. Those who manipulate this unseen mechanism of society constitute an invisible government which is the true ruling power of our country. We are governed, our minds are molded, our tastes formed, and our ideas suggested, largely by men we have never heard of … It is they who pull the wires that control the public mind.[7]

These strategies didn't simply fade into history – they became the blueprint for modern public relations, often weaponised by certain industries to safeguard their interests and sustain their profitability at the expense of human and environmental health.

The tobacco industry is, of course, one of the most famous and systematic adopters of this blueprint. A 1969 memo from cigarette company Brown & Williamson famously declared, 'Doubt is our product, since it is the best means of competing with the "body of fact" that exists in the mind of the general public.'[8]

But numerous industries have dabbled in this playbook. The US National Football League (NFL), for example, borrowed from the tobacco industry's playbook to cast doubt on the link between football and traumatic brain injury, even funding selective research to confuse the public. The issue, they explained with great conviction, wasn't that the NFL had designed a game built almost entirely on skull-to-skull collisions. No, the problem was that the players themselves were stupid. Perhaps they weren't training right. Or maybe they just had rattly brains genetically.

Perhaps nowhere has this playbook been more prominent, persistent and insidious than in the realm of modern-day environmental misinformation. And it is a handful of executives across the fossil fuel industry – comprising oil, coal and natural gas – who have perfected these techniques with alarming precision.*

Fossil fuels are the ghosts of life long gone – ancient burial sites of prehistoric life forms. Hundreds of millions of years ago, long before humans walked the Earth, tiny plants and creatures died. Over the course of their lives, these plants and creatures had been fuelled by the sun, storing this solar energy in their bodies and passing this energy down the food chain. When they died, their

---

* This statement refers specifically to executives of a small number of coal, oil and gas companies who have engaged in misleading marketing and deceptive practices regarding climate. It does not refer to individual fossil fuel workers who have made valuable contributions to society and are not responsible for these corporate strategies.

burial grounds had little or no oxygen, which meant the sun's energy that had fuelled them stayed in their bodies as carbon – the building block of life and the key to their transformation into fossil fuels. Over millennia, layers of mud, sand and rock buried these ancient creatures deeper and deeper in the Earth. Layer upon layer of earth built up over them, its weight creating pressure. A quiet alchemy of nature was underway; millions of years of burial, heat and pressure transformed these ancient plants and creatures. Swamp-dwelling plants were transformed into what we now know as coal. And ancient plankton and algae buried under the sea floor became oil and natural gas. The story of fossil fuels is a tale of ancient sunlight, patient transformation and Earth's quiet magic.

Humanity discovered this primordial bottled sun and set it alight. From the first steam engines to the towering industries of today, these fuels powered revolutions. Fossil fuels ignited our modern world. Coal drove the Industrial Revolution, turning wheels and weaving cloth. Oil and gas moved us further and faster: they powered trains, cars, planes and ships. They lit cities at night, fuelled factories and warmed our homes. They made life more comfortable, prosperous and interconnected for many. From fertilisers to plastics, medicines to electricity, fossil fuels built much of what we rely upon today. We keep using fossil fuels because they are abundant, powerful and deeply embedded in the systems we've built.

Infrastructure worldwide – our power grids, vehicles and industries – is designed around fossil fuels. They've given us a reliable, affordable way to power human life for centuries. Yet the same fuels that advanced humanity now cast a shadow on our future. Every time we use them, we are releasing energy from sunlight that plants captured millions of years ago. Burning them releases greenhouse gases – like carbon dioxide and methane – that warm our atmosphere and alter climates and weather patterns. Their use scars the land, clouds the air and threatens the oceans. Indeed, fossil fuels were a gift from Earth's smallest, earliest life forms – ancient plants and microscopic sea creatures. And then that gift became a curse.

TALES OF DECEIT AND CUNNING: THE PROPAGANDA PLAYBOOK

~

The year was 1959 and a man named Edward Teller stepped to the podium at the 'Energy and Man' symposium, an event organised by the American Petroleum Institute (API) and the Columbia Graduate School of Business. Teller, a theoretical physicist and chemical engineer famously known as the 'father of the hydrogen bomb', stood in front of over 300 government officials, scientists, economists and energy industry executives and delivered a warning. The burning of fossil fuels, he said, was 'contaminating the atmosphere'. This, he said, would have the potential to 'melt the icecap and submerge New York'.[9]

'All the coastal cities would be covered,' he declared with the gravitas of a man accustomed to contemplating the end of the world. 'I think that this chemical contamination is much more serious than most people tend to believe.'[10] His message was unequivocal: humanity needed to start transitioning away from fossil fuels as a source of energy.

This moment would later prove pivotal: it was the first documented warning to an industry that would spend the following decades perfecting the art of ignoring, burying and later outright denying the science of climate change.

By the 1960s, the warnings were gathering steam. In 1965, the API president, Frank Ikard, stood before his industry peers at their annual meeting and delivered a speech. In it, he referenced a report from President Lyndon B Johnson's scientific advisors warning that burning fossil fuels would result in devastating climate change consequences by the year 2000. Ikard quoted the report: 'There is still time to save the world's peoples from the catastrophic consequence of pollution, but time is running out.'[11]

By the late 1970s, a handful of leading fossil fuel companies were deeply immersed in climate science. ExxonMobil, one of the most powerful corporations on Earth, conducted cutting-edge climate research into the effects of burning fossil fuels. Over the subsequent decade, Exxon scientists predicted rising temperatures, melting ice caps and a future defined by climate chaos. Their work showed that humanity was hurtling towards a planetary crisis, and fossil fuels were the driver. Exxon's findings were not just accurate – they were *eerily* precise. Decades later, in 2023, independent scientists reviewed Exxon's original climate models and confirmed that they were astonishing reliable, noting that, 'since the late 1970s and early 1980s, ExxonMobil predicted global warming correctly and skillfully'.[12]

In fact, in 1982, Exxon distributed an internal memo to staff that revealed the company was aware of the link between fossil fuels and climate change. Marked 'not to be distributed externally', the memo acknowledged that avoiding global warming would require 'major reductions in fossil fuel combustion'.[13] Unless that happened, the memo warned, 'there are some potentially catastrophic events that must be considered' as a result of climate change and that 'once the effects are measurable, they might not be reversible'.[14]

It wasn't just Exxon. By the early 1980s, BP, TotalEnergies, Shell and the API had all undertaken or commissioned their own research and had all reached the same sobering conclusion: fossil fuels were warming the planet and driving climate change. Their research even predicted the rising chaos of climate-driven disasters: more frequent and intense floods, droughts and fires. The evidence, they knew, was strong. The burning of fossil fuels was causing irreversible harm to humanity and the planet.

The API and BP had been aware of human-caused climate change as early as the 1950s; the coal industry by the 1960s; electric utilities, TotalEnergies, General Motors and Ford by the 1970s; and Shell by the 1980s.[15]

The industry needed a response to this growing catastrophe. And they quickly agreed that honesty and integrity, while admirable in theory, were entirely unsuited to the task at hand. So, they buried the research. Presumably deciding that things like morality and ethics were time-consuming and tedious, these companies opted

TALES OF DECEIT AND CUNNING: THE PROPAGANDA PLAYBOOK

to stick with the current business model that promised widespread planetary and human harm.

But by the late 1980s, the fossil fuel industry faced growing pressure as global climate science gained wider scientific and public acceptance. This called for a new approach. Inspired by the PR playbook of Edward Bernays and the tobacco industry, the fossil fuel industry didn't just suppress science – they started actively rewriting it. A masterclass in subterfuge soon unfolded across industries.

The playbook for fossil fuels was chillingly straightforward: fund think tanks, pay 'sceptical' scientists, sow seeds of confusion, manufacture doubt and division in public discourse, control the media and information landscape, and undermine efforts to address climate change.

The formation of the UN's Intergovernmental Panel on Climate Change (IPCC) in 1988 marked the start of coordinated international efforts to tackle climate change. This group would dutifully present climate science to the world and report on the impacts of climate change – including how fossil fuels were driving it. In response, the Global Climate Coalition (GCC) was formed in 1989, a lobbying powerhouse made up of members of the automotive, utility, manufacturing, petroleum and mining industries. This group was dedicated to undermining climate science and delaying policy action.[16] It was the veritable who's who of environmental sabotage, including Exxon (a founding member), Chevron, BP, Royal Dutch Shell and the API, among many others. But this campaign against science and climate change warnings was just the beginning.

Exxon alone invested millions of dollars to fund think tanks and organisations that promoted climate scepticism, such as the Competitive Enterprise Institute (CEI) and the Heartland Institute. These organisations often masqueraded as independent research institutions while receiving substantial funding from oil, gas and coal companies, and they all dedicated their might towards questioning the science of climate change, while often misrepresenting data and attacking the credibility of scientists.

According to a report by a US Senate committee, between 2010 and 2020, six fossil fuel companies, including BP, Chevron, Exxon and Shell, spent an estimated $700 million on academic research programs that promoted research favourable to fossil fuels over renewable energy.[17] This wasn't haphazard. It was strategy: a multi-decade, multi-billion-dollar campaign of disinformation, propaganda and lobbying.

Internal strategy memos from Exxon in 1988–89 instructed employees to 'emphasise the uncertainty in scientific conclusions'[18] about climate change and increase 'emphasis on costs/political realities'.[19]

In 1991, a memo by Informed Citizens for the Environment – a front group created by industry bodies including the National Coal Association, the Western Fuels Association and the Edison Electric Institute – outlined their strategy: 'Reposition global warming as theory (not fact).'[20]

In 1998, an 'Action Plan' developed by the API, Chevron, Exxon and a range of mining and utility companies declared that 'it [is] not known for sure whether … climate change actually is occurring',[21] which is odd considering these companies conducted (and buried) their own science that showed that they absolutely knew for sure that climate change was happening. It went on to say that 'Victory will be achieved when average citizens "understand" (recognize) uncertainties in climate science,' and when 'media "understands" (recognizes) uncertainties in climate science'.[22]

Across the world, these organisations were manufacturing doubt about the links between fossil fuels and climate change, tirelessly working at the forefront of climate change denial. They lobbied to block federal and international action to curtail greenhouse gas emissions and 'helped to erect a vast edifice of misinformation that stands to this day'.[23]

The fossil-fuel-backed media campaigns came in fast and furious. Climate advertorials – ads disguised as editorials – blasted headlines like 'Unsettled Science',[24] painting a picture of doubt, deliberately designed to muddy public understanding of climate threats. Between 1996 and 1998, for instance, Exxon ran twelve advertorials timed to coincide with the 1997 UN climate change negotiations in Kyoto, Japan, which sought to engineer doubt about climate change. Ads with headlines read, 'Reset the alarm,' and 'Let's not rush to a decision at Kyoto ... We still don't know what role man-made greenhouse gases might play in warming the planet.'[25]

But as Edward Bernays knew, the key is to appeal to emotion and leverage fear. The most effective way to do this is to convince people that environmental action will make them poor, harm their children and cause widespread hardship. So, they manufactured associations between climate change action and economic hardship while appealing to parents to 'think of the children'.

Through the 1990s and early 2000s, fossil fuel companies and industry groups ran ads in major newspapers like *The New York Times*, manufacturing links between economic hardship, suffering children and climate change action. The saviour? Fossil fuels. Indeed, economic collapse, they warned, was a certainty – unless, of course, you let them keep drilling. Ads took on remarkable diversity, with headlines ranging from 'Lies they tell our children' to 'Oil pumps life', to 'Don't risk our economic future', to 'Apocalypse no', to 'Who told you the earth was warming ... Chicken Little?'[26] But it worked. And it continues to work. It wasn't just denial and manipulation – it was theatre. And these one-liners persist in mainstream media and public consciousness today.

Of course, no PR campaign is complete without a comprehensive credibility attack. Pesky scientists, conservationists and environmentalists were threatening the interests of fossil fuel companies with their science and facts and concern for things like human survival and the environment. This was a problem. To address this problem, they needed to make scientists and environmentalists seem crazy. The fossil fuel industry and their allies leaned heavily into this tactic. By the late-20th century, their PR machine had truly mastered the art of reframing environmental science and advocacy: environmentalists were radicals and extremists, prone to hysteria and hopelessly detached from economic reality.

Thus, the term 'tree hugger' became wielded as a weapon against environmentalists and scientists. This term had once been a symbol of defiance, community spirit and solidarity, emerging from the Chipko movement in 1970s India, in which villagers, often led by women, physically embraced trees to prevent logging.[27]

What began as an earnest badge of honour became, under the deft hand of PR spin, a derogatory term. By the 1980s, 'tree hugger' had become shorthand for anyone supposedly too naive or idealistic to grasp the harsh truths of industrial progress. There is a certain genius in taking people who are quite literally trying to save the world and reducing them to caricatures who, presumably, spend their days sobbing over azaleas. 'Tree hugger' is now a term that manages to be both condescending and whimsical, the verbal equivalent of patting someone on the head before ignoring everything they just said.

But while tree huggers were allegedly naive, they were also masterful enough to be the purveyors of mass harm, out to hurt the working class. The Heartland Institute, the Exxon-funded think tank that has had an illustrious history of denying both the health impacts of cigarettes and climate change, went so far as to sponsor a billboard that compared environmentalists to US mathematics prodigy and domestic terrorist Ted Kaczynski, the Unabomber.[28]

Even as recently as 2024, Australia's Institute of Public Affairs – notably an institution that was funded by the tobacco industry to develop research arguing that passive smoking wasn't bad for you – penned a piece with the headline 'Strangled by Tree Huggers', before stating they had established 'beyond any doubt' that environmental activism is 'preventing Australian industries from operating, jobs from being created, and export revenue from being earned', while also blaming environmentalists for 'harming our regional communities'.[29]

Environmentalists and scientists weren't rational members of society highlighting highly rigorous science and emphasising the need for consideration for the planet and its people. No, they were radicals, prone to hysteria and hopelessly detached from economic reality. 'Tree hugger' became more than a word; it became a cultural marker of impracticality and naivety, and environmentalism became branded as a radical ideology and an enemy to economic stability and the working class.

The long-term consequences of these campaigns have been profound. Public trust in climate and environmental science was eroded, policy paralysis is rife and meaningful legislation continues to be delayed decades later. The labels 'tree hugger' and even 'environmentalist' became a wedge, driving societal splits and framing environmentalism as incompatible with progress.

This is a form of 'othering' wherein one group defines and marginalises another group as fundamentally different, inferior or alien. Othering creates and perpetuates hierarchies, reinforcing power structures by depicting certain groups as 'less than' or as oppositional to the dominant group. It is eerily similar to gender contamination, which drove men away from wearing wristwatches. Both phenomena utilise stereotypes, mockery and social hierarchies to discredit, diminish and often silence voices. Environmentalism, much like the 'contaminated' traits and consumption patterns of women, became coded as soft, emotional and ultimately unserious – a stark contrast to the hard, pragmatic image industries like fossil fuels sought to project.

By the 2000s, it was becoming increasingly difficult to deny or even downplay the evidence of fossil fuel-driven climate change. The science was too strong, the environmental consequences were already evident and consensus had emerged among 97 per cent of scientists. So, major players within the fossil fuel industry matured their strategy, seamlessly transitioning from downplaying climate change to gaslighting the public.

In 2005, BP launched a major advertising campaign popularising the concept of the personal carbon footprint. This deftly reframed climate action as a matter of individual lifestyle tweaks rather than systemic overhaul. Despite BP doing little to address their own footprint – they had made no investments in renewables at the time – the campaign was a PR triumph. By redirecting the spotlight from corporate accountability to personal responsibility, they repositioned the climate crisis as a problem to be solved one lightbulb at a time, conveniently sidestepping their role in sustaining (and expanding) fossil fuel use. The message seemed to be, 'The problem isn't our business model – it is your long showers and morning commutes.'

The personal carbon footprint campaign didn't merely shift blame – it muddied the waters of accountability by ignoring the principle of 'differentiated responsibility'. You see, climate change and nature decline are problems that everybody plays a role in addressing. But not everyone contributed to them in equal measure, nor can they solve them on equal footing. This is differentiated responsibility. It means, simply put, that we cannot believe that an eighteen-year-old girl recycling her Pepsi can carries the same capacity for change as the CEO of a billion-dollar oil and gas company that meets with the US president every month. While individual actions like reducing waste or using public transport are meaningful, even critical, they cannot address the root cause of environmental issues without industry playing their part.

By shifting the focus to personal responsibility, BP and its peers expertly sidestepped the rather inconvenient reality that their own outsized role in driving environmental harm might require a touch more scrutiny. Instead, they created the illusion that climate change can be solved solely through individual consumer choices. This message is heavily entrenched. In fact, you have likely seen evidence of its pervasiveness whenever somebody exclaims, 'But she drives a car!' in response to anybody who advocates for the reduction of fossil fuels.

Despite reasonable public awareness of some of these tactics, BP and its peers had already cast the die and entrenched their message.

Their pollution – and their willingness to continue their pollution despite evidence of widespread harm – was, in fact, your fault.

Coincidentally, plastic pollution is also entirely your fault. Or so the story goes.

The tale of plastics starts with oil – or, more romantically, with those ancient forests and swamps and prehistoric creatures that decomposed into fossil fuels over millennia. From there, it took modern chemistry and slightly diabolical ingenuity to create plastic: that miracle material as light as it is stubborn, as convenient as it is immortal. By the 1970s, plastic was everywhere, from soda bottles to syringes, cheap toys to surgical gloves. It kept your food fresh, your house insulated and your medical procedures sterile. Plastic was progress.

Then progress started choking the oceans, strangling species and slowly killing people. By the 1980s, landfills were swelling, regulations were being discussed, communities were raising concerns and environmental movements were gaining momentum. Plastic was becoming a symbol of pollution. But here's the uncomfortable truth no one likes to say out loud: plastic is a fossil fuel product.

Over 98 per cent of plastic is derived from chemicals sourced from fossil fuels – mainly oil and natural gas.[30]

It's not just plastic, though. Fertiliser, asphalt, synthetic fibres (like nylon and polyester), adhesives (like glue and sticky tape), rubber (like car tyres) and laundry detergents are all derived from fossil fuels. This makes up the petrochemicals industry, represented largely by fossil fuel and chemical companies that have a financial interest in keeping fossil fuel-based products in production and exponential growth.

As the world edges closer to renewable energy and fossil fuels lose their stranglehold on the power grid, fossil fuel companies – and the petrochemical industry they support – are undertaking a strategic shift: to expand the production of fossil fuel-based plastics. This ensures continued profitability while locking in long-term dependence on fossil fuel-derived materials.

As fossil fuel companies began facing growing public pressure to reduce greenhouse gases and move towards cleaner alternatives, plastics emerged as a crucial lifeline for their profitability. According to the International Energy Agency, petrochemicals 'are rapidly becoming the largest driver of global oil demand'.[31] In fact, they will account for more than a third of the growth in oil demand by 2030 and nearly half by 2050, surpassing that of any other sector.[32]

Despite this promising new avenue for profit, the fossil fuel industry found itself in something of a pickle. It was easy enough to deny the existence of invisible greenhouse gases and climate change, which was mainly affecting people who were too poor to be listened to. But plastic waste was irritatingly conspicuous. People began sobbing about trifling matters such as turtles washing up asphyxiated by plastic and human babies having to drink breastmilk laced with microplastics. And then there was the matter of the Great Pacific Garbage Patch – that giant floating mass of plastic trash in the North Pacific Ocean three times the size of France.[33] It was hard to bury the evidence of plastic.

They knew plastic waste wasn't just bad for the environment – it was bad PR. Something had to be done. Not about the problem, of course. About the *optics* of the problem.

Tackling environmental crises, you see, demands solutions that are either bold, clever or creatively corrupt. The fossil fuel industry – joined by some allies in the chemicals industry – in the throes of a valiant fight against science and integrity, settled upon the third option. Their strategy was beautifully simple: plastic wasn't the problem – you were. Or, more precisely, your failure to recycle properly was.

Much like the personal carbon footprint campaign, the narrative around plastic pollution was reframed. Recycling became the focus of the story – not systemic change, not reduced plastics or fossil fuel production, but individual responsibility. Thus began the great plastic recycling ruse, where the answer to a problem created by selling too much plastic was to sell more plastic.

TALES OF DECEIT AND CUNNING: THE PROPAGANDA PLAYBOOK

If the public could be convinced that plastic waste could be successfully recycled, the industry could continue producing plastics at ever-increasing volumes while sidestepping growing calls for regulation. So, in the late 20th century, fossil fuel and plastics companies poured millions into marketing campaigns that painted recycling as the solution to the plastics crisis. This was largely driven by the Society of the Plastics Industry, which later changed its name to the Plastics Industry Association, whose members included Exxon, Chevron, the Dow Chemical Company and DuPont chemicals. The narrative they sold to the public was simple: plastic is recyclable, and recycling is the solution to the mounting waste problem. They pushed for the widespread adoption of recycling labels and systems that would ostensibly give plastic a second life. Reassuring little triangles (the 'chasing arrows') were stamped onto nearly every plastic product and a hopeful tale was told: that discarded plastic would be collected, processed and reborn as something useful.[34] The public, armed with rinsed yoghurt pots and a newfound sense of duty, did their bit, feeling chuffed with themselves as they separated their plastic waste into neat little piles.

What many people didn't realise then – and what we now know – was that this message was a deception, carefully crafted to deflect blame and secure the future of the plastics industry, and with it, the expansion of fossil fuels.

Because plastic recycling was a myth. A fairy tale designed to soothe growing public anger and, crucially, to protect a business model that relied on pumping out billions of tonnes of cheap, disposable plastic while allowing for the continuation of fossil fuel extraction.

Let us begin with a simple truth: plastic is forever. All the plastic ever produced still exists today. It is the cockroach of materials, the undead of packaging. You can bury it, burn it or throw it into the sea and it will persist in some way shape or form.

Less than 10 per cent of the plastic ever produced has been recycled.[35] The rest ends up in landfills, incinerators and the natural environment. But it is far more insidious than that. Even when plastic is recycled, it breaks down into smaller particles. Microplastics – tiny particles that are either produced intentionally or are the result of the breakdown of plastic – have entered the food chain, drinking water and even human bodies. Plastic has been found in our blood, our brains, our breast milk. It has found its way into our kidneys, hearts, livers and lungs; it is found in semen and placenta and breast milk and urine. We have found it in the stomachs of whales, sharks, penguins and polar bears; we have found it tangled around seal bodies and turtle necks. We have seen shorebirds feed plastic pieces to their babies and fish rendered infertile by plastic. We have found it in insect intestines, in fish gills, in the muscle tissue of shellfish. Plastic pollution now contaminates every corner of the globe, from the deepest part of the ocean – in the Mariana Trench – to the highest mountain peak – Mount Everest. And still, plastic production continues to surge, rising from 2 million tonnes per year in the 1950s to over 450 million tonnes per year today.[36]

Internal documents from the 1970s and 1980s reveal that fossil fuel companies and plastic manufacturers promoted recycling to the public while privately acknowledging it would never work at scale. Investigations decades later uncovered industry executives candidly admitting the barriers, one of which was the fact that there was virtually no infrastructure available to recycle the plastic. Added to that, producing virgin plastic from fossil fuels was – and remains – cheaper than recycling, thanks to low oil and gas prices and the high cost of sorting and processing.[37]

In short, the plastics industry – led by petrochemical and fossil fuel companies – knew recycling was a dead-end solution long before your children dutifully started rinsing out peanut butter jars. The promise of plastic recycling was used to produce and sell more plastic.

Recycling is, of course, important and needs to be continuously improved and expanded. Glass bottles and aluminium cans, despite their own unique environmental impacts, can be widely and effectively recycled. Plastic recycling, on the other hand, is fundamentally limited. There are thousands of types of plastic, each with unique chemical compositions, making comprehensive plastic recycling nearly impossible.[38] Unlike metals or glass, plastics degrade in quality each time they are recycled: most plastics can only be 'downcycled' once or twice before becoming waste. Recycling alone cannot address the scale of the plastics crisis – it mitigates only a fraction of the problem and often serves as a scapegoat for systemic failures.

Of course, plastics have revolutionised medicine, made space travel possible and have improved sanitation systems. Plastic straws have been an essential accessibility tool for people with a disability. They have been used to save countless lives: syringes, IV bags, catheters and surgical instruments require plastic, as do defibrillators, ventilators, pacemakers and prosthetics. Plastic piping has enabled clean water systems, while plastic packaging has kept our food fresh and safe. Plastic has provided temporary housing structures in emergency and disaster relief settings; it has allowed for the flow of emergency first-aid kits in war zones and the provision of food and water for refugee communities. It has protected the heads of children as they ride their bikes down the street and constructed the safety features in cars, like airbags.

But it is simultaneously eroding the health of the planet and its people. And the plastics industry has actively undermined efforts to solve this crisis, ramping up production and perpetuating the myth that plastic recycling can 'close the loop'. Lobbying efforts by companies in this industry have systematically blocked legislation aimed at reducing plastic production and have resisted even modest regulatory efforts while simultaneously stymieing efforts to build non-fossil-fuel-based alternatives to plastic.

Today, plastic packaging is used to wrap bananas and avocados and shrink-wrap cucumbers. It is stitched into cheap fast fashion, shedding microplastics with every wash, and covers supermarket multipacks, binding items already wrapped in plastic. It substitutes for grass, carpeting entire playgrounds as artificial turf, it fills the sky in the form of balloons and confetti, it dangles from keychains as decorative trinkets and covers hotel toiletries – tiny shampoo bottles, single-use lotion packets and disposable soap bars. It encases coffee cups and seals their lids, spins into straws, stirrers and cutlery used for mere minutes before being thrown away. It wraps electronics and toys in layers of unnecessary packaging and adds bulk to e-commerce shipments with plastic air pillows. It seals single-serve tomato sauce packets and encases mini soy sauce bottles that hold just 5 millilitres of sauce, and it's crafted into the microbeads that polish our teeth in toothpaste and scrub our bodies in exfoliants. Rather than acknowledging the health and environmental impacts of plastics and making a concerted global effort to find, drive and scale alternatives to (and reductions of) plastics, the plastics industry has doubled down on plastic production to ensure it has become embedded in every aspect of our lives.

The deception around plastic pollution and climate change – largely orchestrated by a handful of fossil fuel and chemical industry players – has left a legacy of misinformation and allowed for the continued production of fossil fuels. Sometimes, the most you can hope for is that people pretend to do good. And so, in this way, the executives of major fossil fuel companies and their allies have truly outdone themselves.

But the foundations of the old are beginning to crack. Increased awareness has sparked a groundswell of action. Investigative reporting and NGO exposés have uncovered decades of environmental deception in startling detail. In 2021, an investigation by Greenpeace revealed ExxonMobil lobbyist Keith McCoy admitting to decades of corporate disinformation. 'Did we aggressively fight against some of the science? Yes,' McCoy admitted, referencing Exxon's role in financing front groups designed to delay climate action.[39] His candid remarks were captured on a hidden camera, exposing calculated attempts to obstruct environmental progress.

TALES OF DECEIT AND CUNNING: THE PROPAGANDA PLAYBOOK

In 2024, a *New York Times* investigation uncovered over 400 leaked documents revealing the plastics industry's behind-the-scenes efforts to counter growing public opposition to plastic pollution.[40] Among their strategies was an ambitious campaign to rebrand PET plastics – used in single-use bottles and containers – as part of a 'zero-waste' future. This campaign included hiring influencers and celebrities to tout the environmental benefits of plastic while also undermining efforts to reduce single-use plastics at major events like the Olympics. The documents also detailed attempts to block plastic reduction regulation and quietly defang a global plastics treaty that was being negotiated to reduce plastic production.

As investigative reports poured in, communities, activists, regulators, lawmakers, governments and NGOs armed with this newfound information started rising en masse to challenge the system of deception driving environmental and human harm. Piece by piece, the system of environmental abuse and propaganda is being dismantled.

In early 2024, the city of Chicago filed a lawsuit against major fossil fuel companies – BP, Chevron, ConocoPhillips, Exxon Mobil and Shell – accusing the companies of discrediting science, misleading the public and causing catastrophic environmental harm, putting its citizens at risk. The lawsuit pulled no punches, saying these companies 'funded, conceived, planned, and carried out a sustained and widespread campaign of denial and disinformation about the existence of climate change and their products' contribution to it'.[41]

By September 2024, the state of California had joined the fray, suing ExxonMobil for what the Attorney-General called 'a decades-long campaign of deception' that unleashed the global plastics pollution crisis.[42]

The dominoes kept falling. In May 2024, Vermont became the first state in the USA to hold fossil fuel companies liable for climate damages. Seven months later, New York followed suit. Fires, floods and hurricanes have become more frequent and devastating, with recovery costs long burdening taxpayers. But no more. Polluting companies are now being forced to foot the bill for climate disaster.*

Los Angeles sued PepsiCo and Coca-Cola for misleading the public about the recyclability of their plastic bottles and downplaying the negative environmental and health impacts of plastic disposal.[43] New York, Kansas and Pennsylvania targeted Exxon, PepsiCo and BP and others for deception over plastics.[44] Puerto Rico and Maine hit Exxon and BP with lawsuits over climate deception. The Makah Indigenous people of Washington state sued Exxon over climate deception. So did the Shoalwater Bay Indian Tribe. Hawaii is suing fossil fuel companies. Alaska is suing. Chicago is suing. Minnesota is suing.

The message is clear: the age of polluting with impunity is ending. This is a reckoning.

It's not just in the USA. These cases are part of a global surge in environmental lawsuits – individuals, communities, governments and Indigenous peoples are holding major fossil fuel companies responsible for environmental deception and damages.

As of September 2024, 2796 environmental lawsuits had been filed across the world.[45] The number of cases filed against fossil fuel companies globally each year has nearly tripled since 2015.

Laws against environmental misinformation are also on the rise. In 2022, France became the first country to ban fossil fuel advertising nationwide.** The Hague has become the first city in the world to pass a law banning advertisements promoting fossil

---

* Climate Analytics estimates that the climate damages attributable to the pollution produced by the largest twenty-five oil and gas companies from 1985 to 2018 amount to a staggering US$20 trillion. ExxonMobil, Shell and BP are estimated to be responsible for climate-related costs of at least US$1 trillion each.

** France's ban was watered down from the initial proposal and doesn't cover the sponsorship of events or ads for natural gas, which is also a fossil fuel.

fuel products, and the European Parliament issued a directive that will outlaw misleading environmental marketing. Meanwhile, regulators in the USA, Australia, Japan and the UK are cracking down on greenwashing.

Shareholder activism and fossil fuel divestment are also shaking up corporate boardrooms. Vast efforts are underway to hold companies accountable for climate change. Campaigns like those led by 350.org have convinced organisations to divest $40.6 trillion from fossil fuel companies, cutting off financial support for those that continue to drive outsized environmental harm. Major institutions like Harvard University, the University of Oxford and Norway's sovereign wealth fund have all pledged to stop investing in oil, gas and coal.[46] Shareholder activism is using investor power to push companies towards change, with investors using their stake in the company to advocate for environmental reform, influencing the trajectory of major fossil fuel companies.

A seismic shift occurred at ExxonMobil in 2021 when a small activist hedge fund, Engine No. 1, ousted fossil fuel-oriented board members, replacing them with climate-focused candidates in a landmark revolt.[47] The trend is spreading. In 2022, climate activist and billionaire Mike Cannon-Brookes spearheaded a shareholder activism campaign at AGL Energy, Australia's largest electricity provider. Determined to put a stop to AGL's continued reliance on coal, he staged a veritable takeover, seizing control of the company in order to push them towards clean energy.[48] The movement is no longer a fringe effort – it is a powerful force of people using their votes and their wallets to drive systemic change.

This has led some companies to pretend to champion environmental action in solidarity with the people. In 2019, BP took to Twitter (X) with a chirpy post, saying, 'The first step to reducing your emissions is to know where you stand. Find out your carbon footprint with our new calculator and share your pledge today!' One user promptly responded, 'I pledge not to spill 4.9 million barrels of oil into the Gulf of Mexico.'[49]

Today, many fossil fuel companies have decided to become human rights organisations. To oppose the expansion of fossil fuels, they insist, is to condemn millions of people to deprivation and poverty. As they explain, increasing fossil fuel production is the *only* reliable means to lift billions out of poverty.

In 2018, a headline in the *Australian Financial Review* read, 'Against Adani? You're Against Lifting the World's Poorest Out of Poverty.'* This message is similarly reflected on the websites of companies like Shell, which is 'addressing global challenges, including those related to poverty and inequality'[50] and Exxon, which states, 'Far too many remain trapped in extreme poverty with no access to electricity or clean cooking fuels.'[51]

Benevolently stepping into the role of human rights defenders, these companies have recast themselves as saviours of the poor, whose products are not just necessary but a moral imperative. It's a curious stance from an industry whose major players seemed perfectly fine smothering Indigenous Nigerian communities in oil, blowing up sacred Indigenous Australian sites, spilling millions of barrels of oil into the ocean, promoting products that increase the risk of childhood cancer, suing and/or intimidating journalists, activists, scientists and charities that condemn their practices, while simultaneously denying (their own) climate change science and blocking efforts to expand renewable energies and transition away from fossil fuels.[52]

Fossil fuels have, of course, been critical in powering industrialisation and providing affordable energy to billions of people. This has facilitated economic growth, created jobs and improved living standards for many. However, these benefits have come with steep environmental, social and economic costs, both immediate and long-term, while at the same time trapping economies in a

---

* In 2024, Indian billionaire Gautam Adani was indicted by US prosecutors for his alleged role in a US$265 million ($407 million) bribery scheme.

cycle of fossil fuel dependence. That may once have seemed justifiable to some, at a time when the damage was less visible and the alternatives less viable. But that is no longer the case. The environmental, social and economic costs are no longer distant or theoretical. And yet, despite this knowledge and the availability of cleaner, cheaper solutions, we continue to uphold a system built on extraction, denial and delay. It is as destructive as it is avoidable. Today, renewable energy systems are increasingly recognised as a more equitable way to extend energy access to underserved and rural populations.[53] If fossil fuel companies were truly committed to alleviating poverty and defending human rights, they could focus on investing in underserved communities while supporting the global transition to clean, renewable energy. Instead, many of them remain focused on expanding the production of fossil fuels globally while actively lobbying against the development of clean energy. In fact, a 2019 report by InfluenceMap revealed that the five largest publicly traded oil companies had been spending over US$200 million per year lobbying to control, delay or block climate change legislation.[54]

While society has become increasingly aware of the impacts of fossil fuels on the environment and human health, the fossil fuel industry has shown an incredible aptitude for reinvention to maintain influence – one that would be inspiring if it wasn't, incidentally, going to do things like end life as we know it.

More recently, certain fossil fuel companies have pivoted to pseudo-solutions, rebranding themselves as allies in the fight against climate change while helpfully ensuring that oil and gas remain firmly embedded in every facet of the economy. Campaigns highlighting investments in 'lower-emissions fuels', natural gas, carbon capture and storage, advanced recycling and (methane-based) hydrogen production paint a picture of corporate sustainability.[55] Some have even rallied around the idea of 'geoengineering' which is, crudely put, a technique that involves putting a shield over the Earth to deflect solar radiation. These initiatives are frequently framed as critical pathways to a cleaner economy. But, for all this rhetoric, clean energy investment by oil and gas companies amounts to just 4 per cent of their overall capital spending.[56]

Such technologies are not without potential, but their promise is frequently overstated, and they tend to be given far more attention in environmental strategies than their real-world effectiveness warrants. Indeed, the primary function of many of these 'solutions' seems to be extending reliance on fossil fuels under the guise of innovation, while sidestepping the actual root cause of environmental harm: fossil fuel production. There's little evidence for these solutions, but their true innovation lies in their ability to generate an entirely new suite of problems while leaving the original problem comfortably undisturbed.

These tactics by many leading fossil fuel companies have led to growing discontent and distrust among the public. Unfortunately, frustration with the environmental impact of fossil fuels sometimes unfairly extends to fossil fuel workers. It's important to separate the decisions of a handful of fossil fuel executives from the contributions of individual workers within this industry. For many years, workers across the coal, oil and gas industries took pride in their work. After all, they were fuelling and feeding people all over the world. Then one morning they woke up to find themselves branded as villains, blamed for the world's environmental crisis. This blame is unjust and misplaced. Fossil fuel workers have been, and continue to be, valuable contributors to society. While we now recognise the urgent need to transition away from fossil fuels, the responsibility for this shift lies with systemic decisions, not individual workers. In fact, it is not unimaginable that, a few decades from now, electric vehicle manufacturers could find themselves in a similar position as fossil fuel workers today, as industries evolve and societal needs shift.

At the same time, it's critical to hold corporate leaders accountable. For decades, many oil and gas executives – arguably possessing some of the most sophisticated understandings of energy systems and climate change – have known about the dangers of fossil fuel dependence. And while we know that transitioning to a fossil fuel-free economy is complex and cannot happen overnight, this transition is being made significantly more difficult by a handful of fossil fuel companies that have spent at least six decades perpetuating disinformation, blocking meaningful environmental policy and delaying the adoption of renewable energy and environmental solutions.

This energy and effort would have been far better spent ensuring a just and equitable transition to clean energy for their workforces. Workers in fossil fuel industries possess highly transferable skills in engineering, project management and energy systems – skills directly relevant to the renewable energy sector.[57] Proper investment in retraining and support could empower these workers to lead the way in building a more sustainable energy future.

Fossil fuels have powered the world and become deeply embedded in modern life. But this level of reliance was not merely a natural progression; it was deliberately engineered to endure. Decades have been spent denying evidence, burying and then denying science and ensuring that every inch of progress is an uphill fight.

In 1954, tobacco companies in the USA released the 'Frank Statement to Cigarette Smokers'. In it, they said:

We believe the products we make are not injurious to health.

We always have and always will cooperate closely with those whose task it is to safeguard the public health.

For more than 300 years tobacco has given solace, relaxation and enjoyment to mankind. At one time or another during those years critics have held it responsible for practically every disease of the human body. One by one these charges have been abandoned for lack of evidence.

Regardless of the record of the past, the fact that cigarette smoking today should even be suspected as a cause of serious disease is a matter of deep concern to us.[58]

Forty years later, in 1994, after decades of denial and burying evidence of the health consequences of cigarettes, seven CEOs of the largest tobacco companies in the USA stood in front of Congress and swore on oath that nicotine wasn't addictive. The debate over tobacco was to continue.

If this sounds familiar, it's because the PR playbook hasn't changed. Indeed, Edward Bernays's PR tactics remain alive and well today. And the true measure of its success is in the chaos, the discord, the division it sows – a strategy designed to ensure that meaningful change is perpetually out of reach.

And if this continues, history may one day credit the fossil fuel industry with two crowning achievements: the birth of industrialisation and the end of humanity.

Chapter 14

The Herpes Agenda:
When Politics and Nature Collide

> Preservation of our environment is not a liberal or
> conservative challenge, it's common sense.
> Ronald Reagan[1]

In 2016, Australia's then deputy prime minister, the Honourable Barnaby Joyce, declared that he wanted to give herpes to a freshwater fish. The object of his devotion was, specifically, the carp.

The European carp is an invasive species that is wreaking havoc on Australia's river ecosystems. Therefore, in an impassioned parliamentary speech, the deputy PM proclaimed, 'We are afflicted in this nation with these disgusting, mud-sucking creatures ...' pausing dramatically, he continued, 'for which the only form of control is a version of herpes.'[2]

The idea had merit, according to scientists. You see, carp have orgies, or, as academics describe it, engage in 'spawning aggregations'.[3] Thus, it was deemed that a gathering of carp in the throes of lovemaking would indeed provide an ideal venue for the mass transmission of herpes. Herpes is deadly to carp, which meant that, theoretically, their populations would reduce. While it is unknown whether the deputy PM would be directly facilitating the transmission of herpes, his emotive speech did prompt a reasonable investment from the Australian federal government.

Later that year, the government formally announced their intent to give carp herpes, pledging $15 million in the budget to infect carp with Cyprinid herpesvirus 3.[4] Yet eight years later, not a single carp has herpes. In fact, their numbers have increased.[5]

Which begs the obvious question: if the deputy prime minister can't give a fish herpes, what hope do we have for mobilising effective political action for nature?

Politics is the stage where humanity's relationship with nature is negotiated – sometimes disastrously, but occasionally with hope and success. It is a messy, imperfect and often exasperating process marked by compromises, conflicts and missteps. Yet it remains one of the most powerful arenas where decisions about the natural world – and our place within it – are made.

On 1 December 1948, José Figueres Ferrer, president of Costa Rica, climbed atop a turret of Fort Bellavista – the headquarters of the Costa Rican army – and looked upon his country in the wake of one of their bloodiest civil wars. In an act that would reverberate through history, Figueres took a sledgehammer and, with a steady swing, smashed a hole in the wall of the fortress.* With the crack of stone, a new era was born. This was the moment that Costa Rica abolished its army. The president, known affectionately as Don Pepe, then handed the keys of the army headquarters to the minister of education, declaring that the fortress would henceforth be known as the National Museum of Costa Rica – a place not of war and weapons but of learning and culture. The budget that had been allocated to the military would now be redirected into education, health and, later, nature.

---

* Figueres initially came to power in 1948, leading a provisional junta – referred to as the Founding Council – that governed for eighteen months following the civil war. True to the council's commitment, he handed over power to the elected leadership in 1949. In the years that followed, he returned to office through democratic means, ultimately serving two additional terms as Costa Rica's president.

THE HERPES AGENDA: WHEN POLITICS AND NATURE COLLIDE

A new constitution was drawn up in 1949. 'The Army as a permanent institution is abolished,' read Article 12.[6] In a region grappling with conflict and political instability, Costa Rica did the unthinkable: it became the first country in the world to constitutionally abolish its standing army. This was the moment that a monumental shift took place in Costa Rica. They were rewriting their story. Piece by piece, the foundations were laid for a society that would come to inspire the world. Soon, Costa Rica would be known world over as a 'pioneer in the protection of peace and nature'.[7]

Following the 1948 civil war and the subsequent abolition of the military, change swept through Costa Rica under the leadership of President Don Pepe. Free education was granted to all citizens and social security was expanded, which laid the groundwork for universal health care. Women were granted the right to vote; so too were African Costa Ricans and the illiterate.[8] Costa Rica's pioneering path was – and is – neither simple nor straightforward, marked by decades of navigating economic, social, political and ecological complexities.

In Costa Rican coffee was the crown jewel of the economy, forming the backbone of its export industry in the 19th century.[9] By the mid-20th century, banana, pineapple and other crops like cocoa, sugarcane and cotton joined the export list, supported by governmental incentives. Soon, rising international demand for beef prompted a rapid expansion of cattle ranching. By the 1980s, Costa Rica had also become one of the leading suppliers of beef to North American fast-food chains.

Agricultural exports were seen as a pathway to economic growth, which meant landowners were incentivised to ranch and farm as much as possible. During this period, agricultural exports made up 72 per cent of the country's total exports.[10] But while the period from 1948 to 1980 saw economic expansion and social reforms, it also came at an immense ecological cost.

By the 1980s, Costa Rica had one of the highest deforestation rates in the world.[11] Costa Rica's forests, once cloaking 75 per cent of the country, had shrunk to a meagre 21 per cent as vast tracts of forest and land were converted to pastures and plantations. Added to that, Costa Rica had become the world's most pesticide-intensive agricultural producer. Toxic chemicals polluted rivers and soil, threatening agriculture, ecosystems and the health of farm workers who faced acute poisoning and long-term health risks, while pesticide exposure drove broader public health issues.[12]

This extractive model brought short-term economic gains but posed long-term threats to the nation's people and its rich biodiversity – though small in size, Costa Rica harboured 5 per cent of the world's biodiversity.[13] This means that 5 per cent of all the known species of plants, animals, fungi and microorganisms on Earth are found within Costa Rica's borders, which is remarkable given Costa Rica covers only about 0.03 per cent of the planet's landmass. Costa Rica was losing its greatest treasures. The very resources that powered the economy were at risk of collapse.

A new awareness began to emerge in Costa Rica, driven by grassroots and Indigenous activism, international environmental movements and forward-thinking policymakers and leaders. The time had come for nature to take centre stage.

Thanks to the social reforms after 1948, by the late 20th century, Costa Rica boasted one of the highest literacy rates in Latin America, provided universal health care and had showed the world that it was possible to live as a pacifist, non-militarised nation. Their institutions were strong, their population was educated, their political environment was relatively stable and their foundations were robust. They were ready for change. What unfolded next was arguably the most extraordinary environmental transformation on the planet: the rebirth of Costa Rica.

Environmental reform swept the country. Inspired by the legacy of President Don Pepe, successive leaders drove ambitious environmental policies that not only reshaped Costa Rica's future but also wove environmental stewardship into the very fabric of its national identity. In 1994, Costa Rica amended its constitution to enshrine environmental protection as a fundamental right. Article 50 was updated to state, 'Every person has the right to a healthy and ecologically balanced environment, being therefore entitled to denounce any acts that may infringe said right and to claim redress for the damage caused. The State shall guarantee, defend, and preserve that right.'[14]

This constitutional recognition of the right to a healthy environment was developed under the leadership of President José María Figueres Olsen. If the name 'Figueres' sounds familiar, it should. Jose María Figueres Olsen was the son of none other than the sledgehammer-wielding Don Pepe. In fact, Don Pepe had two children: his daughter, Christiana Figueres, would become one of the most towering environmental figures in the world – the woman who would unite the world in the name of the environment. She was the architect of the 2015 Paris Agreement, the world's most significant and ambitious climate change treaty.

But Costa Rica didn't stop there. They implemented a raft of groundbreaking initiatives, including the 1996 Forest Law, which prohibited landowners from clearing forests without permits.[15] At the heart of Costa Rica's environmental renaissance was the Payment for Environmental Services (PES) program, launched in 1997.[16] This incentivised landowners to protect forests, restore degraded lands and protect priority ecosystems. Piece by piece, systems that financed environmental destruction were dismantled. To fund this, Costa Rica introduced a fossil fuel tax, redirecting revenue from polluting industries to nature. In Costa Rica, people were being paid to protect nature rather than destroy it.

These legal and policy measures, combined with international cooperation and grassroots activism, led to a remarkable reversal of deforestation trends.

In 2019, Costa Rica received the UN's highest environmental honour, the Champions of the Earth award, in recognition of its conservation efforts.

Today, Costa Rica has reversed its deforestation. Forests cover nearly 60 per cent of its landmass. All of its electricity is generated from renewable energy.[17] A quarter of its landmass and 30 per cent of its marine areas are protected.[18] Every citizen has the right to a healthy environment, and environmental education is integrated into the national curriculum. The country has banned oil exploration and drilling. They have implemented wildlife corridors to connect fragmented habitats, enabling the safe movement of species like the sloth. Once an economy based almost entirely off extractive agricultural industries, Costa Rica is now a global leader in environmental tourism, shifting its economy to one that incentivises and celebrates nature protection.

Costa Rica has consistently placed in the top 5 countries on the Happy Planet Index with the highest happiness score every year from 2006 to 2019.[19] The country has topped the Happiness Index more than any other country on the planet.[20]

Costa Rica is imperfect, as every country, every leader and every transformation inevitably is. Among its most pressing issues is the ongoing struggle to honour the rights of Indigenous communities, whose stewardship has long safeguarded the country's land and seas. Despite landmark legislation in 1977 that declared Indigenous territories 'inalienable' and 'exclusive for the Indigenous communities that inhabit them', significant portions of these lands remain occupied by non-Indigenous ranchers, leaving many Indigenous peoples displaced from their ancestral homes.[21] Efforts to reclaim these lands have often been fraught with violence and insufficient government action. Costa Rica's conservation successes have, at times, come at the expense of Indigenous communities, as protected areas have restricted Indigenous peoples' access to traditional lands and resources. Large-scale development projects, such as hydro-electric dams and infrastructure expansion, have often encroached on Indigenous territories. In more recent years, environmental governance has faced setbacks under the country's current leadership, with the shelving of key agreements like the Escazú Agreement, which sought to empower public participation in environmental decision-making.[22]

Yet the vision of striving for a country centred around people, nature and peace remains strong in Costa Rica. As Costa Rica's former vice minister of environment and energy Patricia Madrigal-Cordero says, 'Nature is in our DNA.'[23] *Pura vida*, Costa Rica's national saying, is the law of the land there: 'pure life'.

Óscar Arias Sánchez was the president of Costa Rica between 1986 and 1990 and from 2006 to 2010. Like many leaders who came before and after him, he sought to carry forward President Figueres's 1948 vision of a country that celebrated people, peace and the planet. Awarded a Nobel Peace Prize for his efforts to promote peace in Central America, Arias's leadership reinforced the idea that Costa Rica's path of peace, social equity and environmental stewardship were interconnected.

When accepting his Nobel Peace Prize in 1987, he said:

My country is a country of teachers. It is therefore a country of peace ... Because our country is a country of teachers, we closed the army camps, and our children go [to school] with books under their arms, not with rifles on their shoulders. We believe in dialogue, in agreement, in reaching a consensus. We reject violence. Because my country is a country of teachers, we believe in convincing our opponents, not defeating them.[24]

Twenty years later, he extended that peace to nature and invited the world to join him and his fellow Costa Ricans in this vision. In 2008, addressing his neighbours at the Fifth Summit of the Americas, he said:

Fifty-nine years ago, Costa Rica declared peace on the world, abolishing its armed forces ... Today, in Costa Rica and around the world, another type of war is raging, and another declaration of peace is necessary: we must declare Peace with Nature and abolish the forces that destroy it ...

It is time to act. We cannot wait for the rest of the world to act, because, although we have not run out of options, we certainly have run out of time.

No one has written the world's last poem yet. No one can call themselves a historian of the end of our species, no one can chronicle our last days. There is still ink left in the well, and we have to decide what we will write with it: will we describe a desert scene, where death has been crowned king, or will we describe life, water, air and the sap of human spirit? Together, we must decide whether we will write the final poem of war, or whether we will write, at last, the poem of our Peace with Nature.[25]

For the past seventy-five years, Costa Rica has charted a path that stands apart from much of the world. They dared to dream of a future where peace and nature were at the core of national identity. Today, that dream remains alive.

Other countries have experimented with their own policy approaches to nature protection and climate action. Australia, for example, is exploring inaction, an approach which they have been fiercely committed to for many decades with consistent results. Most often, this involves political parties spending years bickering over the same topic until the next government is elected and the politicians resume their bickering, while simultaneously approving the expansion of fossil fuel production, clearing vast tracts of forest and dumping pollutants in the ocean.

For example, the Australian government has been arguing for more than seventy years over whether the country should develop nuclear energy, while continuing to expand fossil fuel production.[26] The result is a citizenry that is no more informed on the topic of nuclear energy than they were seven decades ago but who have firmly picked a side and will yell at each other about it in the comments section of a social media post. While Australia has, in recent years, significantly increased its investment in renewable energy and has seen substantial growth across the sector, it still grapples with age-old debates that seek to delay and distract from meaningful environmental action.

In fact, a host of debates have waged in Australia for decades on all manner of topics such as whether the country should continue destroying its remaining native forests, whether polluters should pay for their pollution, whether Indigenous peoples should have decision-making rights over their own land, whether the country should have effective laws to protect its species (or just pretend laws), whether water should remain in rivers, whether the Great Barrier Reef is dying (or if the bleached coral means it is just sleeping), whether fossil fuel companies should pay tax, whether gas companies should be allowed to make environmental laws, and whether it's okay to destroy koala habitat shortly after announcing a koala protection plan.

In this way, environmental debates have been so tedious that by the end of it, even the environmentalists are sick of the environment.

Australia isn't alone in this. Around the world, environmentalists now find themselves oscillating between cautious optimism and existential despair as bold policy reforms are announced before being promptly neutered by the fine print.

Globally, there has been an enormous surge in political attention directed towards environmental issues. Once occupying the fringes of political thought, climate change and, to a lesser degree, nature, have moved into the spotlight. But thus far, environmental policy worldwide is undertaken much like a school group project which tends to involve grand objectives, paired with an unspoken understanding that no one's really going to follow through. Across the world, the journey to protect the environment has been long, slow and annoying. On paper, the global community has rallied around shared visions of a better future. The various UN conventions – on climate change, biodiversity, desertification, plastics and others – are part of a global framework aimed at addressing urgent environmental challenges. Each convention focuses on a specific issue, but they share common aims: to foster international cooperation and catalyse action to protect the planet and its people.

The Paris Agreement, under the UN Framework Convention on Climate Change, aims to limit global warming. The Global Biodiversity Framework, part of the Convention on Biological Diversity, sets targets for ecosystem restoration and species conservation. The Global Plastics Treaty seeks to address plastic pollution. Meanwhile, the UN Convention to Combat Desertification focuses on land degradation. Individual nations sign on to these goals, and every year, world leaders gather to discuss them – often with great ceremony, ambitious speeches and a shared understanding that actual progress will be revisited next year, when they meet to do it all over again. These frameworks reflect a collective recognition of the problem, but they also expose an inconvenient truth: we've become adept at setting goals but are decidedly less skilled at achieving them. The result is that for over thirty years, countries have developed climate and nature targets and then repeatedly failed to meet them.

In October 2024, the United Nations Environment Program released its Emissions Gap Report for that year, a report card on the world's environmental progress.[27] 'You've all tried very hard,' the message to governments effectively said, 'but unfortunately, you're still failing.' In this way, political leaders from around the world received a participation award for their environmental efforts.

You see, despite almost every country in the world announcing sweeping climate and nature initiatives and targets, greenhouse gas emissions keep rising and nature destruction – like deforestation – remains rampant.*[28] The reason for this, according to the UN, is that there is a big gap between what countries promise and what they actually deliver.

---

* In 2023, global greenhouse gas emissions hit a record high of 57.1 gigatonnes. This does not include emissions of militaries, whose figures are largely unreported.

As Australia's former deputy prime minister – and tireless crusader for carp herpes – can attest, it's all well and good to announce your intent, but putting that plan into action and getting it to work are entirely different matters. In short, boldly announcing your ambition just isn't going to cut it. Indeed, the most stubborn obstacle to saving the planet often holds a seat in the halls of government.

Politics, of course, do not exist in a vacuum. They are shaped by a web of vested interests, cultural behaviours and entrenched structural, economic and social systems. Yet despite almost relentless evidence of nature decline and climate catastrophe, political systems remain paralysed, bound by barriers that perpetuate inaction and see environmental protection fall woefully short of budget prioritisation.

Governments worldwide allocate just 0.08 per cent of global GDP (US$165 billion) towards the protection and restoration of nature, with the vast majority coming from just five countries: the US, China, France, Germany and Italy.[29] By comparison, 7 per cent of global GDP is spent on activities that directly harm nature ($7 trillion annually).[30]

The World Economic Forum's *Global Risks Report 2024* places extreme weather events, critical changes to Earth systems, biodiversity loss and ecosystem collapse, and natural resource shortages as the top four global risks over the next ten years.[31] That is, the destruction of nature is the greatest long-term risk to humanity, representing a higher threat than war and conflict.

Spending on nature is vastly disproportionate to the risk it poses. To put this disparity into perspective, in 2023, governments spent US$2.4 trillion on military defence – 2.3 per cent of global GDP.[32] This is a stark contrast to that year's modest investment of US$165 billion directed towards safeguarding the natural systems essential for human survival.

The disconnect between the scale of the risks and the financial and political prioritisation to address them is a result of deep structural and systemic barriers. One such barrier is that politicians, of course, operate on a different timescale to the planet. They think in terms of election cycles, not planetary lifespans and geological ages. This short-term focus often pushes politicians to prioritise policies that deliver quick wins rather than addressing long-term environmental crises.

This myopia is made worse by the outsized influence of vested interests. Well-resourced lobby groups – particularly from the energy, agriculture, pharmaceutical, property and transport industries – have entrenched themselves deeply within political systems through lobbying and donations. While lobbying is a legitimate democratic tool, certain lobby groups have gained enormous influence, shaping policy to favour their interests, often at the expense of the public – and the environment. Comparatively, marginalised voices, communities and consumers remain consistently under-represented, leading to a fundamental imbalance in democratic advocacy.

For example, in the US, well-funded lobbying groups with vested interests have spent millions advocating for car-centric policies, diverting federal investments away from public transport and towards roads and highways. The result is a disproportionate focus on highways and cars over public transport, and cities that are designed in a way that benefits vehicles over people and environments. Public transport systems produce far fewer greenhouse gas emissions per passenger and offer an obvious and affordable solution to climate change, urban congestion and air pollution. Yet US federal funding has allocated four times as much money to roads as to buses, subways and trains.[33] As city planner Brent Toderian notes, 'For as long as there have been cars, car manufacturers have aggressively been seeking to shape our cities for one primary reason: to design cities where that will help them sell more cars.'[34] The result? Urban landscapes designed for vehicles rather than people – at great cost to the environment and public wellbeing.

The influence of vested interests is equally stark in Australia.

Lobbyists outnumber politicians in the nation's capital city by three to one.[35] According to researchers, 'The fossil fuel industry has dominated the Australian economy for decades and has built a strong alliance with governments in various spheres, benefiting from subsidies and influencing policymaking.'[36]

The fossil fuel industry's influence in Australia is so pervasive that it hasn't merely shaped policy – it has toppled leaders. In 2010, Prime Minister Kevin Rudd found himself politically euthanised after daring to propose the Resource Super Profits Tax, a policy targeting excessive mining profits. The mining sector retaliated with a $24 million advertising blitz and lobbying campaign.[37] And with that, the prime minister was unceremoniously dumped from his role as leader of the country.

Such unequal influence erodes democratic fairness. Australian senator Larissa Waters says the Australian government is 'plainly captured by the fossil fuel industry'. According to her, 'coal, gas and oil corporations don't donate hundreds of thousands every year ... because they like democracy' but rather because they're 'buying outcomes'.[38]

Another major barrier to meaningful environmental action is a lack of systems thinking, leading to fragmented and contradictory policies. The environmental crisis isn't just a list of isolated problems – it's a tangled mess of challenges like climate change, biodiversity decline and resource depletion that intersect with healthcare systems, human rights, education, economics and social justice. Yet despite this, governments and institutions often continue to subscribe to the policy that the best way to solve one problem is to ignore the underlying symptoms, prescribe an ineffective treatment and accidentally create two more problems. For example, the development of renewable energy projects to address climate change has, in some cases, proceeded without sufficient consideration of the impacts on wildlife, local communities, the generation and management of waste, or Indigenous land rights. In this way, the mistakes of the past are being transposed to the vision of the future. Added to that, policies themselves are often fragmented and contradictory. For example, while governments roll out ambitious renewable energy programs, they continue to allocate billions of dollars in fossil fuel subsidies annually, undermining their own climate change targets.[39]

Further inflaming these challenges are the forces of ideological division, media concentration and political polarisation. Add social media's amplification of fear, outrage and division, and you create a volatile environment – an unruly mess that complicates and resists the pursuit of transformative and equitable environmental policy.

These problems are significant. But they are not insurmountable. In fact, when faced with clear and immediate danger, history reveals that humanity and political systems are capable of remarkable collective action.

Heroes can come from the most unexpected places. In the 1980s, for example, Margaret Thatcher and Ronald Reagan teamed up and saved the world.* Two of the most polarising conservative leaders of their time, together – as prime minister of the UK and president of the US – did the unthinkable: they orchestrated the most sophisticated, enduring and successful environmental treaty that the world has ever known. It remains to this day the greatest political feat in environmental history.

Their friendship was one for the ages, and their leadership redefined the century. They were political 'soulmates'.[40] She: a blunt,

---

* It took more than two heads of state and a handshake to save the ozone layer. In reality, it required years of quiet, meticulous work by scientists who understood the chemistry, diplomats who understood the assignment, activists, nonprofits and communities who demanded change, and industries that were unafraid to lead the charge. Even so, Reagan and Thatcher are noteworthy – because, against the odds, international politics briefly functioned as intended, and leaders heeded the science.

THE HERPES AGENDA: WHEN POLITICS AND NATURE COLLIDE

uncompromising leader known by the world as the Iron Lady, the longest-serving British prime minister of the 20th century and the first woman to hold the position. He: a charismatic actor turned politician known for his self-deprecating sense of humour and 'warmly ruthless' personality. It was the golden era of conservative politics and Margaret Thatcher and Ronald Reagan sat atop the world. They shared a disdain for government intervention and regulation and a fondness for eliminating communists, fighting with unions and invading other countries for their own good. They believed in military might, free markets and traditional values of personal responsibility, family and faith. They were both resolute, if polarising, leaders of an era marked by ideological and military competition.

And when, in 1984, the planet sat on the brink of catastrophe, it was these two leaders who came to the rescue. And it all started with a great big hole in the sky.

In the 1980s, a group of British scientists – Joe Farman, Brian Gardiner and Jonathan Shanklin – was measuring ozone levels over Antarctica. They noticed something shocking: each spring, a huge chunk of the ozone layer over Antarctica was disappearing, or thinning.[41] The ozone layer is a thin, invisible shield of gas in the stratosphere – Earth's natural sunshield that protects life on Earth from the sun's harmful ultraviolet (UV) rays. And it had been breached. In May 1985, they published their findings in the journal *Nature*. It came to be known as the hole in the ozone. And it was bad. Really bad. Without the ozone layer, complex life on Earth would not exist.

Soon after their findings were published, an atmospheric chemist by the name of Susan Solomon led an expedition to Antarctica to find the cause.[42] She was the sole leader of the expedition and the only woman. Little did she know at the time that she would soon be known as one of the most influential scientists in the world. The cause of the ozone hole, she discovered with her team, was human-made chemicals, specifically chlorofluorocarbons, known as CFCs. CFCs were found in aerosols – like hairspray, foams, refrigerators and air conditioners. And they were destroying the ozone layer.

For the previous decade, scientists had been raising concerns about the destructive effects of CFCs. But a vast global industry had been built around these chemicals. Banning CFCs threatened industries and supply chains that relied heavily on them. Industry giants like chemical company DuPont had enormous financial interest in the production and sale of CFCs and initially lobbied governments and used PR campaigns to cast doubt on the science, maintaining that the data on CFCs was 'inconclusive and didn't warrant drastic action'.[43] Around the world, governments raised concerns about the economic consequences of banning CFCs. Wealthier nations baulked at the cost of restructuring entire industries, while poorer and developing nations feared they lacked the resources to transition to alternatives – assuming alternatives even existed.

The impacts of the ozone hole were also unevenly distributed. The 'ozone hole' had formed over Antarctica, a remote region far from many industrialised or wealthy nations. Such distance made it easier for some governments to dismiss the problem as not being an immediate threat to their people.

To make matters worse, getting people to care about invisible chemicals that were causing an invisible hole in the sky was no easy task, especially when the planetary crisis was conveyed as 'Large losses of total ozone in Antarctica reveal seasonal ClO_x/NO_x interaction'.[44]

But environmental science had two unlikely allies: Ronald Reagan and Margaret Thatcher. Perhaps the most defining feature of the response to the ozone hole was that Margaret Thatcher was, by trade, a scientist. Having trained as a chemist, she understood the science underpinning the crisis. And so, she acted. Wielding her background as a scientist, Thatcher became one of the fiercest advocates for phasing out CFCs. And at her side was none other than Ronald Reagan. Between them, they were two titans of the global economy. And they were about to change the world.

For Ronald Reagan, the matter of the ozone hole was also personal. Three months after the historic study announcing the hole in the ozone, he had undergone surgery to have skin cancer removed from his nose. Without the ozone layer, UV radiation would increase and the future would be blighted by skin cancers, cataracts, dying plants and crops, and damaged ecosystems. Armed with this awareness, as well as his love of the outdoors, his trust in Thatcher and his faith in the science, his advocacy for the ban of CFCs was cemented.

And that was the moment everything changed. Two formidable leaders – neither a friend to regulation – threw their might behind the environment, and with it, they united the world. They were informed, deliberate and unequivocal. CFCs were to be banned. They were going to mend the hole in the ozone.

On 16 September 1987, within two years of the discovery of the ozone hole, the Montreal Protocol – a global United Nations treaty designed to protect the Earth's ozone layer by phasing out the chemicals that deplete it – was agreed upon. Every single country in the world signed it. It was the first universally ratified treaty in UN history.

And it worked.

Today, the hole in the ozone layer is shrinking, and scientists predict it could fully recover by 2066. More than 98 per cent of ozone-depleting substances have been eliminated.[45] Former UN secretary-general Kofi Annan called the Montreal Protocol 'perhaps the single most successful international agreement to date'.[46]

Decades after the signing of the Montreal Protocol, Ronald Reagan's secretary of state, George Shultz, offered a glimpse into the president's mindset on the ozone crisis – and environmental challenges more broadly. Acting decisively to implement the Montreal Protocol and ban environmentally harmful chemicals was, in Reagan's view, an 'insurance policy'.[47] Science, by its nature, is never certain beyond a shadow of a doubt – it is a process of discovery. Yet Reagan knew that even if the science wasn't and isn't ever 100 per cent certain, the risks of inaction were too great. But, as Schultz noted, 'In the case of the Montreal Protocol, the people who were worried were right.'[48]

Today, we are faced with a crisis that is more expansive, more threatening, more destructive than anything we as humanity have ever experienced: climate change and the destruction of nature. Yet decades after scientists first revealed the human-driven nature of these crises, global efforts remain fragmented and insufficient. Replicating the success of the Montreal Protocol for climate change and nature destruction has proven elusive. There are a few reasons for this.

One key difference lies in complexity. The Montreal Protocol tackled a focused problem: CFCs and other ozone-depleting substances, used in specific industries like refrigeration and aerosols. There was a precise, well-defined issue: phasing out a small group of chemicals. The problem was narrow, the solutions were clear and the results (ozone layer recovery) were measurable. Additionally, cost-effective alternatives to CFCs were already being developed, making the transition for industry economically feasible.

In contrast, climate change and nature restoration require a complete overhaul of every facet of industry and the economy: energy systems, transportation, agriculture, property and retail, to name a few. Fossil fuels underpin entire economies, and so too do activities that harm nature, making the transition to sustainable systems more complex and disruptive.

Despite these differences, the Montreal Protocol demonstrated that humanity is capable of extraordinary feats of cooperation when faced with a clear and present danger.

And political leaders should pay heed; there are lessons to be learned.

The Montreal Protocol succeeded because it had powerful advocates at the highest levels of political authority who wielded their full might to confront the problem. They didn't just back the Montreal Protocol – they drove it forward, unapologetically.

Added to that, the Montreal Protocol was legally binding, with real teeth. Countries were held accountable, and those that failed to comply could face penalties like trade restrictions. On the other hand, UN agreements on climate change, biodiversity, desertification and plastic pollution rely mostly on voluntary actions with no real penalties for failing to comply other than a few stern glares and frustrated sighs.

Industries also had clear and compelling financial incentives to comply with the Montreal Protocol. Added to that, nations implemented regulations to restrict the use of ozone-damaging chemicals, meaning that businesses were offered one path: transition away from CFCs or face immediate financial consequences.

The Montreal Protocol also applied the principle of differentiated responsibility, recognising the fact that not all nations had contributed equally to the ozone problem, and nor did they each possess the same resources and capacity to address it. So, rich nations stepped up, providing financing and technical assistance for developing nations when appropriate.[49]

Another factor in the Montreal Protocol's success was that it aimed small to achieve big. By zeroing in on a specific set of ozone-depleting substances, it set clear, tangible goals that were impossible to ignore – and possible to meet. In contrast, agreements addressing climate change, biodiversity loss, plastic pollution and desertification often tackle sprawling, tangled crises that involve long lists of targets and actions, which can often overwhelm and diffuse the focus needed for meaningful progress.

The Montreal Protocol worked because it cut the excuses and stuck to the essentials: power, focus, accountability and enforcement. It didn't try to fix everything – it picked one fight, made it impossible to ignore, and brought everyone to the table with clear rules and incentives. And at its helm stood the Iron Lady herself, who, as luck would have it, was also a scientist.

Today, not a single country in the world has policies strong enough to adequately tackle climate change.[50] This failure persists despite the 2015 Paris Agreement, where nearly every nation pledged to limit global warming to avoid the worst impacts of a hotter planet.

The same is true for nature. In 2010, countries came together under the UN and agreed to halt biodiversity loss by 2020. Not one of the twenty global targets they devised was met.[51] In 2022, they came together again to develop the Global Biodiversity Framework, which set new goals, including protecting 30 per cent of the planet's land and oceans by 2030. But like the Paris Agreement, these targets depend on voluntary commitments.

Perhaps the most profound takeaway from history is this: when governments perceive an existential threat to their people, they act decisively. They regulate, fund and mobilise. They do it for war, and for the spectre of war; they did it for COVID-19; and they did it under the Montreal Protocol. Even the staunchest opponents of regulation – Thatcher and Reagan – found justification for government intervention when the very existence, future and safety of their citizens were at stake. So, the question for today's politicians is this: when faced with overwhelming scientific evidence that climate change and the destruction of nature are driving mass suffering and risking extinction for much of life on Earth, do we perceive this as an existential crisis – and will our response match the gravity of the crisis?

And if we don't perceive it as an existential threat, is waiting to see if we are wrong worth the risk?

PART 5

Solutions: The Tempest, the Fury and the Birdwatcher

Chapter 15

Giants Among Us: The Fight of Our Lives

> None of this will be the fault of nature itself. It will largely be inflicted by the inaction of this generation of adults, in what might fairly be described as the greatest inter-generational injustice ever inflicted by one generation of humans upon the next.
>
> Justice Mordecai Bromberg[1]

We think them fragile, these kids. With their soft hands and fragile bodies. Tender hearts, windswept hair, bright eyes; they're just children. Like saplings and spiderwebs and drops of dew. Like dandelion seeds and butterfly wings. So quiet, so fleeting, we thought these kids were. Just a flicker of a candle in the dark or a whisper in the wind.

But we forget that sometimes, a sapling breaks through the concrete. It defies the weight pressing down on it and threads itself across the pavement. Its tendrils unfurl; it stretches, entwines, reaches for others. Its roots connect, they grip onto one another, communicate, nourish each other. Against all odds, this sapling grows – bold, unyielding – and it pushes through what was once thought immovable. It forms a network and that network becomes a forest and that forest breathes life into the world. And then that forest is enduring even as it dies and is reborn, and then we are the ones that are fleeting.

Like a sapling in the pavement, we underestimated these kids.

And then a fifteen-year-old girl named Greta Thunberg became the most powerful voice in the world to speak on behalf of the planet. And then 7.6 million other kids from 185 countries marched with her for the planet. And then we started to think maybe, just maybe, these kids aren't so fleeting after all.

There are people who dream outrageously and boldly. They fill their minds with stories of hope; chants of joy and pain and rage fall from their lips, and those chants are heard across cultures, across languages and across the world. With their eyes on the skies and the trees and the seas, they write history, together. They're doing it right now.

They march in the streets and the classrooms and the courtrooms and the parliaments and the boardrooms, and we thought they would stop, but they didn't. We thought they would grow tired, but still they didn't. They grew louder and louder until every corner of the world heard their demand: a planet they can live on. That's what they want, and they keep saying it, and they won't stop saying it.

They're too loud, we say. Their dreams too vast. They're unruly and untamed and they shine too bright and burn too strong, these reckless dreamers. These hysterical kids.

But these kids; they waited and waited for the grown-ups to come. But we never did.

So now they paint the world with their rage and their fury and their hope and their passion. The air swells with their shouts, with their laughter, with their joy and with their anger. It is the only thing you can't steal from them, that fire they wield.

They're so young, so delicate but so hardened. We made them hard. They were forged in the fires of our failures.

This burden they carry is so heavy. Too heavy.

But they have learned to weave hope out of shards of fear and wear it as a shield and brandish it as a knife. They march the streets and write history with their footprints; they conjure miracles, these kids. They search and they search for hope and they scramble and they crawl and cry out for something to cling to. It is pain and horror and hope and elation and determination, all rolled into one.

Their footsteps are the revolution; you're watching it. It's happening right now. And you're missing it.

The world will bend and buckle for them. It already is.

They fight and blunder and fail and they get up and they do it again and again and they never stop and they won't stop and they can't stop. We need them, you see – we just don't know it yet.

They're turning the world upside down, these kids. These saplings, they've grown mighty, and they're breathing life into the world again.

In 2021, an eighty-six-year-old Catholic nun named Sister Brigid Arthur walked into the federal court in Melbourne, Australia. With her was sixteen-year-old Anjali Sharma. Together, they carried the hopes of seven other teenagers – and, in fact, all future generations. And they were about to make history.

The year prior, Australian coalmining company Whitehaven Coal submitted a proposal to the minister for the environment outlining their plans to expand one of their coalmines in New South Wales. The project was projected to release 135 million tonnes of carbon dioxide into the atmosphere.[2]

For Sister Brigid, Anjali Sharma and the seven other students, this was unacceptable. So, they did something unprecedented in Australian history. They sued the federal minister for the environment. The teenagers and their legal team argued that the then minister, Sussan Ley, would be breaching a common law duty of care to protect young people against future climate change-related harm if she allowed the mine extension to go ahead.

Specifically, they sought a declaration that the minister owed a duty of care to protect young people from climate harm and called for an injunction to stop the proposed coalmine from being approved. The campaign by these teenagers and Sister Brigid amassed a tidal wave of support. Politicians supported them, and so too did the community – more than $300,000 was poured into the campaign through crowdfunding.

On 27 May 2021, Justice Mordecai Bromberg delivered his ruling.[3]

For the first time in the world, the federal court found that the minister owed a duty of care to protect children from climate change harm caused by greenhouse gas emissions. The judgement sent shockwaves around the world, setting a legal precedent with profound implications for future environmental policy and litigation.

In his ruling, Justice Bromberg said:

> It is difficult to characterise in a single phrase the devastation that the plausible evidence presented in this proceeding forecasts for the Children. As Australian adults know their country, Australia will be lost and the World as we know it gone as well. The physical environment will be harsher, far more extreme and devastatingly brutal when angry. As for the human experience – quality of life, opportunities to partake in nature's treasures, the capacity to grow and prosper – all will be greatly diminished. Lives will be cut short. Trauma will be far more common and good health harder to hold and maintain. None of this will be the fault of nature itself. It will largely be inflicted by the inaction of this generation of adults, in what might fairly be described as the greatest inter-generational injustice ever inflicted by one generation of humans upon the next.

While the court recognised the duty of care, it did not grant an injunction to stop the mine's expansion, citing insufficient evidence to suggest the minister would breach this duty. In September 2021, the minister for the environment approved the Whitehaven coalmine extension.

The following month, the minister appealed the court's decision. And, in a disappointing setback for the teenagers – and all future generations – the Full Court of the Federal Court of Australia overturned the duty of care ruling, stating it was 'unsuitable for judicial determination'.[4]

But the die had already been cast. Young people were already rewriting the rules of the game. And the curious thing is, it might just be working.

Today, Anjali Sharma, who led the landmark case against Australia's environment minister, is twenty-one years old. Now a law student, she is continuing her battle for environmental and youth justice. Her most recent campaign is the Duty of Care Bill. Building on the class-action suit she filed when she was just sixteen, this proposed bill, if passed, would require the Australian Government to consider the impacts of climate change on the health and wellbeing of children and young people when making decisions. Anjali says, 'Doctors owe duties of care to their patients. Teachers owe it to their students. So governments owe it to young people.'

The bill has ignited a wave of support, championed by independent senator David Pocock and embraced by over 25,000 Australians, who have added their signatures to support this urgent call for change.[5]

All over the world, young people are filing lawsuits against governments and companies, demanding greater social and environmental protections – in the US, Canada, India, the Netherlands, Colombia, Pakistan, Germany, Australia, Norway, New Zealand, Brazil, South Korea, South Africa, France, Chile, Mexico, Sweden, Argentina, Italy, Japan and Indonesia, to name just a few. Young people are mobilising in the thousands, setting up powerful networks, creating policy recommendations, engaging with world leaders, advising company CEOs and becoming a force to be reckoned with.

When young people in the Netherlands sued their government for inaction on climate change in 2015, they won. The court ordered the government to reduce its carbon emissions by 25 per cent.*[6]

In Pakistan, it was a seven-year-old girl who established, for the

---

*   At the time, the country had been on track to reduce emissions 17 per cent below 1990 levels. This, the court stated, was 'below the norm of 25% to 40% for developed countries deemed necessary in climate science and international climate policy'.

first time, the rights of a minor to sue in court after she took her government to court over their promotion of fossil fuels in 2016.[7]

In 2018 in Colombia, twenty-five young people between the ages of seven and twenty-six won a lawsuit against their government for failing to protect the Amazon rainforest. The court concluded that deforestation violated the rights of both young people and the rainforest itself, and ordered the government to update its land management plans and reduce deforestation by 2020.[8]

In 2019, young people and NGOs in the Netherlands teamed up and sued Shell for its contribution to climate change. In a historic judgement, the court ordered Shell to slash its net carbon dioxide emissions by 45 per cent by 2030.[9]

In 2020, a First Nations-led group of young people in Australia – some as young as thirteen – made legal history when they won a landmark case against a coalmine project. They argued that the mine would drive climate change and violate their human rights.[10]

And when these kids aren't fighting for the environment and their future in the courtrooms and boardrooms and parliaments and classrooms, they are doing it on the streets. They're showing up in droves: with pickets and placards and banners and T-shirts, with slogans painted on their chests and chants bursting from their throats. And they're not stopping. In fact, they're growing, bigger and bigger. They're paddling into the oceans, they're scaling buildings and climbing cranes. It's an unstoppable tide, converging on streets, coastlines, city squares, forests and farms.

But these kids, they're not alone anymore. People from every walk of life now join them in peaceful protest. Students and teachers. Mums and dads. Grandparents clutching the hands of their grandchildren. Brothers, sisters, cousins, uncles, aunts, friends and allies. Priests, nuns, imams, rabbis, monks, leaders of mosques and temples, churches and synagogues. Indigenous Elders and communities carrying traditions, stories and knowledge of generations. Scientists, nurses, farmers, engineers, artists and tradespeople. They come with wheelchairs and walking sticks and prams and service dogs. They bring drums and megaphones, first-aid kits, water bottles and snacks to share. They're everyone and they're everywhere. And they're not stopping.

So now, we come for them with handcuffs and prison sentences and jail cells. We lock them up, we silence them, and sometimes we kill them.

Indeed, to be an environmental activist in the 21st century is to risk your body and your life. Every two days, one environmental activist is killed, with Indigenous peoples facing a disproportionate share of the attacks.[11] In many countries, including Australia, the UK and the USA, the conviction rate for peaceful environmental protest remains higher than that for rape.[12]

The right to protest is under siege. To lose that right is to lose the right to be heard. And that should terrify us all. After all, it is freedom itself that is at stake. Today, across the globe, protesting environmental inaction carries a harsher penalty than perpetrating environmental abuse.

A crackdown on environmental activism is sweeping the globe in democracies and dictatorships alike, in what human rights organisations describe as a 'violation' of basic human rights.[13] As youth and grassroots environmental activism rises, authorities are responding with new police powers and increasingly harsh new penalties. A raft of new anti-protest legislation has been enacted, particularly in Western democracies – in Australia, the UK, France, Germany and Poland, to name a few. This is a mass silencing.

In just five years, over 7000 people were arrested for environmental activism in the UK alone.[14] Nine hundred of those were arrested for 'slow marching'. In 2024, five environmental activists in the UK were sentenced to four- and five-year prison terms for 'conspiracy to cause a public nuisance'. Their crime? Participating in a private Zoom call to discuss and plan a protest.[15] The severity of the punishment for what amounted to a conversation was condemned by the UN's special rapporteur on the situation of human rights defenders, Michel Forst, who described it as 'shockingly disproportionate'.[16] On the day of judgement, outside the court, eleven more people were arrested for holding signs that reminded jurors of their ancient right to exercise their conscience.[17]

In 2023, police in the Netherlands detained nearly 3000 climate protesters in a single weekend. In Australia, Human Rights Watch has said the states are 'disproportionately punishing climate protesters in violation of their basic rights to peaceful protest'.[18] The penalty for peaceful protest in Australia can incur multi-year prison sentences and fines as high as $50,000.[19]

In 2023, a ninety-seven-year-old minister was arrested for peaceful protest. Hundreds of protesters had gathered in Newcastle, Australia, at an event run by climate action group Rising Tide, to call for an end to fossil fuels. The protesters paddled out on kayaks and surfboards to block the shipping channel at the Port of Newcastle, the world's largest coal port. The protesters were vibrant and varied: teenagers, First Nations peoples, grandparents, coalminers, singers, artists, environmentalists, dogs, members of parliament and students. They had all come together to call on the government to stop new fossil fuel approvals and tax fossil fuel export profits at 75 per cent to fund communities through the energy transition. This was a peaceful protest, bringing a broad cross-section of the community together. One hundred and nine people were arrested, including five children. Rising Tide organiser and spokesperson Alexa Stuart said the youngest protester on the water was fifteen while the oldest was her grandfather, ninety-seven-year-old Reverend Alan Stuart.

In response to his arrest, the reverend said:

I did this for my grandchildren. I did this because I was inspired by my two granddaughters, Jasmine and Alexa, who are working so hard to address the climate crisis we face. I have been so proud watching both girls try so hard to get their voices heard for so many years. They've signed petitions, written letters, organised rallies, visited politicians, and educated the community.

They have tried asking nicely, and their voices have continued to be ignored by our government. And then, in the past year, I have watched both my granddaughters get arrested for climate protests. It was hard to watch. I was scared for their safety and I wished they didn't have to do it.

But more than anything, the fact they were standing up for what they believed in deeply inspired me. So I felt it was up to me to show them I am willing to do all I can to help protect their future and the future of other people.[20]

The legal assault on activism is glaring, but equally insidious are the language and narratives surrounding environmental activism. The rhetoric used to frame environmental activism is deliberately rebranding peaceful protest as criminal insurgency. It has become a weapon, stoking public resentment and justifying draconian punishments.

In 2023 in South Australia, environmental activists disrupted road traffic to protest a fossil fuel conference and advocate for environmental action. In response, the South Australian government introduced new laws increasing the maximum penalty for 'obstructing a public place' from $750 to $50,000 or three months in jail.[21] The leader of the South Australian Liberal Party at the time, David Speirs MP, called the activists 'out-of-touch greenie leftie losers'.[22]

One of the activists had suspended herself from a bridge for thirty minutes to draw attention to the environmental crisis. In response, the state's police commissioner said, 'We can't, as much as we would like to, cut the rope and let them drop.'[23]

Activists are annoying. But we forget, that is the entire point.

By definition, activism is a campaign to bring about political or social change. Few among us are inactive; our choices, actions and voices shape the world around us. And some voices carry more weight than others. Without activism, change would stagnate, injustice would endure and the voices of the marginalised would fade further into obscurity.

Throughout history, protest has been the spark of transformation, the force that bends the arc of history towards justice. Every right we have today was fought for by people who dared to disrupt: the abolition of slavery, civil rights, the fall of apartheid, the dismantling of totalitarian regimes, women's right to vote, the right for non-landowning men to vote, labour movements that secured fair wages and workplace safety standards, the legalisation of same-sex marriage, and even the idea of weekends. These monumental changes didn't arrive quietly or politely; they were fought for with passion, persistence and, yes, activism.

Today, the stakes have never been higher. Environmental activists are among the most persecuted, vilified and criminalised at a time when we need these voices more than ever.

As annoying as an activist may be, it's difficult to ignore the niggling suspicion that perhaps the collapse of Earth's ecosystems and the ensuing widespread calamity might, in the grand scheme of things, prove ever so slightly more annoying.

As journalist George Monbiot points out, 'If you believe a few people sitting in the street is a major impediment to traffic, take a look at what a sea surge, a flash flood, a windstorm, or a rail-buckling, road-melting, bridge-jamming heat event can do to transport infrastructure.'[24]

Activists, far from being the problem, may well be the first responders to a crisis that far too many have been unwilling to confront.

In 1976, sixteen people gathered at the home of a man named Dr Bob Brown in Tasmania, Australia. There was a quiet excitement in the air. For it was here, huddled around the fire on a wintry evening, that a plan was put in place to undertake one of the most extraordinary environmental campaigns in Australian history.

In the rugged wilderness of Tasmania, the Franklin River carves its way through dense forests, winding a sinuous course through a dramatic landscape of mountains, valleys and deep gorges. Branches dip to kiss the river and mists cling to its edges. Its banks cradle lush myrtle beech, delicate mosses, lichens, fungi and towering Huon pines that have stood for millennia. As it weaves its way across the lands, it nourishes the life it touches. For tens of thousands of years, the river has sustained Indigenous communities, providing food, fresh water and materials. It is a place of preternatural beauty, described as mystical, ancient.

But this river was to be dammed. In 1979, the Tasmanian Hydro-Electric Commission (HEC) revealed plans to dam the Franklin River for hydro-electricity. A large barrier would be constructed across the river, halting the flow of its wild waters. This would create a reservoir – an artificial lake – which would be used to power turbines and generate electricity.

The nearby forests would be flooded, and the river's natural rhythm would be stilled, while the delicate ecosystems the river nurtured – teeming with rare and ancient species – would begin to fall.

The cultural destruction would be profound. Among the treasures to be submerged was Kutikina Cave, an archaeological site of immense significance to the Tasmanian Aboriginal community. The cave contains evidence of human habitation dating back more than 20,000 years, an irreplaceable link to Tasmania's Aboriginal history and heritage. Its submersion would erase not just physical artefacts but the spiritual connection of Tasmania's Indigenous peoples to their ancestors and land.

So, it was in the backyard of Dr Bob Brown's farmhouse that this small group of wilderness activists spawned an environmental movement unlike anything Australia had ever seen. This group became the Tasmanian Wilderness Society.[25] And from that moment, the campaign to save the Franklin became a defining battle in Australia.

This grassroots movement swelled, igniting public support and sparking fierce debate. They plastered walls with posters, flooded communities with pamphlets and gathered crowds in impassioned public meetings, educating people about what was at stake: the destruction of a natural and cultural legacy.

What unfolded over the next seven years would forever reshape Australia. This grassroots movement would trigger a referendum, bring down two state premiers in Tasmania, witness the downfall of a prime minister and define the leadership of two Australian national leaders.[26] It would see legal battles between the Tasmanian state government and the federal government. It would see tens of thousands of protesters descend upon Franklin River to physically block the construction of the dam. It would see 1400 protesters arrested in a single day, including Bob Brown and members of the federal parliament.

It was a protest for the ages, and by 1983, there wasn't a person in Australia who didn't know the names Bob Brown or the Franklin River Dam.

The advocacy of the environmental groups involved in this protest set off a series of landmark events that would forever shape Australia's political history: the formation of the Australian Greens political party; a landmark High Court case; the listing of the area as a UNESCO World Heritage site; new regulations under the *National Parks and Wildlife Conservation Act 1975*; the passing of the *World Heritage Properties Conservation Act 1983*; the strengthening of understandings of Indigenous Australian identity and cultural heritage; the emergence of the Wilderness Society, which would become a national charity; and the rise of one of Australia's most influential political and environmental leaders and activists – Dr Bob Brown.

In 1983, as bulldozers descended upon the Franklin River, the High Court case between the Tasmanian Government, who supported the dam, and the federal government of Australia, who supported the protesters, was underway. On 1 July, a final decision was handed down by the High Court: 'There shall be no dam on the Franklin River.'[27]

Whether you agree with their methods or not, one thing is undeniable: activists, protesters and grassroots groups are the backbone of society and have shaped the world in fundamentally positive ways. In every policy, every decision and every bit of freedom you have been given as a citizen, if you look closely, you will undoubtedly find the footprint of these grassroots and community networks that have made that possible.

We stand on the shoulders of invisible giants, and those giants come in the form of young people, grassroots organisations, scientists, activists, nonprofits, Indigenous peoples and communities who have spent centuries laying the groundwork to ensure climate and nature are prioritised.

These individuals and networks are the lifeblood of the progress we enjoy today. These are the people who have marched, fought, bled and begged for the right of everyone to live in a socially equitable, environmentally sustainable society. And rarely have they been celebrated for their sacrifices. Instead, throughout history they have been scorned, belittled, ignored, beaten and imprisoned. Today, these voices are often tokenised; their contributions to climate and nature policy are undermined, while their ideas are rebranded and repackaged in ways that are antithetical to the original intent.

All over the world, grassroots and activist networks have spawned some of the largest policy changes in history: from the original 'tree huggers' movement in 18th-century India, which saw villagers defending forests against commercial logging; to the 1966 Wave Hill protests by the Gurindji people that established First Nations land rights in Australia; to the Greenpeace fleet that took the first ever direct action against whalers in 1975, triggering an international moratorium on commercial whaling; to the small band of scientists in the USA who instigated the banning of the insecticide DDT.

Today, the tradition and legacy of these peripheral but powerful voices are reflected in some of the world's most influential environmental policies and protections. Our bushfire and climate disaster responses, the restoration of habitats and protection of species, the availability of clean air and safe drinking water, the establishment of native title, the protection of oceans and rivers and national parks, the elevation of marginalised voices and the blueprint for clean energy implementation are all thanks to these bands of people we so freely condemn as 'radicals'. They are grassroots organisations, scientists, activists, nonprofits, Indigenous peoples, young people and communities – often volunteers of this thankless work.

These are the people you can thank for taking on the work that nobody else would, work that is considered too 'soft' to be elevated among the most influential circles in Australia. Our young people taking governments to court for the rights of all young people, our wildlife shelters and their voluntary carer networks, our citizen scientists painstakingly identifying and monitoring endangered species, our volunteers restoring coral reefs, our NGOs investigating environmental abuses, our think tanks providing pathways for corporate decarbonisation, First Nations people filing lawsuits against major polluters, our scientists monitoring changes across land and sea and our activists filing shareholder resolutions to get climate on the agenda of the largest companies in the country.

For too long, there has been a false notion that the work and passion of these individuals and networks is inferior, radical – criminal, even. Little recognition has been given to the fact that it is these people who have mobilised millions of others to solve the world's most pressing problems while getting a fraction of the salary (if any at all) and being alienated from key decision-making bodies.

When you're standing on the shoulders of giants, don't forget to keep their fire burning. These grassroots networks are not peripheral to environmental discourse; they are the architects of the progress we enjoy. The privileges we have today were born of advocacy. And advocacy of the past is more palatable to us than the advocacy of the present. But make no mistake, amidst the greatest planetary crisis in human history, never have we needed these voices more.

Today, Tasmania's Franklin River – which represents one of Australia's most iconic environmental victories – would likely be dammed. After all, the success of this campaign rested upon the right for community to show up and advocate – a right that is rapidly being eroded around the world.

In December 2024, the largest court case in history was set in motion.* It was one of the most universally significant moments in environmental history. And it was because of young people.

The ICJ commenced historic hearings to answer two questions: do states have a legal obligation under international law to combat climate change and protect future generations from climate harm? And what are the legal consequences for nations whose actions cause harm?

This was instigated by the youth group Pacific Islands Students Fighting Climate Change. Vanuatu, listening to the students, then lobbied other countries to support their initiative. This legal opinion on climate change could help inform future proceedings and influence domestic legal cases and diplomatic processes. It will likely be cited in thousands of climate-driven lawsuits around the world.[28]

These kids, these activists, these grassroots groups – they're asking us, simply, to let them have a future. To let them imagine a world and a future where the sky is filled with birds and butterflies and the ground is rich with soil and seeds; where the forests are standing, and the ocean breathes life; where the food, oxygen and

---

* Described as the 'largest court case' based on the unprecedented level of global participation: over 130 countries supported the UN resolution requesting the ICJ opinion, making it the most widely endorsed legal action ever brought before the court.

energy provided by nature is plentiful. They're asking not to inherit a world ravaged by floods and droughts and fires and cyclones, a world where they become environmental refugees, where their lungs are poisoned by chemicals and their bodies filled with plastics, a world emptied of wild species. They are asking for something we once took for granted. They want to breathe freely and experience the joy and wonder of nature. They want, simply, to live.

And we are denying them this.

But here's the thing: they aren't asking for our permission anymore. They are claiming their own space as leaders and forging their own path. They are showing us that, regardless of whether they vote or not, they have a major voice in deciding the future. Young people are our customers, our current and future workforce, the shapers of our policy. They are the heirs to the planet, the architects of the climate movement and the power behind the civil rights movement, the Arab Spring, the Occupy Wall Street movement, the Black Lives Matter movement and the #MeToo movement.

To them, protecting their future is not negotiable. So, they are fighting for it – with or without you. And they are just getting started.

These kids, they're turning the world upside down.

Chapter 16

The Delight and Scandal of a Birdwatcher: To Fall in Love

> We're living in the golden age of birding, and like any good cult member, I'm recruiting people to the cause.
> Kate Wong[1]

They are the most curious creature of all. Hyper-focused. Single-minded. Intense. Devoted. Often single. They speak in reverent tones and hushed whispers and can walk with preternatural silence across a bed of leaves. They wield binoculars with the nonchalance of a sommelier sampling a Dom Pérignon. They can crouch in shrubbery for endless hours. They speak in code and use hand signals. They have lists and notebooks and write with lead pencils. They dress with military precision: khaki pants, fitted belt, cedar-brown shirt, wide-brimmed hat, waterproof boots. Their social calendars are governed by migration patterns and their conversations are peppered with whispered phrases like, 'Was that the trill of a reed warbler?'

They are bearers of universal mysteries. Holders of ancient wisdom. They are birdwatchers.

Birding isn't merely a hobby. It's a lifestyle, like wearing Crocs or having a Goop subscription.

Birders are the world's most unwavering optimists. They believe, against all odds, that somewhere out there is a bird they've never seen, just waiting to be discovered. And they'll keep looking, one step (in a pair of sensible shoes) at a time.

They are dealers in wonder and awe, masters of birds and ecosystems, and, despite the many stereotypes we can muster, one should never underestimate the power, joy and influence of the birdwatcher. Indeed, they may just be humanity's greatest source of inspiration and hope.

Birdwatchers, they live among us. In the US, for example, one in every three people is a birdwatcher.[2] That's right – statistically speaking, someone you know is spending their weekends peering into trees and whispering excitedly about warblers.

The birdwatcher could be anyone. The kindly neighbour who seems oddly fixated on their backyard feeders. The co-worker who goes mysteriously quiet during meetings every time a shadow flits past the window. They're everywhere, quietly cataloguing the avian world, one robin or finch at a time.

The practice of observing birds spans a wide spectrum of intensity, from the mildly curious to the dangerously obsessed. Casual enthusiasts are often referred to as 'birdwatchers'. On the other end of the spectrum are the 'birders', a more deliberate, focused and slightly intimidating subset of humanity. Birders are not content to merely watch; they seek to identify, list and photograph as many species as possible – sometimes competitively.

Birdwatching, at first glance, seems like a pursuit designed specifically for people who find stamp collecting too stimulating. But oh, how wrong that is. Behind the courteous nods, the gentle lift of binoculars and the soft pitter-patter of birder footsteps on dew-dappled gum leaves lies a world of unexpected drama – a scandalous underbelly of rivalry, controversy and intrigue. Indeed, what appears to be a peaceful communion with nature masks a tempest of Shakespearean passion.

And so unfolds one of birding's most captivating sagas: the tale of Peter Kaestner, Jason Mann and the frenzied race to find and record 10,000 bird species.

THE DELIGHT AND SCANDAL OF A BIRDWATCHER: TO FALL IN LOVE

The year was 2024 and Peter Kaestner, a retired American diplomat, was on the verge of making history. He was poised to do what no one had done before: log the 10,000th bird on his 'life list'. For the uninitiated, a 'life list' is the ultimate birder's brag book, a meticulous record of every bird species one has ever seen. To put Kaestner's number in perspective, there are around 11,000 bird species on Earth, depending on which ornithological expert you consult. Reaching 10,000 species is akin to climbing Mount Everest, swimming the English Channel and memorising the entire works of Shakespeare – all while blindfolded. It's not just impressive; it's herculean.

In the birding world, Kaestner is a big deal – such a big deal that he's known as the ultimate 'big lister'. Because his list is so big. Big listers are a small but fierce subculture of hypercompetitive birdwatchers, who scour the globe vying to see as many species as possible. Kaestner was about to become the James Bond of birding; after a lifelong quest to find and record as many bird species as possible, he was just one bird shy of 10,000.

On 9 February, Kaestner was standing under the Tinuy-an Falls on Mindanao island in the Philippines when his moment finally arrived. The stage was set, the air electric with possibility. There. On a branch. The feathered emblem of Kaestner's destiny. An orange-tufted spiderhunter. Draped in feathers of muted olive and tawny hues and sporting a tuft of orange atop its head, it's a songbird known for its fondness for banana and its elegantly curved beak. This was it. This was the 10,000th bird. Kaestner was about to etch his name into birding history and claim his rightful place in the annals of birdwatching greatness.

But as every great hero must face a nemesis, so too did Kaestner. Unbeknown to Kaestner at the time, a dark horse had entered the race: Jason Mann.

As the shutter clicked on Kaestner's camera, capturing the orange-tufted spiderhunter in all its glory, chaos was erupting in the birding community. Out of nowhere – or so it seemed – a previously unknown birder named Jason Mann had emerged, armed with claims of his own audacious birding achievements.

While Kaestner was celebrating his 10,000th bird sighting, Jason Mann was also claiming to have reached the elusive 10,000-bird milestone just hours before, snatching the record right out from under Kaestner. The entire birding community was thrown into turmoil.

'Two people break 10,000 species, and on the same day? Can it be?' posted one astonished birder on a birding forum.[3]

'How in the world could this happen? This thing that has never happened in the history of mankind, that two birders get 10,000 on the same day? It's crazy,' said Kaestner.[4]

'It's absolute chaos,' declared another birder.[5] 'The timing of [Mann's] announcement and his lists are just a little too suspicious for me,' one remarked.[6]

A relative unknown in birding circles, Mann had uploaded an astonishing life list of nearly 10,000 species to iGoTerra, a birding platform, just months before the dramatic race to 10,000 reached its climax. Then, on the very same day as Kaestner, Mann conveniently claimed his 10,000th bird in Colombia: a chestnut-bellied flowerpiercer. A feathered fantasy sporting a sleek, iridescent coat of inky blue-black feathers that shimmered like polished obsidian in the sunlight. Mann's ascent to 10,000 birds was meteoric – and downright suspicious to the seasoned birding elite.

On the other hand, Kaestner – the birding equivalent of a national treasure – had spent decades meticulously documenting his sightings, painstakingly recording every bird and branch with utmost precision, all the while sharing his progress with his ever-growing fan base through reputable birding platforms. The entire birding community had been watching his journey towards 10,000 birds with mounting excitement.

To make matters worse, Mann, the enigmatic newcomer, had penned an article congratulating himself on his supposed achievement. He then persuaded a Colombian nature tour company to publish it on their website on 9 February. The headline on the Manakin Nature Tours website read, 'One of the best birdwatchers in the world, Jason Mann breaks world record, passes 10,000 lifetime species in Colombia.'[7] It was published on the very day that Kaestner had celebrated his own 10,000th bird sighting.

Birders world over were positively apoplectic. They descended upon Mann's list with fervour and fury. 'Shamelessly masturbatory,' claimed one birder who saw Mann's announcement. 'Many, many fishy recordings.'[8]

'Not to be trusted,' sniffed another.[9]

'Purple-winged ground-dove raises eyebrows as well,' said one sceptical birder who'd been poring over Mann's list.[10]

The problems with Mann's self-proclaimed triumph went beyond the announcement. His bird list included several species that were long thought extinct, or at least extraordinarily rare, like the New Caledonian nightjar, a bird that hadn't been seen since 1939. Mann had also sidestepped every major birding platform, failing to leave any photos, notes or other evidence of his sightings.

It was the scandal of the century, sending shockwaves through the birding community.

It was only upon internal reflection that Kaestner realised how this series of unfortunate events may have unfolded. He had an Achilles heel, a fatal flaw in his noble pursuit. Just weeks before he reached his 10,000th bird sighting, he had penned an essay for the American Birding Association. In it, he had revealed his plan to be the first to achieve the 10,000 bird milestone.[11] Now the world knew – he was close to becoming the ultimate big lister. 'I realized when I was writing it that I was putting a target on my back,' Kaestner later admitted.[12] Kaestner, in his bid to be as transparent as possible, had inadvertently thrown down the gauntlet, one that Jason Mann had seized.

But Mann's victory was short-lived. Amidst a global outcry from the birding community and a tidal wave of disbelief, Mann folded. He acknowledged the inconsistencies in his records, made his life list private and publicly recognised Kaestner as the rightful record holder. In a post to BirdForum identifying himself, he conceded that he 'made some errors when inputting my sightings into iGoTerra'.[13]

With a magnanimity born of necessity, Mann announced his support for Peter Kaestner as the first birder to reach 10,000 species. Mann said that while he considered over '99% of the list' he uploaded to be accurate, he conceded that, in his haste, 'there were a few oversights'. Given the errors, he said:

I think it best to put my support behind Peter as the first birder to 10,000.

I don't want there to be any question, and in my view he is extremely deserving. I celebrate Peter and other world-class listers, and do not consider myself in competition with any of them.

To me it was a real privilege to have such a close, neck-and-neck race at the end. My hope is that any associated publicity helps move birding forward, raising awareness for more people to enjoy and protect nature.[14]

Manakin Nature Tours, which had published Mann's self-congratulatory article, quickly deleted it and issued a statement in response to the scandal: 'We want to address a recent issue that has sparked considerable controversy within the global birdwatching community,' it read. 'Regarding Mr Mann, we want to clarify that we bear no responsibility for the species records he has observed, whether in our country or elsewhere in the world,' it continued. The website urged impassioned birders to maintain civility during this trying time, noting 'direct attacks against us and accusing us of participating in an alleged "fraud," are unjustified'.[15]

To this day, the truth about Jason Mann remains a mystery. Was he an extraordinary yet reclusive birder or the architect of one of the greatest attempted heists in birding history?

THE DELIGHT AND SCANDAL OF A BIRDWATCHER: TO FALL IN LOVE

What's certain is this: Peter Kaestner is now the uncontested birding champion of the world, officially recognised as the first person in history to log 10,000 bird species and to meticulously document every single one.

The saga of Kaestner versus Mann will, without question, enter birding lore, a story whispered in hushed tones at various binocular-clad gatherings; a cautionary tale of deception and ambition, of passion and determination.

Kaestner's achievement was celebrated not just for its sheer improbability, but its absolute dedication. Decades of fieldwork, patience and a near-encyclopaedic knowledge of birds culminated in a milestone that, for now, seems unassailable. Not just that, he has contributed immensely to the field of environmental science, even discovering a bird species in Colombia: the Cundinamarca antpitta, which was named in his honour with the scientific name *Grallaria kaestneri*.

But birding is far more than a checklist or a competition – it's an invitation to connect with the natural world in profound and unexpected ways. And while it might start as a solitary pursuit, birding is actually an enormous community of care, a movement rooted in the simple yet transformative act of witnessing and safeguarding the world's wonders.

In fact, the quiet attention of birders influences policy, science, culture and the economy. As birder Georgia Angus says, it is the 'gateway drug to environmental awareness'.[16] Birdwatchers are often environmentalists and are more likely to engage in conservation behaviours than the general population.[17]

You see, birdwatchers are the sort of people who notice tiny changes in the world, while the rest of us are busy ignoring the big things. They're the first to notice the dwindling of a bird population, the first to see when a migration pattern shifts or when a habitat shrinks. With their notepads full of tiny, precise observations, they are gathering information, and this information is guiding climate science, ecology and conservation efforts. It has even influenced the design of wind turbines to reduce avian casualties. In their quiet, methodical way, birdwatchers are the khaki-wearing heroes on the frontline of nature conservation.

They redefined a once obscure hobby, turning it into a global network of citizen scientists. Social media has given rise to 'birdfluencers', and viral sightings – like the Steller's sea eagle that appeared in Maine, US – can now mobilise thousands of enthusiastic birdwatchers overnight.

eBird, one of the world's largest biodiversity and citizen science projects, is a global online database powered by the passion and dedication of birdwatchers worldwide. Here, science is democratised – anyone anywhere can contribute their bird sightings. And birdwatchers have contributed millions upon millions of volunteer hours to help build a treasure trove of information on birds and ecosystems.

Through eBird, birders have contributed an astonishing 1.9 billion bird observations from every corner of the globe – and that is just on one platform. This has fuelled efforts to research, understand and protect birds and ecosystems. In fact, researchers have published over 1180 scientific publications incorporating the data from the eBird birding community.[18] It's not just online. The Christmas Bird Count sees more than 80,000 birders flock to thousands of locations to collect data on bird populations. For the Global Big Day, more than 1.3 million birders from 203 countries unite for a twenty-four-hour birding extravaganza, with a challenge to identify as many bird species as possible.[19] Then, of course, there is the 'big year'. To achieve a big year, a participant should expect to identify more than 700 species, travel up to 500,000 kilometres, and spend 270 days away from home in a calendar year. So climactic is this event that a movie was created in its honour, telling the true story of three singularly obsessed men who compete to see who will be the 'best birder in the world'.[20]

THE DELIGHT AND SCANDAL OF A BIRDWATCHER: TO FALL IN LOVE

Birders and bird counting are growing in size and scale all over the world. The Aussie Bird Count. The Global Bird Weekend. Great Backyard Bird Count. There are festivals, competitions and rivalries. There is community. And, importantly, an incredible amount of data is being captured, reshaping how we understand and protect the natural world.

Birds are also big business. Ecotourism centred around birdwatching contributes billions to the global economy each year. Entire towns thrive on seasonal migrations, their economies buoyed by birders who come for flamingos, pelicans and puffins.

In fact, birdwatching tourism is one of the fastest-growing nature-based tourism sectors in the world – in 2023 it was a US$62 billion industry, and rapidly growing.[21] In the US alone, birders spend an impressive $107.6 billion annually on various birding-related activities.[22] This spending supports over 860,000 jobs nationwide.[23]

Colombia has nearly 2000 species of bird and has become a premier destination for birdwatchers. In 2024, a record 6.7 million tourists visited Colombia, with the majority seeking nature experiences, including birdwatching.[24] Birders have poured money into the economy, especially into rural communities, creating nature-based jobs and even prompting the government to rethink the relationship between nature and the economy. In fact, bird-related tourism has the potential to create more than 7500 jobs and contribute $9 million to the Colombian economy.[25] As a result of this, Colombia is ramping up its efforts to protect its birds and their ecosystems, going so far as to develop a strategy to protect its nearly 2000 birds species – one that explicitly incorporates bird appreciation and birdwatching as tools for conservation.[26]

Birders have also descended upon Chile. With over 500 bird species, the country is birder paradise. Chile's birdwatching tourism industry generates US$62.6 million annually.[27] So popular is birding in this country that Chile is preparing to include birds in its GDP.*

Few pastimes inspire more unbridled mockery than birding. It's the Dungeons & Dragons of the natural world. And birders themselves? They occupy the curious cultural space of being both painfully earnest and entirely incomprehensible to outsiders. There are the 'patch birders', fiercely loyal to their local wetland or park, who catalogue every twig and leaf with near-pathological fervour. Then there are 'twitchers' who, upon hearing the faintest whisper of a rare and exotic bird sighting, are conveyed to a fevered delirium; they will abandon weddings, careers and even newborn babies to embark on an intercontinental pursuit for a fleeting glimpse of a golden-winged warbler. There are listers and big listers; backyard birders and urban birders; ear birders, sea birders, competitive birders and social birders.

Birders are everywhere. Birdwatching, as it turns out, is gloriously addictive.

At its heart, birdwatching is an act of quiet rebellion. It is the gentle act of noticing – the willingness to see the world around you. And this simple act is, in itself, revolutionary. In a world that often teaches us to look past the natural world – to pave it over, cut it down or otherwise ignore it – birdwatching demands slowness and attention. It demands that we pause, that we listen, that we care, that we *see*.

And it's never about one single bird. It's about everything that

---

* This does not mean that Chile is issuing tax statements to woodpeckers. What it does mean is that the country is trying to account for the real economic impact of birdwatching – the tours, the binocular sales, the jobs, the local guides with encyclopaedic knowledge of small brown things in bushes – as well as the quieter contributions birds make, like insect control, ecosystem health and encouraging humans to stop, look up and wonder what that strange chirping noise is. It is placing a value on nature.

THE DELIGHT AND SCANDAL OF A BIRDWATCHER: TO FALL IN LOVE

sustains them: the soil. The trees. The rivers. The wind. Soon, it evolves into understanding the delicate interplay of ecosystems – the leaf litter turned over by superb lyrebirds, the seeds dispersed by cassowaries, the flowers kissed into bloom by honeyeaters. A single bird becomes a prism through which to view the entire ecological web, a window into the world, a thread connecting you to nature.

It's no longer possible to see a forest as a commodity or a wetland as an inconvenience when these spaces are home to creatures whose existence depends upon them. Birders understand, perhaps better than most, the fragility of the natural world. They see it in the dwindling numbers of migratory birds, in habitats encroached upon by development, in the ghostly silence where once there was song. Yet they also see hope: in the return of the California condor from the brink of extinction, in the rise of the whooping crane after tireless conservation efforts, in the fierce resilience of nature when given the chance to heal.

To be a birder is to fall in love. Obsessively, irrevocably and, perhaps, foolishly. This love, it creeps up on you. A wedge-tailed eagle slices through the sky. A magpie's liquid song pierces the dawn. A snowy owl gazes unblinking across the tundra.

So, here's to the birdwatchers, those optimistic, slightly eccentric custodians of wonder and joy and passion and love. Because sometimes it is as simple as opening your eyes, stepping outside and looking upon the world around you.

For once you see – once you really see – you cannot unsee. You see the beauty and destruction, the joy and the tragedy of life around you. And you fall in love. And once you are in love, there is nothing you wouldn't do to protect this fragile, extraordinary world.

Epilogue

The Way Forward

Humans are unique. Not in the way that shapeshifting, colour-morphing octopuses are unique – but in the way that a person clipping their toenails on a public bus is unique. The uniqueness lies in the absurdity, the chaos.

For example, humanity is the only species that actively debates whether or not it should maintain the environmental conditions necessary for its own continued existence. Other species do not struggle with this sort of thing. Ensuring the health of things like the atmosphere, forests, oceans, rivers and animals would presumably be an obvious priority for a species that relies entirely on those things to live; something we could all agree on, like 'fire is hot'. But no. Instead, we have managed to convince ourselves that consequences are for other people. We have turned the simple matter of survival into a subject of endless controversy.

We've split the atom, cracked the genetic code and hurtled expensive metal objects into space, sometimes with people inside, mostly just to see if we could. We have built cities that scrape the sky, machines that think and a global network that allows people to yell at each other at any time of day, from any part of the world. But when it comes to the question of whether we should keep our own planet habitable, humans remain curiously undecided.

Some people feel very strongly that we should do something about this, while others insist that everything is probably fine, that the planet is just 'going through a natural cycle' and that if things get bad enough, surely someone – scientists, governments or possibly a billionaire with a rocket – will sort it all out.

I am of the opinion that ensuring the existence (and health) of life on this planet is a reasonably worthwhile pursuit. So, I wrote a book explaining that destroying everything around us is, in fact, bad. That it is not good for the things we are killing, but it is also not good for us. That we really should try to keep more things alive.

In this book, we have traced the tangled lives and fates of humanity and nature – the stories of species and communities on the edge, the machinery of destruction that feels unstoppable, the systems, structures and institutions that enable it.

But we also have also witnessed something else.

We have seen rivers and forests, whales and dolphins granted legal rights. A twelve-year-old girl standing sentinel over an owl. A wildlife rescuer dangling upside down inside a drain on Christmas Eve to save a bird.

We have seen birdwatchers turning wonder into data. Indigenous peoples carrying wisdom through generations. Women knitting sweaters for oil-drenched penguins. Economists designing an economy that puts nature at the centre. Farmers welcoming the wilderness back on their land.

We have seen scientists studying the clitorises of great apes, donning ejaculation helmets to save birds and reading the secrets of the sea in whale poo. Journalists exposing destruction against the tide of media monopolies. Diplomats spending years in airless rooms so that environmental policy may still breathe.

We have seen companies investing in nature. Lawyers defending nature. Billionaires directing their fortunes towards the environment. Nuns linking arms with teenagers to stop coalmines in the courts.

We have seen otters regenerating sea forests and helping other marine life thrive. Vultures halting the spread of disease and cleansing the land of decay. Geckos inspiring space technologies.

Pandas bridging nations in times of conflict. Plants and trees weaving medicine from sunlight and alleviating pain from the world.

We have seen a world being rebuilt. Messily, with mistakes and missteps. Stumbling, mending, remaking. But it is happening. The world as we know it is fracturing. In its place, something new is taking shape – not a world bent on dominating nature, but one learning to coexist with it.

This future is being written in pieces: in the decisions we make, in the fights we refuse to abandon, in the status quo we refuse to accept, in the communities we forge. It is being written in courtrooms and classrooms, in forests and oceans, in laboratories and streets, in economic models and corporate strategies.

You see, the destruction of nature is not accidental – it is embedded in the very systems that shape our world: our dominant economies, institutions, laws, politics, technologies, media and even our shared cultural psyche. These are the systems we live by, guided by rules of our own making – rules that we conjured up, declared immutable, built belief systems around and now follow with devotion. For centuries, these systems have dictated how we define progress, power and civilisation itself.

Dominant economic models prioritise endless production, extraction and consumption on a planet with finite resources. Political systems, driven by short-term election cycles, neglect long-term ecological and social stability while remaining vulnerable to corporate influence, aggressive lobbying and a concentrated media landscape that shapes public perception. Polarisation becomes political strategy as division, misinformation and distortion are wielded to win votes, shape policy and dictate public discourse.

Cultural narratives deepen our disconnect with nature, feeding the illusion that we can dominate the living world without consequence. Curiosity is often replaced by certainty, public debate framed as a battle to be won rather than a conversation to deepen understanding and connection.

Legal frameworks often uphold and reinforce these systems,

prioritising a narrow set of human interests at the expense of ecological wellbeing, while legitimising harm to nature and communities: destroying an ancient forest is considered development; protesting its destruction is often a criminal offence.

These are the human systems that shape our world order. And they were sometimes inadvertently – built on a logic of dominance over both nature and communities. A logic that we continue to uphold today.

Yet there is nothing inevitable about the way we run our economies, draft our laws or govern our societies. These systems have no independent existence; they exist because we will them to. They are human constructs.

Unlike the laws of nature – like gravity or thermodynamics, which apply equally to everyone – human-made systems like economics, law and politics are often structured to the advantage of a select group of people. This is why gravity applies to everyone but tax loopholes do not.

And these human constructs don't just allow environmental destruction – they normalise it. They make it seem inevitable, even necessary. In this way, environmental destruction persists not because we lack alternatives but because we allow it to remain profitable, socially tolerated and politically viable through these systems.

But the world doesn't have to work this way. We made it this way. And that means we can make it work differently. Like all human-made systems, they can be rewritten – assuming, of course, we want to.

There is no single way forward, no silver-bullet solution to end the destruction of nature and communities. But there are thousands of ways forward. What we are witnessing – and what we have witnessed in this book – is a collision of grassroots movements, political transformations, economic shifts, corporate upheavals, scientific breakthroughs, social and cultural evolutions and legal battles that together are changing the course of history.

This is how we end the destruction of nature and communities: by rethinking the fundamental forces driving this exploitation. This means addressing the root cause of harm – not just treating the symptoms, but changing the conditions that created them in the first place.

It is not enough to make small adjustments within a system that is designed to extract, consume and discard at an ever-increasing pace. As long as deforestation, pollution, over-consumption and inequality remain embedded in the logic of our economic, legal, cultural, social and political systems, even the most well-intentioned efforts will struggle to create lasting change.

Planting trees, for example, is often seen as a simple, tangible solution. Tree planting is noble – essential even. But planting a tree within an economic system that rewards excessive deforestation, within a political system that allows an extractive company to have outsized influence over environmental policy, within a cultural system that tells us to take more than we need and within a legal system that sanctions this is a bit like trying to fill a bucket with water while someone else is enthusiastically poking holes in the bottom.

Planting trees is necessary. Recycling your plastic is responsible. Choosing a bike over a car is admirable. But changing the economic and political logic that makes mass deforestation, excessive plastic production and fossil fuel production highly profitable is world-changing.

This is how systems shift – through deliberate and collective efforts to rethink and reshape the dominant institutions, economic structures, political systems and social and cultural narratives that we have convinced ourselves are inevitable and the only way forward. Piece by piece.

This may sound impossible – fanciful, even. But this is precisely where the individual becomes powerful – not (solely) in the singular act of planting a tree or meticulously sorting recyclables but in something far older, far more intrinsic to our species: our ability to cooperate. To form communities. To mobilise.

Indeed, the real impact of individual action is not in its ability to single-handedly shift the course of history, but in our ability to cooperate with others and leverage the unique and diverse strengths of individuals within a community.

Change very rarely comes about from people standing in opposing corners, shouting at each other. But when individuals come together, things happen. Sometimes it's bad things, like reality television, but sometimes it's rather useful things, like libraries and the legal recognition of nature.

Change is shaped by how we as individuals choose to navigate the world – by the ideas we engage with, the people we connect with, the choices we make, the actions we take and the way we engage with the world around us. After all, cultural shifts precede policy shifts, and the greatest impact is achieved when individual responsibility is linked to community and collective action.

But collective action (and impact) is not just a matter of effort; it is a matter of direction. It does not come from retreating into ideological strongholds or treating every disagreement as a war to be won. It happens in the harder, messier work of finding common ground in unlikely places, of speaking and listening with the intent to understand rather than to defeat, of recognising that progress is not the product of perfectly crafted arguments but of people – imperfect, conflicted and often stubborn – choosing, against all odds, to build something better together.

We know that individual influence is not equal. The power to effect change is indeed distributed unevenly. A consumer making ethical choices has some impact, but a corporate buyer who chooses sustainable suppliers can shift entire markets. A voter supporting an environmental policy is important, but a politician drafting that policy can determine the course of nations.

But the power of the individual is immense in its ability to mobilise, connect and amplify change within a community. The key isn't just doing *something* but doing the right thing in the right place, where it has the most leverage. Change happens when individuals recognise their unique strengths and work within a community, directing those strengths towards the places where they have the most impact. When this individual action is woven into collective momentum, when people pool their diversity of strengths as part of a community, systems begin to shift.

This is how history moves. And this is how the future is shaped. Not just in grand, sweeping gestures but in the steady accumulation of small actions that ripple outward – shifting perspectives, influencing policies and rewriting the boundaries of what is possible.

It begins with curiosity.

Someone wonders whether trees should have rights. An economist asks what the value of nature is. A birdwatcher logs a bird sighting into an app. A politician takes a sledgehammer to a building. A philosopher asks whether happiness should replace GDP. A teenager skips school to protest climate inaction. A teacher rewrites a lesson plan.

And then, one day, a river becomes a person. An economy is rebuilt to protect birds. A nation is founded on the principles of peace and nature. A government rewrites its constitution to give future generations a legal voice.

Change does not announce itself. It does not arrive all at once. But piece by piece, voice by voice, decision by decision, it takes shape.

And before we know it, we are living in a world that once seemed impossible.

Make a Difference

This book doesn't offer a comprehensive plan to save the world, and it doesn't pretend that individual action alone can undo decades of systemic harm. But change does, at the very least, require participation – from everyone.

Our world is shaped by systems – legal, economic, political, cultural – often built on short-term goals and unequal benefits. The destruction of nature isn't a glitch in these systems. It's a feature. But systems can be changed. They are, after all, made by us.

Below are nine levers of change. They overlap, interact and reinforce one another – and they are by no means exhaustive. Yet together they play a large role in shaping the operating logic of our societies. Shifting any one of them matters. Shifting several of them at once? That's how history moves.

Although the broader themes and frameworks here apply globally, some examples and resources in this book reflect a British context.

1. Political and Policy Transformation

How we make rules, allocate power and set national priorities. Politics is where priorities are set, legislation is made and funding decisions are determined. Shifting political systems towards long-term, ecological and community-centred outcomes is foundational.

Practical Actions:
- **Vote:** Support candidates who prioritise long-term environmental and social justice.
- **Read beyond political slogans:** Review candidates' voting records, party manifestos and policy proposals. Verify with independent sources. Always check the data supports the claim.
- **Write to your MP or local councillors:** Call for stronger environmental laws, increased nature investment, enhanced community consultation, integrity reforms, etc.
- **Push for transparency and integrity reforms:** Support real-time political donation disclosure, limits on corporate lobbying, stronger anti-corruption measures and putting a stop to deceptive political advertising.
- **Support civic education programs:** Political literacy is uneven and declining. Support programmes that strengthen public

understanding of democratic processes, media literacy and civic engagement.
- **Follow global agreements shaping global policy:** This includes the Paris Agreement, the Global Biodiversity Framework, the UN Convention to Combat Desertification, the Global Plastics Treaty, and initiatives like CITES and the Global Initiative to End Wildlife Crime.

Examples:
- Track MP voting records: theyworkforyou.com
- Transparency and anti-corruption movement: openDemocracy and Transparency International UK
- Independent research on democracy and public integrity: Institute for Government and UK in a Changing Europe
- Engage in public consultations on critical legislation: gov.uk/government/consultations
- Environmental scorecards and policy analysis: Friends of the Earth UK, Green Alliance, UK Climate Change Committee
- Fact-check political claims: Full Fact
- Track climate policy performance: climateactiontracker.org and OECD's Climate Action Database.
- Track government progress on biodiversity and nature: WWF's NBSAP Tracker.
- Corporate lobbying on climate policy: influencemap.org

2. Economic Logic and Redesign

How we value, produce, consume, trade and measure.
Our economies are structured around endless extraction, consumption and growth – on a finite planet. Rethinking these foundations is essential.

Practical Actions:
- **Advocate for wellbeing budgets:** Alternative measures of

progress that go beyond GDP growth as the primary policy driver.
- **Push for subsidy reform:** End public subsidies for activities that harm ecosystems, drive emissions and accelerate biodiversity loss.
- **Support regenerative, circular, and care-based economies:** Design waste and harm out of the system from the start.
- **Promote post-growth and degrowth models:** Support economic frameworks that recognise ecological limits and prioritise social wellbeing over perpetual extraction.
- **Engage with nature-positive economic frameworks:** Explore models that integrate ecosystem health, climate stability and biodiversity protection into economic planning and investment.

Examples:
- Understand how nature harm is subsidised: Research by Global Witness and fossilfuelsubsidytracker.org documents fossil fuel and nature-harming subsidies
- Explore economic redesign models: doughnuteconomics.org
- Engage with 'degrowth' thinking: Less is More: How Degrowth Will Save the World by Jason Hickel
- Learn about integrating nature into economic systems: UK Natural Capital Committee (NCC)
- Design waste out of the economy: WRAP UK, Ellen MacArthur Foundation
- Join or follow Wellbeing Economy Alliance (WEAll).
- Explore roadmaps for a nature-positive economy: Read WWF's 'Global Roadmap for a Nature-Positive Economy'

3. Industry and Production Systems

How goods and services are created – and at what cost.
Industry underpins our physical world. Shifting production systems towards regeneration rather than extraction is critical.

Practical Actions:
- **Push for green manufacturing hubs:** This includes those with zero-emissions, biodiversity strategies and just transition plans for workers and communities.
- **Demand mandatory supply chain due diligence laws:** Companies must be required to identify, disclose and eliminate environmental and human rights harms across their operations and supply chains.
- **Support corporate initiatives for regenerative supply chains:** Engage with those working towards full traceability and ecosystem restoration.
- **Support companies adopting science-based targets for nature:** Including commitments to eliminate deforestation, biodiversity loss and ecosystem degradation from their supply chains.
- **Engage your workplace or industry sector** in initiatives that embed nature, climate and human rights considerations into procurement, production and reporting systems.

Examples:
- Mobilise business support for nature-positive economies: businessfornature.org
- Scalable environmental restoration solutions: drawdown.org
- Roadmaps for companies to prioritise nature: World Business Council for Sustainable Development (WBCSD) Roadmaps to Nature Positive.
- Science Based Targets Network (SBTN): sciencebasedtargets.org
- Read Minderoo Foundation's Plastic Waste Makers Index.

4. Finance, Investment and Capital Flows

Where money flows – and who decides.
Capital determines what gets built, protected or destroyed. Shifting financial flows is crucial for systemic change.

Practical Actions:
- **Divest personal savings, superannuation and banking** away from fossil fuels, deforestation, ocean degradation and other destructive industries.
- **Screen investments** for links to ecosystem destruction, ocean degradation and violations of Indigenous land rights and human rights.
- **Choose investment managers** who actively use their shareholder influence to engage companies on environmental standards, Indigenous rights and human rights – and drive measurable change.
- **Support shareholder activism:** Vote at AGMs and back resolutions demanding stronger action on climate, biodiversity and social justice.
- **Invest in regenerative enterprises,** ethical investment funds and nature-based restoration projects.
- **Advocate for robust implementation of nature and climate-related financial risks,** including frameworks like the Taskforce on Nature-related Financial Disclosures (TNFD).
- **Support the development of nature-positive investment frameworks,** not only for climate, but for biodiversity, water systems and ocean health.
- **Advocate for strong global climate finance mechanisms,** including the Loss and Damage Fund to support vulnerable nations impacted by climate change and nature destruction.

Examples:
- Join shareholder activism campaigns: ShareAction, Follow This, marketforces.org.uk and ClientEarth
- Responsible investing and a sustainable financial system: UK Sustainable Investment and Finance Association (UKSIF) and Principles for Responsible Investment (PRI)
- Learn about the Finance for Biodiversity Pledge: financeforbiodiversity.org
- Support the Taskforce on Nature-related Financial Disclosures

(TNFD), a global framework for nature-related risk disclosure: tnfd. global
- Build financial systems that value nature: naturefinance.net
- Engage with Barbados prime minister Mia Mottley's Bridgetown Initiative to undertake systemic reform of how capital flows globally: bridgetown-initiative.org

5. Legal Systems and the Law

What the law protects, prioritises and permits.
Laws shape what is permissible and whose interests are upheld. Current legal frameworks often protect corporations better than ecosystems or communities.

Practical Actions:
- **Support environmental litigation:** This includes legal centres that defend ecosystems, communities and climate stability.
- **Advocate for the Rights of Nature:** Legal recognition of ecosystems as rights-bearing entities, not mere resources. Push for the criminalisation of ecocide: The recognition of severe environmental destruction as an international crime alongside genocide and crimes against humanity.
- **Defend protest rights:** The UK has seen increasing restrictions on environmental protest and civil disobedience through recent legislation and policing powers. Support efforts to protect civil liberties and democratic rights.
- **Support environmental law reform efforts in the UK:** This includes stronger environmental impact assessment standards, independent enforcement and accountability mechanisms.

Examples:
- Friends of the Earth
- Stop Ecocide UK

- UK Rights of Nature Network
- UK Environmental Law Association
- Global Alliance for the Rights of Nature: garn.org

6. Knowledge, Culture, Communication and Misinformation

How we know, tell and believe the story of the world.
The stories we believe shape the actions we take. Challenging misinformation and rebuilding ecological literacy are essential for reclaiming a future that includes nature.

Practical Actions:
- **Actively support independent, science-informed journalism and public broadcasters:** Especially those covering environmental, Indigenous and human rights issues.
- **Challenge misinformation calmly but persistently:** Correct falsehoods when you encounter them, and model critical inquiry.
- **Familiarise yourself with disinformation tactics:** Understand how fear, division and doubt are manufactured to serve vested interests.
- **Ask critical questions about information sources:** Who produced this? Who funded it? Who benefits from its spread?
- **Strengthen your ecological literacy:** Engage directly with scientific sources on climate, biodiversity and planetary systems.
- **Support cultural storytelling and education initiatives** that reconnect people to nature and community.

Examples:
- Institute for Public Policy Research (IPPR): Fact-based public research
- Media Bias Chart: adfontesmedia.com
- DeSmog Climate Disinformation Database: desmog.com/climate-disinformation-database

- Support independent news outlets: The Conversation UK, The Guardian, openDemocracy
- Understand the science of climate change: ipcc.ch
- Understand the science of biodiversity: ipbes.net
- Play 'Bad News', an online game by the University of Cambridge aimed at strengthening misinformation resistance.
- Report greenwashing: Advertising Standards Authority (ASA) and/or the Competition and Markets Authority (CMA)

7. Technology and Innovation

The tools we build and the futures they shape.
Technological development must be guided by ecological and social principles, not just profit motives.

Practical Actions:
- **Support public-good technological innovation:** Technologies that restore ecosystems, strengthen communities and serve the public interest under democratic oversight.
- **Demand strong ethical standards and governance** for emerging environmental technologies – including robust regulation and transparent reporting.
- **Fund or donate to organisations** advancing nature-based technologies, Indigenous-led innovation and community-driven tech solutions.
- **Invest in technology development led by women and communities:** Innovation that prioritises social justice, ecological resilience and equitable futures.
- **Challenge techno-fixes** that externalise harm, bypass community consultation or deepen systemic inequalities.

Examples:
- Engage with conservation technology initiatives: UK Nature Impact Fund and WWF.

- ETC Group: Independent watchdog monitoring geoengineering, synthetic biology, and risky technological interventions in nature: etcgroup.org
- Ada Lovelace Institute: an independent research institute with a mission to ensure that data and AI work for people and society.

8. Community Mobilisation, Organising and Greening

Where real change begins.
Change is not built by lone heroes. It is built by communities moving together.

Practical Actions:
- **Join or start local voluntary environmental groups:** Building capacity for nature regeneration and community-building.
- **Participate in local renewable energy projects:** Support rooftop solar cooperatives, community-owned renewables, local energy schemes, battery storage and EV infrastructure initiatives.
- **Support urban greening projects:** Advocate for community gardens, food-growing spaces, composting schemes, beach and river clean-ups, tree planting, and rewilding efforts.
- **Advocate to your local council:** Champion ambitious tree canopy targets, biodiversity corridors, wetland protections and public access to green spaces.
- **Engage in citizen science:** Contribute to biodiversity monitoring, wildlife rescue, and community-led conservation efforts.

Examples:
- Find and support community groups, including Transition Network International, Friends of the Earth, UK Youth Climate Coalition, The Wildlife Trusts (local branches across the UK) and Nature Neighbourhoods (People's Plan for Nature).
- Become a citizen scientist: Big Butterfly Count, RSPB Big Garden Birdwatch and inaturalist.org

- Become a wildlife carer or rescuer: British Wildlife Rehabilitation Council (BWRC), RSPCA and Hessilhead Wildlife Rescue

9. Personal Behaviour, Individual and Lifestyle Decisions

The everyday choices we make – and the systems we reinforce or resist.
Lifestyle choices matter most when they are linked to collective action and cultural shifts.

Practical Actions:
- **Shift habits consciously:** Walk or bike where possible, minimise meat consumption, buy less and buy local, prioritise durability and repairability, and reduce waste – but always link personal shifts to broader systemic change.
- **Align your information ecosystem:** Curate your newsfeeds and social media to amplify science-based voices, Indigenous knowledge holders, conservationists, independent journalists, human rights experts and grassroots organisers.
- **Speak up against extractive cultural norms:** Challenge narratives that glorify over-consumption, luxury emissions, environmental exploitation or extreme individualism.
- **Normalise conversations about systems change:** Bring ecological and justice issues into workplaces, homes and social spaces.

Examples:
- Update your social media settings. Review and adjust social media and news settings to avoid algorithm-driven misinformation and engagement traps. Platform settings can strongly shape what information is surfaced – or suppressed.
- Follow the work of NGOs, scientists, environmentalists, human rights experts, and independent journalists and researchers on social media.

To follow Natalie Kyriacou's work, connect on:
LinkedIn: https://www.linkedin.com/in/nataliekyriacou/
Instagram: @nat_kyriacou
Website: nataliekyriacou.com

If you loved the book (or even mildly tolerated it), consider leaving a short review on Goodreads, or wherever you buy your books. It helps other curious humans stumble upon it.

Acknowledgements

This is the section where I thank people, mostly to avoid uncomfortable conversations later.

Presumably, the only people reading this are my friends, who will be making sure that I have adequately acknowledged them. You see, I am both blessed and burdened with friends and loved ones who are fierce, intelligent, generous and kind – and petty enough to check whether they've been named and in what specific order. With that in mind, I would like to thank Andy, Antoinette, Arik, Catherine, Charlotte, Chrissy, Clare, Dilan, Elyse, Jo, Judy, Keeshia, Laura, Lucas, Miriam, Nicole, Nisha, Peter, Talita, Thomas and Tracey (in alphabetical order). Genuinely, I am so grateful for you. I don't deserve you, but I sure am glad to have you. Thank you for your love, your neuroses, your unsolicited opinions and your pathological belief in me. This book honestly *would* have been possible without you. It just would have been worse.

Thank you to the people who have proofread my work, supported me, tolerated me, mentored me and reminded me where to find joy in times of tragedy. Thank you to Kelly and the team at Affirm Press (and Simon & Schuster), to Tara and the Curtis Brown team, to my readers (hopefully there are some), to the brave souls who entrusted me with their stories and to my big, odd, loving family – you are the best people I know.

To everyone else who offered support, thank you. You may not be named, but you are known and loved. Possibly not by me, but by someone, I'm sure.

Finally, to the true heroes of this book – environmental

ACKNOWLEDGEMENTS

scientists, rangers, conservationists, activists, wildlife carers and vets and rescuers, community leaders, volunteers, wisdom-keepers, journalists and NGO workers: you are the heartbeat of this book. I will never stop amplifying your stories. I will never stop talking about you. And I will never stop fighting with you and for you.

Sources

Introduction

1. McConaghy, Charlotte, *Migrations*. Penguin Books Australia, Melbourne 2021
2. Fuller, Errol, *The Great Auk: The Extinction of the Original Penguin*, Errol Fuller Publications, Southborough, Kent, 1999
3. Rose, Laurence, *The Long Spring: Tracking the Arrival of Spring Through Europe*, Bloomsbury Publishing, London, 2018
4. Bradshaw, Corey J. A., and Giovanni Strona, 'Children born today will see literally thousands of animals disappear in their lifetime, as global food webs collapse' *The Conversation*, 17 Dec 2022, https://theconversation.com/children-born-today-will-see-literally-thousands-of-animals-disappear-in-their-lifetime-as-global-food-webs-collapse-196286
5. Carson, Rachel, *Lost Woods: The Discovered Writing of Rachel Carson*, Beacon Press, Boston, 1999

Chapter 1

1. Swift, Graham, *Shuttlecock*, United Kingdom, Scribner UK, London, 2019
2. Dussex, Nicolas et al., 2021, 'Population genomics of the critically endangered Kākāpō', *Cell Genomics*, Vol. 1, doi.org/10.1016/j.xgen.2021.100002
3. Department of Conservation, 'Kākāpō Research', https://www.doc.govt.nz/our-work/kakapo-recovery/what-we-do/research-for-the-future/
4. Clout, Mick N., Graeme P. Elliot and Bruce C Robertson, 2002, 'Effects of supplementary feeding on the offspring sex ratio of kakapo:

a dilemma for the conservation of a polygynous parrot', *Biological Conservation*, Vol. 107, doi:10.1016/S0006-3207(01)00267-1

5. Department of Conservation, 'Sounds of Science archive.' https://doc.govt.nz/news/podcast/sounds-of-science-archive/
6. Ibid.
7. Ibid.
8. *Stuff*, 'The kākāpō "ejaculation helmet" and efforts to save the bird population' 6 Mar. 2018, https://www.stuff.co.nz/environment/102033870/the-kkp-ejaculation-helmet-and-efforts-to-save-the-bird-population
9. Department of Conservation, 'Sirocco the kākāpō conservation superstar' https://www.doc.govt.nz/nature/native-animals/birds/birds-a-z/kakapo/sirocco/
10. Hiraiwa-Hasegawa, Mariko, 2000, 'The sight of the peacock's tail makes me sick: The early arguments on sexual selection', *Journal of Biosciences*, Vol. 25, doi:10.1007/BF02985176
11. Schilthuizen, Menno, 2014, *Nature's Nether Regions: What the Sex Lives of Bugs, Birds and Beasts Tell Us About Evolution, Biodiversity and Ourselves*, Penguin Books, New York, 2015
12. Lu, Donna 'Taronga zoo lyrebird perfectly mimics the ear-splitting wail of a crying baby' *The Guardian*, 2 September 2021, https://www.theguardian.com/environment/2021/sep/02/targona-zoo-lyrebird-perfectly-mimics-the-ear-splitting-wail-of-a-crying-baby
13. *National Geographic*, 'Argonaut (Paper Nautilus)' https://www.nationalgeographic.com/animals/invertebrates/facts/argonaut
14. Falk, Dan 'How Darwin's "Descent of Man" Holds Up 150 Years After Publication' *Smithsonian Magazine*, 24 February 2021, https://www.smithsonianmag.com/science-nature/how-darwins-descent-man-holds-150-years-after-publication-180977091/
15. Darwin, Charles, et al., 'The Descent of Man, and Selection in Relation to Sex, Princeton University Press, Princeton, 1981
16. Rosenthal, Gil G., and Michael J. Ryan, 2022, 'Sexual selection and the ascent of women: Mate choice research since Darwin.' *Science*, vol. 375, https://doi.org/10.1126/science.abi6308

17. Darwin, Charles, et al., 'The Descent of Man, and Selection in Relation to Sex, Princeton University Press, Princeton, 1981
18. Darwin Correspondence Project, 'Darwin in letters, 1882: Nothing too great or too small', https://www.darwinproject.ac.uk/letters/darwins-life-letters/darwin-letters-1882-nothing-too-great-or-too-small
19. Fuentes, Agustín, 2021, "The Descent of Man" 150 Years On, *Science*, Vol. 371, https://doi.org/10.1126/science.abj4606
20. Rosenthal, Gil G., and Michael J. Ryan, 2022, 'Sexual selection and the ascent of women: mate choice research since Darwin', *Science*, Vol. 375, https://doi.org/10.1126/science.abi63082024
21. Horgan, John, 'Darwin Was Sexist, and So Are Many Modern Scientists', *Scientific American*, 18 December 2017, https://www.scientificamerican.com/blog/cross-check/darwin-was-sexist-and-so-are-many-modern-scientists/
22. Mayr, Ernst, 'Darwin's influence on modern thought', *Scientific American*, 283.1 (2000): 78-83.
23. Ah-King, Malin, et al., 'Genital Evolution: Why Are Females Still Understudied?', *PLOS Biology*, Vol. 12, 6 May 2014, https://doi.org/10.1371/journal.pbio.1001851
24. Small, Meredith F., 'Casual Sex Play Common Among Bonobos', *Discover Magazine*, 21 May 2025, http://discovermagazine.com/mind/casual-sex-play-common-among-bonobos
25. Angier, Natalie, *Woman: An Intimate Geography*, Houghton Mifflin Harcourt, Boston, 1999
26. Bateman, A. J., 1948, 'Intra-Sexual Selection in Drosophila.' *Heredity*, Vol. 2, https://doi.org/10.1038/hdy.1948.21
27. Loehnen, Elise, 'Angela Saini: How Science Got Women Wrong', *Elise Loehnen*, 10 February 2021, https://www.eliseloehnen.com/episodes/angela-saini-how-science-got-women-wrong
28. Fine, Cordelia and Mark A. Elgar, 'Promiscuous Men, Chaste Women, and Other Gender Myths', *Scientific American*, 1 September 2017, https://www.scientificamerican.com/article/promiscuous-men-chaste-women-and-other-gender-myths

29. Scelza, Brooke A., 2013, 'Choosy but Not Chaste: Multiple Mating in Human Females', *Evolutionary Anthropology*, Vol. 22, https://doi.org/10.1002/evan.21373
30. Smith, Jennifer E. et al., 2020, 'Obstacles and opportunities for female leadership in mammalian societies: A comparative perspective', *The Leadership Quarterly*, Vol. 31, doi: :10.1016/j. leaqua.2018.09.005
31. Frank, Laurence G., 1997, 'Evolution of genital masculinization: why do female hyenas have such a large 'penis'?', *Trends in Ecology & Evolution*, Vol. 22, https://doi.org/10.1016/s0169-5347(96)10063-x
32. Smith, Jennifer E. et al., 2020, 'Obstacles and opportunities for female leadership in mammalian societies: A comparative perspective', *The Leadership Quarterly*, Vol. 31, doi: 10.1016/j. leaqua.2018.09.005
33. Ibid.

Chapter 2

1. Leaky, Richard E. and Roger Lewin, *The sixth extinction: Patterns of life and the future of humankind*, Knopf Doubleday Publishing Group, New York, 1995
2. Møller, Anders Pape, 2019, 'Parallel declines in abundance of insects and insectivorous birds in Denmark over 22 years', *Ecology and Evolution*, Vol. 9, https://doi.org/10.1002/ece3.5236
3. World Wide Fund for Nature (WWF). *Living Planet Report 2022: Building a Nature-Positive Society*. WWF International, 2022
4. Ceballos, Gerardo, et al., 2017, 'Biological annihilation via the ongoing sixth mass extinction signaled by vertebrate population losses and declines', *Proceedings of the National Academy of Sciences*, Vol. 114, https://doi.org/10.1073/pnas.1704949114
5. Barnsky, A. et al., 2011, 'Has the Earth's sixth mass extinction already arrived?' *Nature*, https://doi.org/10.1038/nature09678
6. Cowie, Robert H., Philippe Bouchet and Benoît Fontaine, 2022, 'The Sixth Mass Extinction: fact, fiction or speculation?', *Biological Reviews*, Vol. 97, https://doi.org/10.1111/brv.12816
7. Leaky, Richard E. and Roger Lewin, *The sixth extinction: Patterns*

of life and the future of humankind, Knopf Doubleday Publishing Group, New York, 1995

8. Grant, Richard, 'How Scientists Tracked the Movements of a 17,000-Year-Old Woolly Mammoth', *Smithsonian Magazine*, November 2023, https://www.smithsonianmag.com/science-nature/scientists-tracked-movements-17000-year-old-woolly-mammoth-180983064/; Brassey, Charlotte A. and James D. Gardiner, 2015, 'An advanced shape-fitting algorithm applied to quadrupedal mammals: improving volumetric mass estimates', *Royal Society Open Science*, Vol. 2, https://doi.org/10.1098/rsos.150302 American Museum of Natural History, 'How the Smilodon Got Its Teeth', *1 July 2015,3* https://www.amnh.org/explore/news-blogs/research-posts/how-the-smilodon-got-its-teeth

9. Siebert, Charles, *NRDC: The Secret World of Whales*, Chronicle Books, San Francisco, 2011

10. Lindsey, Rebecca, 'Climate change: atmospheric carbon dioxide', *Climate.gov*, 9 April 2024, https://www.climate.gov/news-features/understanding-climate/climate-change-atmospheric-carbon-dioxide

11. Kovarik, Bill, 'Changing views of extinction in history', *Environmental history*, https://environmentalhistory.org/changing-views-of-extinction-in-history/

12. Met Office, 'The Great Smog of London 1952', https://weather.metoffice.gov.uk/learn-about/weather/case-studies/great-smog#:~:text=A%20fog%20so%20thick%20and

13. Ripple, William J. et al., 2017, 'World Scientist's Warning to Humanity: A Second Notice', *BioScience*, Vol. 67, https://www.researchgate.net/publication/322251387_World_Scientists'_Warning_to_Humanity_A_Second_Notice

14. International Union of Geological Sciences, 'The Anthropocene', 20 March 2024, https://www.iugs.org/_files/ugd/f1fc07_40d1a7ed58de458c9f8f24de5e739663.pdf?index=true

15. Zhong, Reymond, 'Geologists Make It Official: We're Not in an 'Anthropocene' Epoch', *The New York Times*, https://www.nytimes.com/2024/03/20/climate/anthropocene-vote-upheld.html

16. Davidson, Nicola, 'The Antropocene epoch: have we entered a new phase of planetary history?', *The Guardian*, 30 May 2019, https://www.theguardian.com/environment/2019/may/30/anthropocene-epoch-have-we-entered-a-new-phase-of-planetary-history
17. Davidson, Nicola, 'The Antropocene epoch: have we entered a new phase of planetary history?', *The Guardian*, 30 May 2019, https://www.theguardian.com/environment/2019/may/30/anthropocene-epoch-have-we-entered-a-new-phase-of-planetary-history
18. International Union of Geological Sciences, 'The Anthropocene', 20 March 2024, https://www.iugs.org/_files/ugd/f1fc07_40d1a7ed58de458c9f8f24de5e739663.pdf?index=true
19. Carrington, Damian, 'Quest to declare Antropocene an epoch descends into epic row', *The Guardian*, 8 March 2024, https://www.theguardian.com/science/2024/mar/07/quest-to-declare-anthropocene-an-epoch-descends-into-epic-row
20. Pyron, Alexander R., ' We Don't Need to Save Endangered Species. Extinction Is Part of Evolution', *The Washington Post*, 22 November 2017, https://www.washingtonpost.com/outlook/we-dont-need-to-save-endangered-species-extinction-is-part-of-evolution/2017/11/21/57fc5658-cdb4-11e7-a1a3-0d1e45a6de3d_story.html
21. Safina, Carl, 'In Defense of Biodiversity: Why Protecting Species from Extinction Matters, *Yale Environment 360*, 12 February 2018, https://e360.yale.edu/features/in-defense-of-biodiversity-why-protecting-species-from-extinction-matters
22. De Vos, Jurriaan M. et al., 2015, 'Estimating the normal background rate of species extinction', *Conservation Biology*, Vol. 29, https://doi.org/10.1111/cobi.12380
23. Hitler, Adolf, *Mein Kampf*, Houghton Mifflin, Boston, 1943
24. Zubrin, Robert, *The New World on Mars: What We Can Create on the Red Planet*, Diversion Books, New York, 2024
25. Martin, Newell, *Studies from the Biological Laboratory*, Oxford University, 1883 European Libraries Collection, https://archive.org/details/studiesfrombiol00martgoog/page/386/mode/2up

26. Hall, G. Stanley and Yuzero Motora, 1887, 'Dermal sensitiveness to gradual pressure changes', *The American Journal of Psychology*, Vol. 1, https://www.jstor.org/stable/1411232

Chapter 3

1. *The Sydney Morning Herald*, 'New Strategy in a War on the Emu', 5 July 1953
2. Elton-Pym, James, 'Harald Holt: The Australian prime minister who disappeared', *SBS News*, 17 December 2017, https://www.sbs.com.au/news/article/harold-holt-the-australian-prime-minister-who-disappeared/scs448smz.
3. Ameer, Zahid, *The Great Emu War: How Flightless Birds Bested the Australian Army*, 2024
4. McManus, Sam, 'Australia's Emu War spawns feature film, jokes and memes 90 years on', *ABC News*, 10 December 2022, https://www.abc.net.au/news/2022-12-10/great-emu-war-90-years-on-army-wheatbelt-battle-history/101752238
5. *Sydney Morning Herald*, 'Emu war: won every round so far', 9 November 1932
6. Majerus, Michael EN, and Naomi R. Stevens, 2006, 'The peppered moth: a problem not to be sneezed at', *Biologist* 53.1
7. *Butterfly Conservation*, 'Peppered Moth and natural selection', https://butterfly-conservation.org/moths/why-moths-matter/amazing-moths/peppered-moth-and-natural-selection
8. Miller, Michael E., 'Behold, the bin chicken: Sydney's stinky, grimy but (mostly) beloved bird', *The Washington Post*, 5 February 2024
9. Facebook, International Glare at Ibises Day Event, 21 December 2016, https://www.facebook.com/events/hyde-park-local/international-glare-at-ibises-day/1580074002019864/
10. Fischer, Sascha, 2018, 'From Bin Chicken to Totem: The Unlikely Rise of the Ibis as a Symbol of Modern Australia', *ABC News*, 7 September, https://www.abc.net.au/news/2018-09-07/ibis-bin-chicken-rise-totem-for-modern-australia/10209332

11. Smith, Gregory C., 2008, 'White Ibis: The Urban Survivor', *Corella*, Vol. 32, https://absa.asn.au/wp-content/uploads/2021/04/Cor-Vol-32-Pg58–65-White-Ibis.pdf.
12. *The Week*, 'Humans 'worse for natur than world's worst nuclear accident',' 6 October 2015, https://theweek.com/65617/humans-worse-for-nature-than-worlds-worst-nuclear-accident
13. *World Nuclear Association*, 'Chernobyl Accident 1986', 17 February 2025, https://world-nuclear.org/information-library/safety-and-security/safety-of-plants/chernobyl-accident
14. Ibid.
15. Mousseau, Timothy A. and Anders P. Møller, 2016, 'The Animals of Chernobyl and Fukushima', *Genetics, Evolution and Radiation*, doi:10.1007/978-3-319-48838-7_21
16. Ibid.
17. Galván, Ismael, et al., 2014, 'Chronic exposure to low-dose radiation at Chernobyl favours adaptation to oxidative stress in birds', *Functional Ecology*, https://doi.org/10.1111/1365-2435.12283
18. Orizaola, Germán, 'Chernobyl has become a refuge for wildlife 33 years after the nuclear accident', *The Conversation*, 8 May 2019, https://theconversation.com/chernobyl-has-become-a-refuge-for-wildlife-33-years-after-the-nuclear-accident-116303
19. Coghlan, Andy, 'Wildlife is thriving around Chernobyl since the people left' *New Scientist*, 5 October 2015, https://www.newscientist.com/article/dn28281-wildlife-is-thriving-around-chernobyl-since-the-people-left/

Chapter 4

1. Carson, Rachel, *Silent Spring*, Houghton Mifflin, Boston, 1962
2. Penly, Taylor, 'Hippos from hell: Pablo Escobar's private zoo leaves nation racing against time to save ecosystem 40 years on', *Fox News*, 4 February 2024, https://www.foxnews.com/media/hippos-hell-pablo-escobar-private-zoo-nation-racing-time-save-ecosystem-40-years
3. Dutton, Christopher L., 2018, 'Organic matter loading by hippopotami causes subsidy overload resulting in downstream hypoxia and

fish kills, *Nature Communications*, Vol. 9, https://doi.org/10.1038/s41467-018-04391-6

4. Shurin, Jonathan B. et al., 2020, 'Ecosystem effects of the world's largest invasive animal', *Ecology*, Vol. 101, doi:10.1002/ecy.2991
5. Subalusky, Amanda L. et al., 2023, 'Rapid population growth and high management costs have created a narrow window for control of introduced hippos in Colombia', *Scientific Reports*, Vol. 13, doi:10.1038/s41598-023-33028-y
6. In recent years, conservationists have debated whether it's always feasible or beneficial to fight non-native species, especially in ecosystems already heavily altered by human activity.
7. IPBES, 'The thematic assessment report on invasive alien species and their control', 2023, https://www.ipbes.net/ias
8. Latombe, Guillaume, 'Invasive species risk a biodiversity disaster – but there is still time to stop it', *The Conversation*, 6 September 2023, https://theconversation.com/invasive-species-risk-a-biodiversity-disaster-but-there-is-still-time-to-stop-it-212818
9. Ikeda, Tohru et al., 2004, 'Present status of invasive alien raccoon and its impact in Japan, *Global Environmental Research*, Vol. 8, https://www.researchgate.net/publication/228486361_Present_status_of_invasive_alien_raccoon_and_its_impact_in_Japan
10. Suzuki, Takaaki, and Tohru Ikeda. 'Challenges in Managing Invasive Raccoons in Japan.' *Pacific Conservation Biology*, vol. 25, no. 3, 2019, pp. 318–325. https://doi.org/10.1071/PC18024.
11. Bellard, Céline, et al. 'Invasive Species Are a Leading Cause of Animal Extinctions.' *Frontiers in Ecology and the Environment*, vol. 18, no. 5, 2020, pp. 206–212. https://doi.org/10.1002/fee.2195.
12. *Kyodo News*, 'Raccoon numbers surge in Tokyo, causing damage to crops', 29 April 2024, https://english.kyodonews.net/news/2024/04/9a3ccf062656-raccoon-numbers-surge-in-tokyo-causing-damage-to-crops.html
13. Watari, Yuya, et al., 2021, 'First synthesis of the economic costs of biological invasions in Japan', *NeoBiota*, Vol. 67, https://doi.org/10.3897/neobiota.67.59186

14. Cuthbert, Ross N., 2021, 'Global economic costs of aquatic invasive alien species', *Science of the Total Environment*, Vol 775, https://doi.org/10.1016/j.scitotenv.2021.145238
15. Sarill, Michael, 'Burmese Pythons in the Everglades', B*erkeley International & Executive Programs Rausser College of Natural Resources*, 2016, https://iep.berkeley.edu/content/burmese-pythons-everglades
16. DeSantis, Ron, 'The Focus on Fiscal Responsibility Budget', 2025, https://flgov.com/eog/sites/default/files/shared/2025/02/FY%202025-26%20Governor%20Budget%20Highlights.pdf
17. Fantle-Lepczyk, Jean E., et al., 2022, 'Economic costs of biological invasions in the United States, *Science of the Total Environment*, Vol. 806, https://doi.org/10.1016/j.scitotenv.2021.151318
18. *The Carroll Herald*, 'Otter Fur is of a Most Exquisite Fineness and Richness in Both Colour and Texture', 15 August 1883
19. Badelt, Brad, 'The otter, the urchin and the Haida', *Canadian Geographic*, 1 March 2024, https://canadiangeographic.ca/articles/the-otter-the-urchin-and-the-haida/
20. *Hartford Weekly Times*, 'Exterminating Sea Otters', 21 July 1892
21. *Arizona Weekly Journal-Miner*, 19 April 1899
22. *The Carroll Herald*, 'North Pacific Otters', 15 August 1883
23. Preston, Christopher J., 'The far-reaching influence of Alaska's sea otters', B*BC Future*, 1 March 2023, https://www.bbc.com/future/article/20230228-how-alaskas-sea-otters-came-back
24. Larson, Shawn E., Glenn R VanBlaricom and James L. Bodkin (eds.), *Sea Otter Conservation*, Academic Press, Cambridge (MA), 2015
25. Nicholson, Teri E., et al., 2024, 'Sea otter recovery buffers century-scale declines in California kelp forests, *PLOS Climate*, doi:10.1371/journal.pclm.0000290
26. Shil, Susmita, et al., 'KELP FOREST: ECOLOGICAL SIGNIFICANCE AND RESTORATION PROSPECTS.' PLANTA 7 (2023): 1419-1427. https://www.researchgate.net/publication/377150541_KELP_ FOREST_ECOLOGICAL_ SIGNIFICANCE_AND_ RESTORATION_PROSPECTS

Chapter 5

1. Stanford, Craig B., The Last Tortoise: *A Tale of Extinction in Our Lifetime*, Belknap Press, Cambridge (MA), 2010
2. International Rhino Foundation, 'One rhino poached is too many – every rhino counts', https://rhinos.org/blog/one-rhino-poached-is-too-many-every-rhino-counts/
3. Gilchrist, Jason, 'South Africa's 70,000 kg rhino horn stockpile must be burnt to prevent illegal trading' *The Conversation*, 3 July 2024, https://theconversation.com/south-africas-70-000kg-rhino-horn-stockpile-must-be-burnt-to-prevent-illegal-trading-232030, Ripple, William J., et al., 2015, 'Collapse of the world's largest herbivores', *Science advances*, Vol. 1, doi: 10.1126/sciadv.1400103
4. Ol Pejeta Conservancy, 'Goodbye Sudan, the world's last male Northern White Rhino', https://www.olpejetaconservancy.org/2024/06/11/goodbye-sudan-the-worlds-last-male-northern-white-rhino/
5. Aubudon, 'Billions to none... the extinction of the Passenger Pigeon', https://johnjames.audubon.org/conservation/billions-none-extinction-passenger-pigeon
6. Shell, Hanna Rose, 'The Face of Extinction', *Natural History*, Vol. 113, no. 4, May 2004
7. Smithsonian Institution. (n.d.). *The Passenger Pigeon*. Biodiversity Heritage Library. Retrieved from https://www.biodiversitylibrary.org/bibliography/39140
8. National Museum of Australia, 'Extinction of thylacine', https://www.nma.gov.au/defining-moments/resources/extinction-of-thylacine
9. Middleton, Nathalie, 'The Man Who Made Lonesome George Less Lonely', *Orion Magazine*, 29 November 2023, https://orionmagazine.org/article/lonesome-george-tortoise-endlings-extinction-fausto-llerena/
10. IPBES, 'Global Assessment Report on Biodiversity and Ecosystem Services', 2019, https://ipbes.net/global-assessment
11. Ritchie, Hannah, 'How many species are there?', Our World in Data, 30 November 2022, https://ourworldindata.org/how-many-species-are-there2025.
12. Duthé, Vanessa, 2023, 'Reductions in home-range size and social

interactions among dehorned black rhinoceroses (Diceros bicornis)', *PNAS*, Vol. 120, https://doi.org/10.1073/pnas.2301727120

13. Ibid.
14. Duthé, Vanssa, 'How dehorning affects rhino behavior', *PNAS*, 31 July 2023, https://www.pnas.org/post/podcast/dehorning-affects-rhino-behavior
15. Hausheer, Justine E., ' Can Helicopter-deployed Toad Sausages Save Australia's Northern Quoll?', *Cool Green Science,* 18 September 2022, https://blog.nature.org/2015/11/09/can-helicopter-deployed-toad-sausages-save-australias-northern-quoll/
16. Gilchrist, Jason, 'Rhinos: scientists are hanging them upside-down from helicopters – here's why', *The Conversation*, 14 September 2021, https://theconversation.com/rhinos-scientists-are-hanging-them-upside-down-from-helicopters-heres-why-167832
17. *ABC News*, 'Scientists seek deodorant for smelly kiwis', 24 September 2010, https://www.abc.net.au/news/2010-09-24/scientists-seek-deodorant-for-smelly-kiwis/2273452
18. Smith, Jodie, 'Colchester Zoo use mirrors to help flamingos to breed', *BBC*, 26 July 2010, http://news.bbc.co.uk/local/essex/hi/people_and_places/nature/newsid_8854000/8854531.stm
19. Nicholls, Henry, *Lonesome George: The Life and Loves of a Conservation Icon*, Macmillan Science, New York, 2010
20. Scott, Robert F., 'Sumatran Rhino Conservation – SSC Endorsement of Singapore Proposals', *Rhino Resource Center*, 19 December 1984, http://www.rhinoresourcecenter.com/index.php?s=1&act=pdfviewer&id=1651846992&folder=165
21. Scott, Robert F., 'Sumatran Rhino Conservation – SSC Endorsement of Singapore Proposals', *Rhino Resource Center*, 19 December 1984, http://www.rhinoresourcecenter.com/index.php?s=1&act=pdfviewer&id=1651846992&folder=165
22. Rookmaaker, Kees, 'The history of the captive breeding programs of the Sumatran Rhinoceros 1984-2019', *The International Rhino Foundation*, October 2019, http://www.rhinoresourcecenter.com/pdf_files/163/1634651531.pdf

23. Kolbert, Elizabeth, *The Sixth Extinction: An Unnatural History*, Henry Holt and Co, New York, 2014
24. Rabinowitz, Alan, 1995, 'Helping a species go extinct: the Sumatran Rhino in Borneo', *Conservation Biology*, Vol. 9, https://doi.org/10.1046/j.1523-1739.1995.09030482.x
25. Hancy, Jeremy, 'A herd of dead rhinos', *Mongabay*, 24 September 2018, https://news.mongabay.com/2018/09/a-herd-of-dead-rhinos/
26. Li, Tan Cheng, 'Saving rhinos: Our fatal blunder', *Borneo Rhino Alliance*, 2 June 2014, https://www.borneorhinoalliance.org/resources/articles/saving-rhinos-our-fatal-blunders/
27. Cincinnati Zoo, 'A history of the zoo's Sumatran Rhino breeding program', 1 December 2015, https://cincinnatizoo.org/a-history-of-the-zoos-sumatran-rhino-breeding-program/
28. Kolbert, Elizabeth, *The Sixth Extinction: An Unnatural History*, Henry Holt and Co, New York, 2014

Chapter 6

1. Lamott, Anne, *Bird by Bird: Some Instructions on Writing and Life*, Anchor Books, New York, 1995
2. DeNapoli, Dyan, *The Great Penguin Rescue: 40,000 Penguins, a Devastation Oil Spill, and the Inspiring Story of the World's Largest Animal Rescue*, Free Press, New York, 2010
3. Ibid.
4. Ibid.
5. Ibid.
6. Curtis, Russ, '2000 – Treasure Spill – South Africa', *International Bird Rescue*, 30 June 2000, https://www.birdrescue.org/2000-treasure-spill-south-africa-2/
7. DeNapoli, Dyan, *The Great Penguin Rescue: 40,000 Penguins, a Devastation Oil Spill, and the Inspiring Story of the World's Largest Animal Rescue*, Free Press, New York, 2010
8. Englefield, Bruce, et al., 2019, 'The Demography and Practice of Australians Caring for Native Wildlife and the Psychological, Physical and Financial Effects of Rescue, Rehabilitation and Release

of Wildlife on the Welfare of Carers', *Animals*, Vol. 9, doi: 10.3390/ani9121127

9. Maathai, Wangari, *Unbowed: A Memoir*, Anchor Books, New York, 2006
10. Hayanga, Jeremiah A., 2006, 'Wangaru Mathai: an African woman's environmental and geopolitical landscape', *International Journal of Environmental Studies*, Vol. 63, doi:10.1080/00207230600948024
11. Maathai, Wangari, *Unbowed: A Memoir*, Anchor Books, New York, 2006
12. Ibid.
13. Vince, Gaia, 'Africa's 'tree lady',' *Nature Climate Change*, 23 October, 2011, https://www.nature.com/articles/nclimate1275
14. Maathai, Wangari, *Unbowed: A Memoir*, Anchor Books, New York, 2006
15. Ibid.
16. Kabiru, Joseph, 'Farewell Wangari Maathai, you were a global inspiration – and my heroine', *The Guardian*, 26 September 2011, https://www.theguardian.com/global-development/poverty-matters/2011/sep/26/farewell-wangari-maathai-my-heroine
17. Ibid.
18. Ighobor, Kingsley, 'Wangari Maathai, the woman of trees, dies', *United Nations*, 17 January 2012, https://africarenewal.un.org/en/magazine/wangari-maathai-woman-trees-dies
19. Ibid.
20. Ibid.
21. *United Nations*, ' Secretary-General, Seddened by Wangari Maathai's Death, Hails Nobel Laureate's Linking of Human Rights, Poverty, Environmental Protection, Security', 26 September 2011, https://press.un.org/en/2011/sgsm13846.doc.htm
22. Ighobor, Kingsley, 'Wangari Maathai, the woman of trees, dies', *United Nations*, 17 January 2012, https://africarenewal.un.org/en/magazine/wangari-maathai-woman-trees-dies
23. Ibid.
24. *UCL Institute for Global Prosperity*, 'Citizen science and botanic

knowledge among herders and farmers in Kenya', https://www.ucl.ac.uk/bartlett/news/2019/jul/citizen-science-and-botanic-knowledge-among-herders-and-farmers-kenya

25. Hoyte, Simon and Jerome Lewis, 'Forest peoples and Extreme Citizen Science in Cameroon, *Development Studies Association*, https://nomadit.co.uk/conference/dsa2018/paper/43005

26. Fagan-Jeffries, Erinn P., et al., 2024, 'Hymenopteran parasitoids of fall armyworm (Spodoptera frugiperda (J.E. Smith) (Lepidoptera: Noctuidae)) in Australia, with the description of five new species in the families Braconidae and Eulophidae', *Austral Etnomology*, Vol. 63, https://doi.org/10.1111/aen.12682

27. Pensoft Publishers, 'Tiny, beautiful, and completely unknown animals': Citizen scientists discover new beetles from the Borneo forest', *Phys.org*, 20 March 2024, http://phys.org/news/2024-03-tiny-beautiful-unknown-animals-citizen.html

28. Pensoft Publishers, 'Citizen Scientists Help Discover a New, Giant Slut from Europe', *Lab Manager*, 22 July 2022, https://www.labmanager.com/citizen-scientists-help-discover-a-new-giant-slug-from-europe-28319

29. Bittel, Jason, 'First-ever photos show humpback whales mating–and they're males', National Geographic, 29 February 2024, https://www.nationalgeographic.com/animals/article/humpback-whale-mating-recorded-first-time-males

30. Wegener, Alfred, 'Citizen scientists help discover microplastics along the entire German coastline', *Phys.org*, 25 September 2024, https://phys.org/news/2024-09-citizen-scientists-microplastics-entire-german.html

31. *Atlas of Living Australia*, 'An iNaturalist Australia milestone', 15 October 2024, https://www.ala.org.au/blogs-news/an-inaturalist-australia-milestone/

Chapter 7

1. Lambertini, Marco, 'Technology can help us save the planet. But more than anything, we must learn to value nature', *World Economic*

Forum, 23 August 2023, https://www.weforum.org/ stories/2018/08/here-s-how-technology-can-help-us-save-the-planet/

2. *Al Jazeera*, 'China angry over Obama-Dalai Lama meeting', 17 July 2011, https://www.aljazeera.com/news/2011/7/17/china-angry-over-obama-dalai-lama-meeting

3. *Native Knowledge 360°*, 'Pacific Coast Region' https://americanindian.si.edu/nk360/pnw-history-culture-regions/pacific-coast

4. *Columbia River Inter-Tribal Fish Commission*, 'We are all Salmon People', https://critfc.org/salmon-culture/we-are-all-salmon-people/

5. Hall, Loura, 'New Commercial Robot Copies Gecko's Toes', *NASA*, https://www.nasa.gov/technology/tech-transfer-spinoffs/new-commercial-robot-copies-geckos-toes/

6. Frauenhofer-Gesellschaft, 'Robotic Arm Inspired by Elephants', *Science Daily*, 9 July 2007, https://www.sciencedaily.com/releases/2007/07/070706140906.htm

7. Hamilton, Tyler, 'Whale-Inspired Wind Turbines: Mimicking the bumps on humpback-whale fins could lead to more efficient wind turbines', *MIT Technology Review*, 6 March 2008, https://www.technologyreview.com/2008/03/06/221447/whale-inspired-wind-turbines/

8. Linic, Suzana, 2020, 'Experimental and Numerical Methods for Concept Design and Flow Transition Prediction on the Example of the Bionic High-Speed Train', *Experimental and Computational Investigations in Engineering, Proceedings of the International Conference of Experimental and Numerical Investigations and New Technologies*, CNNTech 2020, doi:10.1007/978-3-030-58362-0_5

9. Lai, Yubo et al., 2024, 'Medical ultrasound application beyond imaging: insights from ultrasound sensing and biological response', *Authorea*, doi:10.22541/au.172629798.83925659/v1

10. Malsbury, Erin, 'How We Lifted Flight from Bird Evolution', *Nation Museum of Natural History*, 17 December 2020, https:// www.smithsonianmag.com/blogs/national-museum-of-natural-history/2020/12/17/how-we-lifted-flight-bird-evolution/

11. *Convention on Biological Diversity*, 'Pharmaceuticals and Biodiversity: Securing the Health of the Planet', 29 June 2021, https://www.cbd.int/article/pharmaceuticals-biodiversity-planet
12. Whiting, Kate, 'This is how biodiversity loss impacts medicine and human health', *World Economic Forum*, 23 November 2023, https:// www.weforum.org/stories/2023/11/biodiversity-nature-loss-health-medicine/
13. Newman, David and Gordon M. Cragg, 2020, 'Natural Products as Sources of New Drugs over the Nearly Four Decades from 01/1981 to 09/2019', *Journal of Natural Products*, Vol. 83, https://doi.org/10.1021/acs.jnatprod.9b01285
14. Piper, Ross et al., 'Nature is a rich source of medicine – if we can protect it', *The Conversation*, 14 December 2018, https://theconversation.com/nature-is-a-rich-source-of-medicine-if-we-can-protect-it-107471
15. *Convention on Biological Diversity*, 'Pharmaceuticals and Biodiversity: Securing the Health of the Planet', 29 June 2021, https://www.cbd.int/article/pharmaceuticals-biodiversity-planet
16. Greenstone, Michael, et al. 'The Economic and Health Costs of the Vulture Crisis in India.' *American Economic Review*, vol. 113, no. 6, 2023, pp. 1904–1951, www.aeaweb.org/articles?id=10.1257/aer.20230016&from=f. Accessed 6 Dec. 2024.
17. Ibid.
18. Menéndez Frenández, Pelayo et al., 2020, 'The Global Flood Protection Benefits of Mangroves', *Scientific Reports*, Vol. 10, doi:10.1038/s41598-020-61136-6
19. *Tourism and Events Queensland*, 'Great Barrier Reef', https://teq.queensland.com/au/en/industry/industry-resources/great-barrier-reef-resources
20. Agence France-Press, 'Eager beavers: rodents engineer Czech wetland project after years of human delay', *The Guardian*, 12 February 2025, https://www.theguardian.com/world/2025/feb/11/beavers-save-czech-taxpayers-by-flooding-ex-army-training-site
21. *BBC*, 'Drunk Swedish elk found in apple tree near Gothenburg' 8 September 2011, https://www.bbc.com/news/world-europe-14842999

SOURCES

22. Nyberg, Per, 'Drunken moose ends up stuck in Swedish apple tree.' *Center for Biological Diversity*, 9 Sept. 2011, https://www.biologicaldiversity.org/news/center/articles/2011/CNN-09-09-2011.html
23. Hanson, Hilary, 'Drunken Moose Terrorize Sweden: Public Left with Questions, Antlers', *Huff Post*, 16 September 2013, https://www.huffpost.com/entry/drunk-elk-sweden_n_3936590
24. Welsh, Jennifer, 'A Gang Of Alcoholic Moose Terrorized A Swedish Homeowner', *Business Insider*, 11 September 2013, https://www.businessinsider.com/alcohol-imbibing-animals-2013-9
25. Wynick, Alex, 'Gang of deer 'drunk' from rotten apples stop man from entering his own home', *Mirror*, 2 September 2013, https://www.mirror.co.uk/news/weird-news/drunk-deer-stop-man-entering-2245311
26. Sveriges Radio, 'Drunken elk threaten home owner', 28 August 2013, https://www.sverigesradio.se/artikel/5629237
27. *'Drunken Moose Ends Up Stuck in Swedish Apple Tree.'* Sveriges Radio, 9 Sept. 2013, https://sverigesradio.se/artikel/5629237. Accessed 6 Dec. 2024
28. Sveriges Radio, 'Another drunk moose destroys swing set' 15 September 2011, https://www.sverigesradio.se/artikel/4697489.
29. Halpin, James, 'Buzzwinkle, moose who wandered downtown Anchorage, killed Ihumanely,' *The Seattle Times*, 6 April 2008, www.seattletimes.com/seattle-news/buzzwinkle-moose-who-wandered-downtown-anchorage-killed-humanely/
30. Ditmer, Mark A. et al., 2017 'Moose at their bioclimatic edge alter their behavior based on weather, landscape, and predators' *Current Zoology*, Vol. 64, doi:10.1093/cz/zox047
31. Ibid.
32. IKEA, 'COP25: IKEA on what climate actions are needed to improve land use', 27 November 2019, https://www.ikea.com/global/en/newsroom/sustainability/cop25-ikea-on-what-climate-actions-are-needed-to-improve-land-use-191127/
33. IKEA, 'IKEA launches 2030 forest agenda', 25 January 2021, https://www.ikea.com/global/en/newsroom/sustainability/

ikea-launches-new-2030-forest-agenda-to-push-for-improved-forest-management-and-biodiversity-globally-210125/

34. *Statista*, 'Chocolate Confectionary – Worldwide', https://www.statista.com/outlook/cmo/food/confectionery-snacks/confectionery/chocolate-confectionery/worldwide

35. Bhutada, Govind, 'Cocoa's bittersweet supply chain in one visualization', *World Economic Forum*, 4 November 2020, https://www.weforum.org/stories/2020/11/cocoa-chocolate-supply-chain-business-bar-africa-exports/

36. Vidal, John, 'The Real Price of a Chocolate Bar: West Africa's Rainforests', Yale Environment 360, 21 Feb. 2019, Yale School of the Environment, https://e360.yale.edu/features/the-real-price-of-a-chocolate-bar-west-africas-rainforests

37. Beg, Mohd Shavez, et al., 2017, 'Status, supply chain and processing of cocoa - A review', *Trends in Food Science & Technology*, Vol. 66, https://doi.org/10.1016/j.tifs.2017.06.007

38. Asante, Paulina A. et al., 2025, 'Climate change impacts on cocoa production in the major producing countries of West and Central Africa by mid-century', *Agricultural and Forest Meteorology*, Vol. 362, https://doi.org/10.1016/j.agrformet.2025.110393

39. Mars Incorporated, 'Sustainability Reports', https://www.mars.com/sustainability-plan/sustainability-reports

40. Costanza, Robert et al., 1997, 'The value of the world's ecosystem services and natural capital', *Nature*, Vol. 387, https://www.nature.com/articles/387253a0#auth-Robert-Costanza-Aff1-Aff2

41. Sutton, Paul, 'Can you put a dollar value on nature', *World Economic Forum*, 6 March 2015, https://www.weforum.org/stories/2015/03/can-you-put-a-dollar-value-on-nature/

42. Greenfield, Patrick, 'How much is an elephant worth? Meet the ecologists doing the sums', *The Guardian*, 28 January 2021, https://www.theguardian.com/environment/2021/jan/28/how-much-is-an-elephant-worth-meet-the-ecologists-doing-the-sums-aoe

43. Farmer, Brian, 'Prof says businesses should measure cost to planet.' *BBC*, 3 July 2024, https://www.bbc.com/news/articles/c047gdm8r3xo

44. Stevens, William K, 'How Much Is Nature Worth? For You, $33 Trillion', *The New York Times*, 20 May 1997, https://www.nytimes.com/1997/05/20/science/how-much-is-nature-worth-for-you-33-trillion.html
45. *Chicago Tribune*, 'An economist is a man who states...', 26 December 1993, https://www.chicagotribune.com/1993/12/26/an-economist-is-a-man-who-states/
46. Avery, Helen, 'Conservation Finance: Can Banks Embrace Natural Capital?', *Euromoney*, 8 October, 2019, https://www.euromoney.com/article/27bjsstsqxhkmh1y5f4ts/sustainability/conservation-finance-can-banks-embrace-natural-capital/
47. Bokat-Lindell, Spencer, 'Do We Need to Shrink the Economy to Stop Climate Change?', *The New York Times*, 16 September 2021, https://www.nytimes.com/2021/09/16/opinion/degrowth-cllimate-change.html
48. Stevens, William K, 'How Much Is Nature Worth? For You, $33 Trillion', *The New York Times*, 20 May 1997, https://www.nytimes.com/1997/05/20/science/how-much-is-nature-worth-for-you-33-trillion.html
49. Monbiot, George, 'The UK government wants to put a price on nature – but that will destroy it', *The Guardian*, 16 May 2018, https://www.theguardian.com/commentisfree/2018/may/15/price-natural-world-destruction-natural-capital
50. Monbiot, George, 'Put a price on nature? We must stop this neoliberal road to ruin', *The Guardian*, 24 July 2014, https://www.theguardian.com/environment/georgemonbiot/2014/jul/24/price-nature-neoliberal-capital-road-ruin
51. Ibid.
52. Hemming, Polly, 'The Climate Crisis Is an Integrity Crisis', *The Australia Institute*, 21 March 2024, https://australiainstitute.org.au/post/the-climate-crisis-is-an-integrity-crisis-polly-hemming/
53. Sparrow, Jeff, 'The call to 'put a price on nature' can be appealing – but It misunderstands what's at stake', *The Guardian*, 16 December 2022, www.theguardian.com/commentisfree/2022/dec/16/the-call-to-put-a-price-on-nature-can-be-appealing-but-it-misunderstands-whats-at-stake

54. Monbiot, George. 'The Call to Put a Price on Nature Can Be Appealing – but It Misunderstands What's at Stake.' *The Guardian*, 16 Dec. 2022, www.theguardian.com/commentisfree/2022/dec/16/the-call-to-put-a-price-on-nature-can-be-appealing-but-it-misunderstands-whats-at-stake.
55. United Nations, 'Secretary-General Calls on States to Tackle Climate Change 'Time Bomb' through New Solidarity Pact, Acceleration Agenda, at Launch of Intergovernmental Panel Report', 20 March 2023, https://press.un.org/en/2023/sgsm21730.doc.htm

Chapter 8

1. Planetary Business Podcast, 'Biodiversity: the ultimate global challenge – with scientist Dr. Frauke Fischer', 7 July 2022, https://podcasts.apple.com/au/podcast/biodiversity-the-ultimate-global-challenge-with/id1572270495?i=1000569068726
2. Placani, Adriana and Stearns Broadhead., 2022, 'Moral Dimensions of Offsetting Luxury Emissions.' *Ethics, Policy & Environment*, Vol. 25, https://doi.org/10.1080/21550085.2022.2104099
3. Grant, Madeline, 'Rich 'eco-sinners' can't buy environmental absolution through carbon offsetting', *The Telegraph*, 21 August 2019, www.telegraph.co.uk/environment/2019/08/21/rich-eco-sinners-cant-buy-environmental-absolution-carbon-offsetting/ Hunt, Christian and Alex Randall, 'Cheat Neutral' *Actipedia*, 19 November 2009, https://actipedia.org/project/cheat-neutral
4. IPCC, 'Climate Change: The IPCC 1990 and 1992 Assessments', https://www.ipcc.ch/site/assets/uploads/2018/05/ipcc_90_92_assessments_far_full_report.pdf
5. The United Nations Framework Convention on Climate Change, 'Article 2 Objective', *United Nations*, https://unfccc.int/resource/ccsites/zimbab/conven/text/art02.htm
6. Xu, Chi et al., 2020, 'Future of the human climate niche', *Proceedings of the National Academy of Sciences*, Vol. 1117, https://doi.org/10.1073/pnas.1910114117
7. Greenfield, Patrick and Nyasha Chingono, ''We don't know

where the money is going': the 'carbon cowboys' making millions from credit schemes', *The Guardian*, 15 March 2024, https://www. theguardian.com/environment/2024/mar/15/ money-carbon-credits-zimbabwe-conservation-aoe.
8. McCoy, Terrence et al., 'How 'carbon cowboys' are cashing in on protected Amazon forest', *The Washington Post*, 24 July 2024,
9. https://www.washingtonpost.com/world/interactive/2024/brazil-amazon-carbon-credit-offsets/
10. Cannon, John, "Cowboys' and intermediaries thrive in Wild West of the carbon market', *Mongabay*, 9 January 2024, https://news.mongabay.com/2024/01/cowboys-and-intermediaries-thrive-in-wild-west-of-the-carbon-market/
11. Rathi, Akshat, 'Inside the billion-dollar market for junk carbon offsets', *Australian Financial Review*, 21 November 2022, https:// www.afr.com/markets/debt-markets/inside-the-billion-dollar-market-for-junk-carbon-offsets-20221121-p5c053
12. Greenfield, Patrick, 'Revealed: more than 90% of rainforest carbon offsets by biggest provider are worthless, analysis shows', *The Guardian*, 19 January 2023, https://www.theguardian.com/ environment/2023/jan/18/revealed-forest-carbon-offsets-biggest-provider-worthless-verra-aoe
13. Lethal Humidity Global Council, 'Real Zero vs Net Zero', https://www.lethalhumidity.org/real-zero/
14. Marshall, Claire, 'Kenya's Ogiek people being evicted for carbon credits - lawyers', *BBC*, 9 November 2023, https://www.bbc.com/ news/world-africa-67352067 Greenfield, Patrick, "Nowhere else to go': forest communities of Alto Mayo, Peru, at centre of offsetting row', *The Guardian*, 19 January 2023, https://www.theguardian.com/environment/2023/ jan/18/forest-communities-alto-mayo-peru-carbon-offsetting-aoe
15. Bryan, Kenza and Clara Murray, 'Shell plant reported millions of 'phantom' carbon credits', *Financial Times*, 5 May 2024, https://www.ft.com/content/93938a1b-dc36-4ea6-9308-170189be0cb0
16. Tegel, Simeon, 'Timber companies claim carbon credits for

trees they don't cut down', *The Washington Post*, 21 April 2024, https://www.washingtonpost.com/world/2024/04/21/redd-carbon-credits-amazon-peru/

17. Dolmetsch, C., 2024, Former C-Quest CEO Kenneth Newcombe Charged With Carbon Credit Fraud. [online] Bloomberg.
18. com. Available at: https://www.bloomberg.com/news/articles/2024-10-02/former-c-quest-ceo-newcombe-charged-with-carbon-credit-fraud
19. Greenfield, Patrick, "'Nowhere else to go': forest communities of Alto Mayo, Peru, at centre of offsetting row', *The Guardian*, 19 January 2023, https://www.theguardian.com/environment/2023/jan/18/forest-communities-alto-mayo-peru-carbon-offsetting-aoe; Wenzel, Fernanda, 'Verra suspends carbon credit projects following police raid in Brazil', *Mongabay*, 11 June 2024, https://news.mongabay.com/2024/06/verra-suspends-carbon-credit-projects-following-police-raid-in-brazil/; SourceMaterial, 'The Carbon Con', 18 January 2023, https://www.source-material.org/vercompanies-carbon-offsetting-claims-inflated-methodologies-flawed/
20. Sabin Center for Climate Change Law, Berrin v. Delta Air Lines Inc., 2023, https://climatecasechart.com/case/berrin-v-delta-air-lines-inc/
21. Sabin Center for Climate Change Law, Australian Parents for Climate Action v. EnergyAustralia, 2023, https://climatecasechart.com/non-us-case/australian-parents-for-climate-action-v-energyaustralia/
22. Sabin Center for Climate Change Law, Dorris v. Danone Waters of America, 2022, https://climatecasechart.com/case/dorris-v-danone-waters-of-america/
23. United Nations, 'Bogus Net-Zero Pledges 'Rank Deception', Sham Must End, Secretary-General Stresses at Launch of Report by High-Level Expert Group on Non-State Actors' Commitments', 8 November 2022, https://press.un.org/en/2022/sgsm21576.doc.htm
24. Marsh, Alastair, 'Market for Carbon Credits Faces Fresh Blow as Offsets Slammed', *Bloomberg*, https://www.bloomberg.com/news/articles/2024-07-02/carbon-credits-face-fresh-blowback-as-ngos-unite-to-slam-offsets?embedded-checkout=true

25. Morton, Adam, 'Australia's carbon credits system a failure on global scale, study finds', *The Guardian*, 27 March 2024, https://theguardian.com/environment/2024/mar/27/australias-carbon-credits-system-a-failure-on-global-scale-study-finds
26. Garside, Ben, 'Pope Francis encyclical warns on use of carbon credits', *Carbon Pulse*, 4 Juli 2015, https://carbon-pulse.com/5149/
27. Hagelberg, Niklas, 'Carbon offsets are not our get-out-of-jail free card', *UN Environment Program*, 12 June 2019, https://www.unep.org/news-and-stories/story/carbon-offsets-are-not-our-get-out-jail-free-card
28. Setzer, Joana and Higham, Catherine, 'Global trends in climate change litigation: 2024 snapshot', *Grantham Research Institute on Climate Change and the Environment*, 27 June 2024, https://www.lse.ac.uk/granthaminstitute/publication/global-trends-in-climate-change-litigation-2024-snapshot/
29. Chami, Ralph et al., 'Nature's Solution to Climate Change', *Finance & Development*, December 2019, https://www.imf.org/external/pubs/ft/fandd/2019/12/pdf/natures-solution-to-climate-change-chami.pdf
30. Hogg, Carolyn et al., 'NSW biodiversity reforms signal progress, but gaps still exist', *The University of Sydney*, 24 July 2024, https://www.sydney.edu.au/news-opinion/news/2024/07/24/nsw-biodiversity-reforms-signal-progress-but-gaps-still-exist-environment-conservation-expert.html
31. Slezak, Michael, "'Nature credits' could make Australia the 'Green Wall Street' for the world, Tanya Plibersek says', *ABC News*, 1 September 2022, https://www.abc.net.au/news/2022-09-01/australia-hopes-to-create-green-wall-street-with-credit-scheme/101392808
32. Australian Government Clean Energy Regulator, 'ACCU project and contract register', https://cer.gov.au/markets/reports-and-data/accu-project-and-contract-register?view=Projects
33. Sangha, Kamaljit K. et al., 2021, 'Assessing the value of ecosystem services delivered by prescribed fire management in Australian tropical

savannas', *Ecosystem Services*, Vol. 51, https://doi.org/10.1016/j.ecoser.2021.101343

34. Australian Government Department of Agriculture, Fisheries and Forestry, 'Australia's Indigenous land and forest estate (2024)', 28 October 2024, https://www.agriculture.gov.au/abares/forestsaustralia/forest-data-maps-and-tools/spatial-data/indigenous-land-and-forest
35. UN Environment Programme, 'Global annual finance flows of $7 trillion fueling climate, biodiversity, and land degradation crises', 9 December 2023, https://www.unep.org/news-and-stories/press-release/global-annual-finance-flows-7-trillion-fueling-climate-biodiversity
36. Ibid.
37. *Earth Track*, 'Environmentally harmful subsidies update: $2.6 trillion/year and a continuing threat to nature', 17 September 2024, https://www.earthtrack.net/blog/environmentally-harmful-subsidies-update-26-trillionyear-and-continuing-threat-nature
38. Hemming, Polly, 'What is Climate Integrity?', *The Australia Institute*, 16 February 2023, https://australiainstitute.org.au/post/what-is-climate-integrity/

Chapter 9

1. Bayly, Sami, 'Most people avoid ugly animals. I'm obsessed with them', *The Guardian*, 1 October 2019, https://www.theguardian.com/commentisfree/2019/oct/01/most-people-avoid-ugly-animals-im-obsessed-with-them
2. Angier, Natalie, 'A Masterpiece of Nature? Yuck!', *The New York Times*, 9 August 2010, https://www.nytimes.com/2010/08/10/science/10ugly.html
3. Curtin, Polly and Sarah Papworth, 2020 'Coloring and size influence preferences for imaginary animals, and can predict actual donations to species-specific conservation charities', *Conservation Letters*, Vol. 13, https://doi.org/10.1111/conl.12723
4. Thomas-Walters, Laura and Nichola J Raihani, 2017, 'Supporting conservation: The roles of flagship species and identifiable victims', *Conservation Letters*, Vol. 10, https://doi.org/10.1111/conl.12319

SOURCES

5. Caldwell, Iain R. et al., 2024 'Global trends and biases in biodiversity conservation research', *Cell Reports Sustainability*, Vol. 1, doi:10.1016/j.crsus.2024.100082
6. Caldwell, Iain R. et al., 2024 'Global trends and biases in biodiversity conservation research', *Cell Reports Sustainability*, Vol. 1, doi:10.1016/j.crsus.2024.100082
7. IUCN, 'Barometer of Life', https://www.iucnredlist.org/about/barometer-of-life
8. WWF, 'Living Planet Report 2022' https://livingplanet.panda.org/nature-loss-crisis/
9. IUCN, 'Barometer of Life', https://www.iucnredlist.org/about/barometer-of-life
10. Matthew, Christopher, 'The cult of the eagles in the Roman Republic', in *Religion and Classical Warfare*, Pen and Sword Books, South Yorkshire, 2020
11. Verderame, Lorenze, 'The Seven Attendants of Hendursaĝa: A study of animal symbolism in Mesopotamian cultures', in *The First Ninety Years: A Sumerian Celebration in Honor of Miguel Civil*, De Gruyter, 2017
12. Robinson, Tom, 'Native bats should be celebrated like other Australian wildlife, ecologist says', 28 April 2024, https://www.abc.net.au/news/2024-04-28/australia-bats-poor-reputation-covid-links-and-vampire-myths/103669082
13. Ibid.
14. Celley, Courtney, 'Bats are one of the most important misunderstood animals' *U.S. Fish & Wildlife Service*, www.fws.gov. Available at: https://www.fws.gov/story/bats-are-one-most-important-misunderstood-animals
15. Cohn, Jeffery P., 2012, 'Bats and White-Nose Syndrome Still a Conundrum', *BioScience*, Vol. 62, https://doi.org/10.1525/bio.2012.62.4.19
16. Cheng, Tina L. et al., 2021, 'The scope and severity of white-nose syndrome on hibernating bats in North America', *Conservation Biology*, Vol. 35, https://doi.org/10.1111/cobi.13739
17. LaCapra, Veronica, 'White-Nose Syndrome: A Scourge In The Bat

Caves', *NPR*, 5 April 2012, https://www.npr.org/2012/04/05/150000682/white-nose-syndrome-a-scourge-in-the-bat-caves

18. Hadlock, Jordan and Brent Z. Kaup, 2023, 'Locking-in white-nose syndrome? The limits of the endangered species act & non-charismatic megafauna', *Environmental Sociology*, Vol. 10, 10.1080/23251042.2023.2261684

19. Tobin, Ben, 'Sharks' *Australian Institute of Marine Science*, https://www.aims.gov.au/docs/projectnet/sharks-02.html

20. Keartes, Sarah, 'Why Do Sharks Bite People?', *Save Our Seas Foundation*, https://saveourseas.com/worldofsharks/why-do-sharks-bite-people

21. Bansal, Agam, et al., 2018, 'Selfies: A boon or bane?', *Journal of Family Medicine and Primary Care*, Vol. 7, doi:10.4103/jfmpc.jfmpc_109_18; Johns Hopkins Medicin, 'A Dangerously Tasty Treat: The Hot Dog Is a Choking Hazard', https://www.hopkinsmedicine.org/health/wellness-and-prevention/a-dangerously-tasty-treat-the-hot-dog-is-a-choking-hazard

22. International Fund for Animal Welfare (IFAW), 'FAQ About Sharks', 20 June 2024 https://www.ifaw.org/au/journal/faq-about-sharks

23. Elflein, John, 'Deadliest animals worldwide by annual number of human deaths as of 2022', *Statista*, 22 May 2024, https://www.statista.com/statistics/448169/deadliest-creatures-in-the-world-by-number-of-human-deaths

24. Byard, Roger W., 2025, 'Death and injuries caused by cattle: A forensic overview', *Forensic Science Medicine and Pathology*, Vol. 21, doi: 10.1007/s12024-024-00786-8

25. Florida Museum, 'The ISAF 2024 shark attack report', https://www.floridamuseum.ufl.edu/shark-attacks/yearly-worldwide-summary/

26. Pacoureau, Nathan, et al., 2021, 'Half a century of global decline in oceanic sharks and rays', *Nature*, Vol. 589, doi:10.1038/s41586-020-03173-9

27. Gallagher, Austin J. et al., 2014, 'Evolved for Extinction: The Cost and Conservation Implications of Specialization in Hammerhead Sharks', *Bioscience*, Vol. 64, https://doi.org/10.1093/biosci/biu071

SOURCES

28. National Parks and Wildlife Service South Australia, 'Neptune Islands Group (Ron and Valerie Taylor) Marine Park, https://www.marineparks.sa.gov.au/find-a-park/eyre-peninsula/neptune-islands
29. 'Conservation Groups Prioritize Iconic Species', *The Earth Times*, https://earthtimes.org/blogs/conservation/conservation-groups-prioritize-iconic-species?srsltid=AfmBOopoudDkZFgWdSnKxqRy74qhRN2NT-VdEKdDcgQ4YHv5L9S5a5PH
30. Ibid.
31. Colléony, Agathe et al., 2017, 'Human preferences for species conservation: Animal charisma trumps endangered status', *Biological Conservation*, Vol. 206, https://doi.org/10.1016/j.biocon.2016.11.035
32. Black, Madeline, 'An Ode to the Blobfish', *Ocean Conservancy*, 10 January 2020, https://oceanconservancy.org/blog/2020/01/10/an-ode-to-the-blobfish/
33. Kiberd, Roisin, 'It's Hard Out There for a Blobfish: Ugly Animals Need Conservation Too', *Vice*, 14 August 2015, https://www.vice.com/en/article/the-blobfish-needs-love-too/
34. Schultz, Colin and Sonja, 'In Defense of the Blobfish: The 'World's Ugliest Animal' Is Our Fault', *Smithsonian Magazine*, 16 January 2024, https://www.smithsonianmag.com/smart-news/worlds-ugliest-animal-blobfish-6676336/
35. Gannon, Megan, 'Blobfish Named World's Ugliest Animal', *Live Science*, 13 September 2013, https://www.livescience.com/39621-blobfish-worlds-ugliest-animal.html
36. Hance, Jeremy, "'Cute' umbrella video of slow loris threatens primate', *Mongabay*, 13 March 2011, https://news.mongabay.com/2011/03/cute-umbrella-video-of-slow-loris-threatens-primate/
37. Cowan, Carolyn, 'When "cute" is cruel: social media videos stoke loris pet trade, study says', *Mongabay*, 27 June 2023, https://news.mongabay.com/2023/06/when-cute-is-cruel-social-media-videos-stoke-loris-pet-trade-study-says
38. Cowan, Carolyn, 'When "cute" is cruel: social media videos stoke loris pet trade, study says', *Mongabay*, 27 June 2023, https://news.mongabay.com/2023/06/

when-cute-is-cruel-social-media-videos-stoke-loris-pet-trade-study-says
39. Kvizhashvili, Lika, 'The endearing giant panda: a symbol of conservation', *European Wilderness Society*, https://wilderness-society.org/the-endearing-giant-panda-a-symbol-of-conservation/
40. Hvistendahl, Mara and Joy Dong, 'The Panda Factories', *The New York Times*, 15 October 2024, https://www.nytimes.com/interactive/2024/10/15/world/asia/pandas-zoo-breeding-death-captivity.html
41. Ibid.
42. Ibid.
43. Ibid.
44. Goodall, Jane, 'Chimpanzees – Bridging the Gap' in *The Great Ape Project*, St. Martin's Griffin, New York, 1993
45. History, 'Jane Goodall observes a chimpanzee making and using tools', 27 May 2025, https://www.history.com/this-day-in-history/jane-goodall-observes-a-chimpanzee-making-and-using-tools

Chapter 10

1. Discover Whanganui, 'I am the river, the river is me', http://discoverwhanganui.nz/guides/the-river-is-me/
2. Evans, Edward Payson, The Criminal Prosecution and Capital Punishment of Animals, W. Heinemann, London, 1906
3. Ibid.
4. Ibid.
5. Ibid.
6. Ibid.
7. Ibid.
8. Ibid.
9. Girgen, Jen, 2003 'The historical and contemporary prosecution and punishment of animals' https://www.animallaw.info/sites/default/files/lralvol9_p97.pdf
10. Evans, Edward Payson, *The Criminal Prosecution and Capital Punishment of Animals*, W. Heinemann, London, 1906

11. Ibid.
12. Roach, Mary, *Fuzz: When Nature Breaks the Law*, W.W. Norton & Company, New York, 2021
13. Evans, Edward Payson, *The Criminal Prosecution and Capital Punishment of Animals*, W. Heinemann, London, 1906
14. Vatomsky, Sonya, 'When Societies Put Animals on Trial', *JSTOR Daily*, 13 September 2017, https://daily.jstor.org/when-societies-put-animals-on-trial/
15. Evans, Edward Payson, *The Criminal Prosecution and Capital Punishment of Animals*, W. Heinemann, London, 1906
16. Gallica, 'Le Progrès des Glaciers et les Phénomènes Glaciaires Actuels', *Bibliothèque Nationale de France*, https://gallica.bnf.fr/ark:/12148/bpt6k58330777/f59.image.r=glacier?rk=21459;2
17. Dearing, Aissa, 'Legal Personhood: Extending Rights to Nature?', *JSTOR Daily*, 11 July 2024, https://daily.jstor.org/legal-personhood-extending-rights-to-nature/
18. O'Donnell, Erin and Anna Arstein-Kerslake, 2021, 'Recognising personhood: the evolving relationship between the legal person and the state', *Griffith Law Review*, Vol. 30, doi:10.1080/10383441.202 1.2044438
19. Stone, Christopher D., 'Should trees have standing? – Toward legal rights for natural objects' in *Environmental Right*, Routledge, 2017
20. Ibid.
21. Ibid.
22. Ibid.
23. Isherwood, Aaron and Ross Macfarlane, 'River as Plaintiff', *Sierra Club*, 25 January 2023, https://www.sierraclub.org/sierra/river-plaintiff-william-o-douglas.
24. Barkham, Patrick, 'Should rivers have the same rights as people?', *The Guardian*, 25 July 2021, https://www.theguardian.com/environment/2021/jul/25/rivers-around-the-world-rivers-are-gaining-the-same-legal-rights-as-people
25. BBC Radio 4, 'Rivers and the Rights of Nature' https://www.bbc.co.uk/programmes/m001gwz2
26. Ibid.

27. The Whanganui River Report, Ngā Tāngata Tiaki o Whanganui, 1999, https://ngatangatatiaki.co.nz/assets/Uploads/Important-Documents/Whanganui-River-Report-1999.pdf
28. Te Awa Tupua (Whanganui River Claims Settlement) Act 2017', New Zealand Legislation, https://www.legislation.govt.nz/act/public/2017/0007/latest/whole.html
29. *BBC Radio* 4, 'Rivers and the Rights of Nature' https://www.bbc.co.uk/programmes/m001gwz2
30. *The Economist*, 'New Zealand declares a river a person', 25 March 2017, https://www.economist.com/asia/2017/03/25/new-zealand-declares-a-river-a-person
31. Boyd, David R., 2018 'Recognizing the Rights of Nature: Lofty Rhetoric or Legal Revolution?', *Natural Resources & Environment*, Vol. 32, https://www.jstor.org/stable/26418846
32. Ibid.
33. Sen, Sudipta, 'Of Holy Rivers and Human Rights: Protecting the Ganges by Law', *Yale University Press*, 25 April 2019, https://yalebooks. yale. edu/2019/04/25/of-holy-rivers-and-human-rights-protecting-the-ganges-by-law
34. Australian Earth Laws Centre, 'Rights of Nature in Ecuador', https://www.earthlaws.org.au/aelc/rights-of-nature/ecuador/
35. Global Alliance for the Rights of Nature, 'Rights of Nature timeline', https://www.garn.org/rights-of-nature-timeline
36. Law of the Rights of Mother Earth, translated by Ministry of Foreign Affairs of Bolivia, 2010
37. Climate Change Litigation Chart, 'Atrato River Decision T-622/16 of November 10, 2016', https://climatecasechart.com/non-us-case/atrato-river-decision-t-622–16-of-november-10–2016/
38. Putzer, Alex et al., 2022, 'The Rights of Nature as a Bridge between Land-Ownership Regimes: The Potential of Institutionalized Interplay in Post-Colonial Societies', *Transnational Environmental Law*, Vol. 11, doi:10.1017/S2047102522000334
39. Earth Law Center, 'Panama Passes National Rights of Nature Law', https://www.earthlawcenter.org/panama

40. Eco Jurisprudence Monitor, 'Colombia Court Decision Protecting Bees', https://ecojurisprudence.org/initiatives/court-decision-protecting-bees-in-colombia/
41. Marshall, Candic,. 'The Australian river legally recognised as a 'living entity'', *Australian Geographic*, 19 September 2024, https://www.australiangeographic.com.au/topics/history-culture/2024/09/yarra-river-legally-recognised-as-a-living-entity/
42. Ibid.
43. Greenfield, Patrick, '"Sweet City': the Costa Rica suburb that gave citizenship to bees, plants and trees', *The Guardian*, 29 April 2020, https://www.theguardian.com/environment/2020/apr/29/sweet-city-the-costa-rica-suburb-that-gave-citizenship-to-bees-plants-and-trees-aoe
44. Putzer, Alex et al., 2022 'Putting the rights of nature on the map. A quantitative analysis of rights of nature initiatives across the world', *Journal of Maps*, Vol. 18, https://doi.org/10.1080/17445647.2022.2079432
45. Takoko, Mere, 'A Descendant's Call for Whale Legal Personhood', *Atmos*, 26 March 2024, https://atmos.earth/a-descendants-call-for-whale-legal-personhood/
46. Kowhai, Te Rina, 'Māori king and other indigenous Pacific leaders sign up to granting whales legal personhood', *Te Ao Māori News*, 28 March 2024, https://www.teaonews.co.nz/2024/03/28/maori-king-and-other-indigenous-pacific-leaders-sign-up-to-granting-whales-legal-personhood/
47. Kennedy, Rónán, 'Trees, rivers and mountains are gaining legal status – but it's not been a quick fix for environmental problems', *The Conversation*, 17 August 2023, https://theconversation.com/trees-rivers-and-mountains-are-gaining-legal-status-but-its-not-been-a-quick-fix-for-environmental-problems-211542
48. Kennedy, Rónán, 'Trees, rivers and mountains are gaining legal status – is this a promising approach to protect nature?', *World Economic Forum*, 29 August 2023, https://www.weforum.org/stories/2023/08/legal-rights-nature-environmental-protection/

49. Kauffman, Craig M. and Pamela L. Martin, 2023, 'How Ecuador's Courts Are Giving Form and Force to Rights of Nature Norms', *Transnational Environmental Law*, Vol. 12, doi:10.1017/S2047102523000080
50. Marshall, Virginia, 2020, 'Removing the Veil from the 'Rights of Nature': The Dichotomy between First Nations Customary Rights and Environmental Legal Personhood', *Australian Feminist Law Journal*, Vol. 45, https://doi.org/10.1080/13200968.2019.1802154
51. Ibid.
52. Human Rights Law Centre, 'New Evidence Shows Right to Protest in Peril in Australia', 3 July 2024, https://www.hrlc.org.au/news/2024/07/03/protest-peril
53. UCLA School of Law, "Elephants in the jungle': Stephen Bainbridge on his new book about corporate purpose', 15 March 2023, https://law.ucla.edu/news/elephants-jungle-stephen-bainbridge-his-new-book-about-corporate-purpose
54. Eco-Jurisprudence Monitor, 'Ecuador court case on rights of nature violations from mining in the Los Cedros Protected Forest', 2021, https://ecojurisprudence.org/initiatives/los-cedros/; Surma, Katie, 'Landmark Ruling on Uncontacted Indigenous Peoples' Rights Strikes at Oil Industry', *Inside Climate News*, 13 March 2025, https://insideclimatenews.org/news/13032025/landmark-ruling-uncontacted-indigenous-peoples-rights-ecuador-oil-industry/
55. Stilt, Kristen, 2021, 'Rights of Nature, Rights of Animals', *Harvard Law Review*, Vol. 134, https://harvardlawreview.org/forum/vol-134/rights-of-nature-rights-of-animals/
56. Invasive Species Council, 'Our Work' https://invasives.org.au/our-work/feral-animals/cats-in-australia/feral-cats
57. Stilt, Kristen, 2021, 'Rights of Nature, Rights of Animals', *Harvard Law Review*, Vol. 134, https://harvardlawreview.org/forum/vol-134/rights-of-nature-rights-of-animals/
58. Ibid.
59. Smith, Justin E. H., 'Nature Is Becoming a Person', *Foreign*

Policy, 24 November 2021, https://foreignpolicy.com/2021/11/24/nature-person-rights-environment-climate-philosophy-law/

60. Pardo, Michelle C., 2023, 'Legal Personhood for Animals: Has Science Made Its Case?', *Animals*, Vol. 13, https://doi.org/10.3390/ani13142339
61. Gulatsi, Taylor, 'Legal Personality for Animals in India and Pakistan', *Library of Congress*, 30 August 2023, https://blogs.loc.gov/law/2023/08/legal-personality-for-animals-in-india-and-pakistan/
62. The Nonhuman Rights Project, 'Challenging the rightlessness of nonhuman animals', https://www.nonhumanrights.org/litigation/
63. ABC News, 'Monkey selfie: PETA settles in fight over rights to macaque's photos', 12 September 2017, https://www.abc.net.au/news/2017-09-12/monkey-selfie-lawsuit-ends-in-settlement/8894906
64. PETA, 'PETA Sues SeaWorld for Violating Orcas' Constitutional Rights', 25 October 2011, https://www.peta.org/blog/peta-sues-seaworld-violating-orcas-constitutional-rights/
65. 'Wild Animals in Brazil.' Superior Justice Tribunal, https://ecojurisprudence.org/wp-content/uploads/2022/02/Brazil_Wild-Animals-in-Brazil_164-1.pdf. Accessed 14 Jan. 2025.
66. Doughty, Steve, 'Brown bear who was 'sentenced' to time in a human jail has been freed to live in a zoo', *Daily Mail*, https://www.dailymail.co.uk/news/article-7696233/Brown-bear-sentenced-time-human-jail-freed-live-zoo.html
67. Salter, Jessica, 'Dog appears as witness in murder trial', *The Telegraph*, 10 September 2008, www.telegraph.co.uk/news/worldnews/europe/france/2775597/Dog-appears-as-witness-in-murder-trial.html
68. Reuters, 'Nigerian police detain goat over armed robbery', 24 January 2009, https://www.reuters.com/article/lifestyle/nigerian-police-detain-goat-over-armed-robbery-idUSTRE50M4BM/
69. Sukheja, Bhavya. 'Sheep Sentenced to Three Years in Jail for Killing a Woman in Africa', *NDTV*, 24 May 2022, https://www.ndtv.com/offbeat/sheep-sentenced-to-three-years-in-jail-for-killing-a-woman-in-africa-3003935

Chapter 11

1. World Bank, 'Selina Neirok Leem: 'a small island girl with big dreams'', 8 March 2016, www.worldbank.org/en/country/pacificislands/brief/selina-neirok-leem-a-small-island-girl-with-big-dreams
2. Fava, Marta, 'Ocean plastic pollution: an overview – data and statistics', *UNESCO Ocean Literacy Portal*, 9 May 2022, https://oceanliteracy.unesco.org/plastic-pollution-ocean/
3. Institute of Environmental Management and Assessment (IEMA), 'Plastic pollution leading to huge death toll in developing world', https://www.iema.net/articles/plastic-pollution-leading-to-huge-death-toll-in-developing-world
4. Fuller, Richard et al., 2022, 'Pollution and health: a progress update' *The Lancet Planetary Health*, Vol. 6, doi:10.1016/S2542-5196(22)00090-0
5. Miles, Tom, 'China says it won't take any more foreign garbage', *Reuters*, 19 July 2017, https://www.reuters.com/article/us-china-environment-idUSKBN1A31JI/?il=0
6. Cameron, Shaun, 'China: rejecting rubbish.' *The Interpreter*, 10 November 2021, https://www.lowyinstitute.org/the-interpreter/china-rejecting-rubbish
7. NSW Environment Protection Authority, 'Response to the enforcement of the China National Sword Policy', https://www.epa.nsw.gov.au/Your-environment/Recycling-and-reuse/Response-to-China-National-Sword
8. The World Counts, 'World Waste Facts', https://www.theworldcounts.com/challenges/planet-earth/state-of-the-planet/world-waste-facts
9. Kellenberg, Derek, 2015 'The Economics of the International Trade of Waste', *Annual Review of Resource Economics*, Vol. 7, https://doi.org/10.1146/annurev-resource-100913-012639
10. Katz, Cheryl, 'Piling Up: How China's Ban on Importing Waste Has Stalled Global Recycling', *Yale Environment 360*, 7 March 2019, https://e360.yale.edu/features/piling-up-how-chinas-ban-on-importing-waste-has-stalled-global-recycling

11. Jain, Aarushi, 'Trash Trade Wars: Southeast Asia's Problem With the World's Waste', *Council on Foreign Relations*, 8 May 2020, https://www.cfr.org/in-brief/trash-trade-wars-southeast-asias-problem-worlds-waste
12. Gutberlet, Jutta, 'Global Plastic Pollution and Informal Waste Pickers', *Cambridge Prisms: Plastics*, Vol. 1, doi:10.1017/plc.2023.10
13. Chen, Sulan, 'Unsung heroes: Four things policymakers can do to empower informal waste workers', *United Nations Development Programme (UNDP)*, 28 December 2023, https://www.undp.org/blog/unsung-heroes-four-things-policymakers-can-do-empower-informal-waste-workers
14. Haddad, Aja Barber. *Consumed: The Need for Collective Change – Colonialism, Climate Change, and Consumerism.* 1st ed., Octopus, 2021.
15. Ibid.
16. Gündoğdu, Sedat, *Plastic Waste Trade: A New Colonialist Means of Pollution Transfer*, Springer Nature, 2024.
17. Michaelson, Ruth, ''Waste colonialism': world grapples with west's unwanted plastic', *The Guardian*, 31 December 2021, https://www. theguardian.com/environment/2021/dec/31/waste-colonialism-countries-grapple-with-wests-unwanted-plastic
18. Sembiring, Margareth, *Global Waste Trade Chaos: Rising Environmentalism or Cost-Benefit Analysis?*, S. Rajaratnam School of International Studies, 2022, https://www.jstor.org/stable/resrep26804
19. ABC News, 'Malaysia to send plastic waste back to Australia and other developed nations', 29 May 2019, https://www.abc.net.au/news/2019-05-29/malaysia-to-send-tonnes-of-plastic-waste-back-to-foreign-nations/11159208
20. BBC, 'Rodrigo Duterte: Philippines not a 'dump site' for Canadian waste', 24 April 2019, https://www.bbc.com/news/world-asia-47901709
21. UN Environment Programme, 'China's trash ban lifts lid on global recycling woes but also offers opportunity', 6 July 2018, https://

22. www.unep.org/news-and-stories/story/chinas-trash-ban-lifts-lid-global-recycling-woes-also-offers-opportunity
22. Ibid.
23. European Commission, 'Plastics', https://environment.ec.europa.eu/topics/plastics_en
24. Boyd, David R., 'Additional Sacrifice Zones: Supplementary Information to the Report of the Special Rapporteur on Human Rights and the Environment', *United Nations Human Rights Special Procedures*, 2022, https://www.ohchr.org/sites/default/files/2022-03/Annex1_to_A_HRC_49_53.pdf
25. Ibid.
26. Ibid.
27. Landrigan, Philip J. et al., 2017, 'The Lancet Commission on pollution and health', The Lancet, Vol. 391, doi:10.1016/S0140-6736(17)32345-0
28. Boyd, David R., 'Additional Sacrifice Zones: Supplementary Information to the Report of the Special Rapporteur on Human Rights and the Environment', *United Nations Human Rights Special Procedures*, 2022, https://www.ohchr.org/sites/default/files/2022-03/Annex1_to_A_HRC_49_53.pdf
29. Ibid.
30. Maitland Parker and Yaara Bou Melhem, *Yurlu Country*, Illuminate Films, 2025
31. Melville, Kirsti, 'How mesothelioma devastated this Indigenous community in the Pilbara', *ABC News*, 7 February 2019, https://www.abc.net.au/news/2019-02-07/how-asbestos-devastated-wittenoom-indigenous-community/10781312
32. Boyd, David R., 'Additional Sacrifice Zones: Supplementary Information to the Report of the Special Rapporteur on Human Rights and the Environment', *United Nations Human Rights Special Procedures*, 2022, https://www.ohchr.org/sites/default/files/2022-03/Annex1_to_A_HRC_49_53.pdf
33. '"We're Dying Here": The Fight for Life in a Louisiana Fossil Fuel Sacrifice Zone'. Human Rights Watch, 25 Jan. 2024, https://www.

hrw.org/report/2024/01/25/were-dying-here/fight-life-louisiana-fossil-fuel-sacrifice-zone. Accessed 6 Jan. 2025.

34. Boyd, David R., 'Additional Sacrifice Zones: Supplementary Information to the Report of the Special Rapporteur on Human Rights and the Environment', United Nations Human Rights Special Procedures, 2022, https://www.ohchr.org/sites/default/files/2022-03/Annex1_to_A_HRC_49_53.pdf

35. Ibid.

36. United Nations Conference on Trade and Development, 'The Least Developed Countries Report 2023', https://unctad.org/publication/least-developed-countries-report-2023

37. Rouquette, Pauline, 'Australia offersrRefuge to Tuvaluans as rising sea levels threaten Pacific archipelago', *France 24*, 11 November 2023, https://www.france24.com/en/asia-pacific/20231111-australia-offers-refuge-to-tuvaluans-as-rising-sea-levels-threaten-pacific-archipelago

38. Tuvalu, 'The First Digital Nation', https://www.tuvalu.tv/about/

39. Ibid.

40. Boyd, David R., 'Additional Sacrifice Zones: Supplementary Information to the Report of the Special Rapporteur on Human Rights and the Environment', *United Nations Human Rights Special Procedures*, 2022, https://www.ohchr.org/sites/default/files/2022-03/Annex1_to_A_HRC_49_53.pdf

41. Office of the Prime Minister, 'Standard Operating Procedure for Planned Relocation in the Republic of Fiji', 2023, https://fijiclimatechangeportal.gov.fj/ppss/standard-operating-procedures-for-planned-relocation-in-the-republic-of-fiji/

42. United Nations, 'Why women are key to climate action', https://www.un.org/en/climatechange/science/climate-issues/women

43. MacDonald, Rhona, 2005, 'How Women Were Affected by the Tsunami: A Perspective from Oxfam', PLoS Medicine, Vol. 2, https://doi.org/10.1371/journal.pmed.0020178

44. United Nations OHCHR, 'Protecting the Rights of Internally Displaced Persons in Natural Disasters', 2011, https://pacific.ohchr.org/docs/idp_report.pdf

45. True, Jacqui and Yolanda Riveros-Morales, 2018, 'Towards inclusive peace: Analysing gender-sensitive peace agreements 2000–2016', *International Political Science Review*, Vol. 40, https://doi.org/10.1177/0192512118808608
46. UN Women, 'Facts and figures: Women's leadership and political participation', 11 March 2025, https://www.unwomen.org/en/articles/facts-and-figures/facts-and-figures-womens-leadership-and-political-participation
47. International Union for Conservation of Nature, 'Gender equality for greener and bluer futures: Why women's leadership matters for realising environmental goals', 2024, https://iucn.org/sites/default/files/2024-03/2024-gender-equality-for-greener-and-bluer-futures.pdf
48. Hinchliffe, Emma and Nina Ajemian, 'The share of women running Global 500 companies falls to just 5.6%', *Fortune*, 5 August 2024, https://fortune.com/2024/08/05/the-share-of-women-running-global-500-companies-falls-to-just-5-6/
49. Parson, Laurie, *Carbon Colonialism: How Rich Countries Export Climate Breakdown*, Manchester University Press, Manchester, 2023
50. United Nations Environment Programme, 'Rich countries use six times more resources, generate 10 times the climate impacts than low-income ones', 1 March 2024, https://www.unep.org/news-and-stories/press-release/rich-countries-use-six-times-more-resources-generate-10-times
51. Hickel, Jason et al., 2022 'Imperialist appropriation in the world economy: Drain from the global South through unequal exchange, 1990 – 2015', *Global Environmental Change*, Vol. 73, https://doi.org/10.1016/j.gloenvcha.2022.102467
52. Ibid.
53. World Bank, 'Poverty, Prosperity, and Planet Report 2024: Pathways Out of the Polycrisis', 2024, http://hdl.handle.net/10986/42211
54. Barich, Anthony, 'China, UK expand mining presence in Africa; US, Canada, Australia lose ground', *S&P Global*, 16 September 2024, https://www.spglobal.com/marketintelligence/en/news-insights/

latest-news-headlines/china-uk-expand-mining-presence-in-africa-us-canada-australia-lose-ground-83154484

55. Nigerian Upstream Petroleum Regulatory Commission, 'Nigeria: Leading Crude Oil Producer in Africa', 25 July 2024, https://www.nuprc.gov.ng/nigeria-leading-crude-oil-producer-in-africa
56. International Energy Agency, 'Nigeria' https://www.iea.org/countries/nigeria/oil
57. Bjerde, Anna, 'Lighting Up Africa: Nigeria can show the way', *World Bank Blogs*, 29 February 2024, https://blogs.worldbank.org/en/africacan/lighting-up-africa-nigeria-can-show-the-way
58. World Bank, 'Nigeria: Overview', https://www.worldbank.org/en/country/nigeria/overview
59. Ndimele, Prince E., *The Political Ecology of Oil and Gas Activities in the Nigerian Aquatic Ecosystem*, Academic Press, Cambridge (MA) 2018
60. Steyn, Phia, 2009, 'Oil exploration in colonial Nigeria, c. 1903 – 58', *The Journal of Imperial and Commonwealth History*, Vol. 37, https://doi.org/10.1080/03086530903010376
61. Frynas, Jêdrzej George et al., 2000, 'Maintaining Corporate Dominance after Decolonization: the 'First Mover Advantage' of Shell-BP in Nigeria', *Review of African Political Economy*, Vol. 27, doi:10.1080/03056240008704475
62. Ordinioha, Best and Seiyefa Brisibe, 2013, 'The human health implications of crude oil spills in the Niger delta, Nigeria: An interpretation of published studies', *Nigerian Medical Journal*, Vol. 54, doi:10.4103/0300-1652.108887
63. Greenpeace, 'Confronting injustice: Racism and the environmental emergency', 2022, https://www.greenpeace.org.uk/wp-content/uploads/2022/09/Confronting-Injustice-2022-web.pdf
64. Amnesty International, 'Petroleum, Pollution, and Poverty in the Niger Delta', 2009, https://www.es.amnesty.org/fileadmin/noticias/Niger_Delta_ Campaign_Digest_01.pdf
65. Ansah, Christabel Edena et al., 2022, 'Environmental Contamination of a Biodiversity Hotspot—Action Needed for Nature Conservation

in the Niger Delta, Nigeria', *Sustainability*, Vol. 14, https://doi.org/10.3390/su142114256

66. Greenpeace, 'Confronting injustice: Racism and the environmental emergency', 2022, https://www.greenpeace.org.uk/wp-content/uploads/2022/09/Confronting-Injustice-2022-web.pdf
67. Amnesty International, 'Investigate Shell for complicity in murder, rape and torture', 28 November 2017, https://www.amnesty.org/en/latest/press-release/2017/11/investigate-shell-for-complicity-in-murder-rape-and-torture/
68. Amnesty International, 'Nigeria: Shell complicit in the arbitrary executions of Ogoni Nine as writ served in Dutch court', 29 June 2017, https://www.amnesty.org/en/latest/press-release/2017/06/shell-complicit-arbitrary-executions-ogoni-nine-writ-dutch-court/
69. Craig, Jess 'The village that stood up to big oil – and won', *The Guardian*, 1 June 2022, https://www.theguardian.com/environment/ng-interactive/2022/jun/01/oil-pollution-spill-nigeria-shell-lawsuit
70. Cavcic, Melisa, 'Shell offloads its Nigerian onshore business to focus on deepwater and integrated gas', *Offshore Energy*, 17 January 2024, https://www.offshore-energy.biz/shell-offloads-its-nigerian-onshore-business-to-focus-on-deepwater-and-integrated-gas/
71. Rowell, Andy and James Marriot, 'Is Shell's exit from Nigeria a front to dodge legal responsibilities?', *openDemocracy*, 6 June 2024, https://www.opendemocracy.net/en/shell-sell-nigeria-subsidiary-niger-delta-oil-spills-ogoni-nine-renaissance-front-legal-responsibilities/
72. Amnesty International, 'Nigeria: Shell must clean up devastating oil spills in the Niger Delta', 2 February 2023, https://www.amnesty.org/en/latest/news/2023/02/nigeria-shell-oil-spill-trial/
73. Shell Nigeria, 'Shell and Milieudefensie settle long-running case over oil spills in Nigeria', 23 December 2022, https://www.shell.com.ng/media/_jcr_content/root/main/section_1819883623/list/list_item.multi.stream/1743157389457/88e34b4ff8ea5d0ba808aa8ba87432918d2edc50/2022-shell-nigeria-press-releases-archive.pdf
74. Shell, 'Powering Progress: Annual Reports and Accounts', 2022,

https://www.shell.com/investors/results-and-reporting/annual-report-archive/_jcr_content/root/main/section_812377294/tabs/tab_1219767661/text.multi.stream/1742905354181/24ff673614687df7a13f18ce0cbd126cd73bc788/shell-annual-report-2022.pdf

75. Amnesty International, 'Nigeria: Shell must clean up devastating oil spills in the Niger Delta', 2 February 2023, https://www.amnesty.org/en/latest/news/2023/02/nigeria-shell-oil-spill-trial/
76. Obia, Christopher B. et al., 2021, 'Counting the cost of the Niger Delta's largest oil spills: Satellite remote sensing reveals extensive environmental damage with >1million people in the impact zone', *Science of The Total Environment*, Vol. 775, https://doi.org/10.1016/j.scitotenv.2021.145854
77. Boyd, David R., 'Business, planetary boundaries, and the right to a clean, healthy, and sustainable environment' *United Nations General Assembly*, 2024, https://documents.un.org/access.nsf/get?OpenAgent&DS=A/HRC/55/43&Lang=E
78. Okonjo-Iweala, Ngozi, 'Mia Mottley', *Time*, 23 May 2022, https://time.com/collection/100-most-influential-people-2022/6177695/mia-mottley/
79. Oliver Velez, Denise, 'Caribbean Matters: The world celebrates Barbados PM Mia Mottley's response to reporter's question', *Daily Kos*, 2 March 2023, https://www.dailykos.com/stories/2023/3/2/2155424/-Caribbean-Matters-The-world-celebrates-Barbados-PM-Mia-Mottley-s-response-to-reporter-s-question
80. Mansoor, Sanya, 'Prime Minister Mia Amor Mottley: The Plight of Barbados Is Also a Dire Climate Warning for the U.S.', *Time*, 7 June 2022, https://time.com/6184996/mia-mottley-barbados-time100-summit-2022/
81. United Nations Framework Convention on Climate Change, 'WLS - Opening Ceremony: Remarks by Mia Amor Mottley, Prime Minister of Barbados', 1 November 2021, https://unfccc.int/sites/default/files/resource/BARBADOS_cop26cmp16cma3_HLS_EN.pdf
82. The Bridgetown Initiative. https://www.bridgetown-initiative.org/

83. BBC, 'My day at COP26: 'I told world leaders: We're not drowning, We're fighting'' BBC News, 2 November 2021, https://www.bbc.com/news/science-environment-59121480
84. "Tuvalu Minister to Address COP26 Knee Deep in Water to Highlight Climate Crisis and Sea Level Rise." The Guardian, 8 Nov. 2021, www.theguardian.com/environment/2021/nov/08/tuvalu-minister-to-address-cop26-knee-deep-in-seawater-to-highlight-climate-crisis. Accessed 6 Apr. 2025.
85. Whiting, Kate, 'COP27: The top quotes from climate and world leaders at the UN Summit' World Economic Forum, 11 November 2022, https://www.weforum.org/stories/2022/11/cop27-quotes-climate-leaders/; United Nations, 'Warning Time Is Running Out, Small Island Developing States Demand Urgent Action to Address Climate Crisis They Did Not Create, as General Debate Continues', 27 September 2024, https://press.un.org/en/2024/ga12638.doc.htm
86. His Excellency Terrance Michael Drew, Prime Minister of Saint Kitts and Nevis, 27 September 2024, *Statement at the General Debate of the 79th Session of the United Nations General Assembly*, gadebate.un.org/en/79/saint-kitts-and-nevis
87. His Excellency Philip Edward Davis, Prime Minister of the Bahamas, 27 September 2024, *Statement at the General Debate of the 79th Session of the United Nations General Assembly*, gadebate.un.org/en/79/bahamas
88. United Nations Framework Convention on Climate Change, 'Fund for responding to Loss and Damage' https://unfccc.int/loss-and-damage-fund-joint-interim-secretariat

Chapter 12

1. Twain, Mark, 1905, 'The Czar's Soliloquy', *The North American Review*, Vol. 180, http://www.jstor.org/stable/25105366
2. Sandison, A. T., 2012, 'The Madness of the Emperor Caligula' *Medical History*, Vol. 2, https://doi.org/10.1017/S0025727300023759
3. Graves, Robert I, Claudius: From the Autobiography of Tiberius Claudius, Born 10 B.C., Murdered and Deified A.D. 54. Vintage

Books, 1989; 'Caligula.' Empires: The Roman Empire in the First Century, PBS, https://www.pbs.org/empires/romans/empire/caligula.html.

4. History.com, 'Caligula', 2 March 2025, https://www.history.com/topics/ancient-rome/caligula
5. Suetonius Tranquillus, Gaius, *The Lives of the Twelve Caesars*, translated by J. C. Rolfe, Loeb, Harvard University Press, 1913
6. Carlson, Deborah N., 2002 'Caligula's Floating Palaces- Archaeologists And Shipwrights Resurrect One Of The Emperor's Sumptuous Pleasure Boats' *Archaeology*, Vol. 55, https://www.jstor.org/stable/41779576
7. Tanner, Jeremy and Andrew Gardner, *Materialising the Roman Empire*, UCL Press, London, 2024
8. Ibid.
9. Ibid.
10. Pliny, Natural History, XIV, 1: ' laxitas mundi et rerum amplitudo damno fuit. ' Both quotes translated by this author. See Lao 2011 for a discussion of Pliny and luxury.
11. Pliny the Elder. The Natural History, translated by W. Thayer, 2002, https://www.perseus.tufts.edu/hopper/text?doc=Perseus%3Aabo%3Aphi%2C0978%2C001%3A33
12. Vassy, Aurélie, 'The Top 6 Most Expensive Hermès Birkin Bags', Sotheby's, 16 December 2024, https://www.sothebys.com/en/articles/the-top-6-most-expensive-hermes-birkin-bags
13. Stuart Hughes, 'iPhone 5 Black Diamond', https://stuarthughes.com/shop/luxury-mobile-phones/apple/iphone-5-black-diamond/
14. Darwin, Liza, 'This Is What a $1 Million Fur Coat with Silver Tips Looks Like', *Refinery29*, 18 July 2015, https://www.refinery29.com/en-us/2015/07/90934/fendi-one-million-dollar-fur
15. The Jewellery Editor, 'The $55m Hallucination watch: Graff Diamonds hits unprecedented heights at Baselworld 2014', 28 March 2014, ww.thejewelleryeditor.com/watches/article/the-55m-hallucination-watch-graff-diamonds-hits-unprecedented-heights-at-baselworld-2014/

16. Ensar Oud, 'Mélange Privé SQ', https://www.ensaroud.com/product/melange-privee-sq/
17. Thompson, I. D. et al., 'Expensive, Exploited and Endangered: A Review of the Agarwood-Producing Genera Aquilaria and Gyrinops—CITES Considerations, Trade Patterns, Conservation, and Management', *International Tropical Timber Organization*, 2022, https://cites.org/sites/default/files/documents/E-CoP19-Inf-12-R1.pdf
18. Guinness World Records, 'Most Valuable Gold-Plated Car', https:// www.guinnessworldrecords.com/world-records/452819-most-valuable-gold-plated-car
19. '7,000 cars, over 1,700-room palace, and 4,300-cr golden jet: This ultra rich sultan once hosted PM Modi', *The Economic Times*, 29 October 2024, https://economictimes.indiatimes.com/ news/international/global-trends/when-pm-modi-was-hosted-by-one-of-worlds-wealthiest-monarch-bruneis-sultan-hassanal-bolkiah-who-owns-7000-cars-palace-with-over-1700-rooms-/ articleshow/114725414.cms?from=mdr
20. BBC, 'Jeff Bezos' superyacht will see historic bridge dismantled', 4 February 2022, https://www.bbc.com/news/world- europe-60241145
21. Architectural Digest, 'Inside Antilia, Mukesh Ambani's $2 billion Mumbai mansion.', 23 May 2024, https://www.architecturaldigest.in/content/mukesh-ambani-antilia-home-mumbai/
22. 22The New Indian Express, 'Mukesh Ambani built Antilia on orphanage land illegally sold in 2005: Maharashtra State Board of Wakfs', 29 November 2017, https://www.newindianexpress.com/nation/2017/Nov/28/mukesh-ambani-built-antilia-on-orphanage-land-illegally-sold-in-2005-maharashtra-state-board-of-wak-1712977.html
23. D'Arpizio, Claudia et al., 'Luxury in Transition: Securing Future Growth', *Bain & Company*, 2024, https://www.bain.com/insights/luxury-in-transition-securing-future-growth/
24. Veblen, Thorstein, *Conspicuous Leisure. The Theory of the Leisure Class: An Economic Study in the Evolution of Institutions*, Macmillan, London, 1899

25. Baudrillard, Jean, *For a Critique of the Political Economy of the Sign*, translated by Charles Levin, Verso Books, London; New York 2019
26. Sundie, James M. et al., 2011, 'Peacocks, Porsches, and Thorstein Veblen: Conspicuous consumption as a sexual signaling system', *Journal of Personality and Social Psychology*, vol. 100, https://doi.org/10.1037/a0021669
27. Rice University, 'Does driving a Porsche make a man more desirable to women?' *ScienceDaily*, 16 June 2011, https://www.sciencedaily.com/releases/2011/06/110616092647.htm
28. Ibid.
29. Miller, Geoffrey. 'Must-Have: The Hidden Instincts Behind Everything We Buy'. Vintage, 2010.
30. Our World in Data, 'Per capita GHG emissions vs. per capita CO_2 emissions, 2023', https://ourworldindata.org/grapher/per-capita-ghg-co2-including-land-use?tab=table
31. World Population Review, 'Ecological Footprint by Country', https://worldpopulationreview.com/country-rankings/ecological-footprint-by-country
32. Earth Overshoot Day, 'How Many Earths? How Many Countries?', https://overshoot.footprintnetwork.org/how-many-earths-or-countries-do-we-need/
33. Alestig, Mira et al., 'Carbon Inequality Kills: Why curbing the excessive emissions of an elite few can create a sustainable planet for all', *Oxfam Policy & Practice*, 28 October 2024, https://policy-practice.oxfam.org/resources/carbon-inequality-kills-why-curbing-the-excessive-emissions-of-an-elite-few-can-621656/
34. Ibid.
35. Wilk, Richard and Beatriz Barros, 'Private planes, mansions and superyachts: What gives billionaires like Musk and Abramovich such a massive carbon footprint', *The Conversation*, 17 February 2021, https://theconversation.com/private-planes-mansions-and-superyachts-what-gives-billionaires-like-musk-and-abramovich-such-a-massive-carbon-footprint-152514

36. Khalfan, Ashfaq et al., 'Climate Equality: A Planet for the 99%', *Oxfam International*, 2023. https://www.oxfam.org.au/wp-content/uploads/2023/11/Climate-Equality-A-Planet-for-the-99.pdf
37. Oxfam international, 'Billionaires emit more carbon pollution in 90 Minutes than the average person does in a lifetime', 28 October 2024, https://www.oxfam.org/en/press-releases/billionaires-emit-more-carbon-pollution-90-minutes-average-person-does-lifetime
38. Cozzi, Laura et al., 'The world's top 1% of emitters produce over 1000 times more CO_2 than the bottom 1%', *International Energy Agency*, 2023, https://www.iea.org/commentaries/the-world-s-top-1-of-emitters-produce-over-1000-times-more-co2-than-the-bottom-1
39. Ibid.
40. Rosendo-Rios, Veronica and Paurav Shukla, 2023, 'When luxury democratizes: Exploring the effects of luxury democratization, hedonic value and instrumental self-presentation on traditional luxury consumers' behavioral intentions', *Journal of Business Research*, Vol. 155, https://doi.org/10.1016/j.jbusres.2022.113448
41. Niinimäki, Kirsi et al., 2020, 'The environmental price of fast fashion', *Nature Reviews Earth & Environment*, Vol. 1, doi:10.1038/s43017-020-0039-9
42. Gall, Melanie, 'Why insurance companies are pulling out of California and Florida, and how to fix some of the underlying problems', *The Conversation*, 7 June 2023, https://theconversation.com/why-insurance-companies-are-pulling-out-of-california-and-florida-and-how-to-fix-some-of-the-underlying-problems-207172
43. Thallinger, Günther, 'Climate, Risk, Insurance: The Future of Capitalism', *LinkedIn*, 25 March 2025, www.linkedin.com/pulse/climate-risk-insurance-future-capitalism-g%C3%BCnther-thallinger-smw5f/
44. Moody's, '2023 Catastrophe Review Report' https://www.moodys.com/web/en/us/insights/insurance/catastrophe-modeling-report.html
45. Welch, Craig, 'First study of all Amazon greenhouse gases suggests the damaged forest is now worsening climate change', *National*

Geographic, 12 March 2021, https://www.nationalgeographic.com/environment/article/amazon-rainforest-now-appears-to-be-contributing-to-climate-change

46. Haworth, Jon, '4 charged after fully functional solid gold toilet called 'America' stolen from Winston Churchill's birthplace', *ABC News*, 8 November 2023, https://abcnews.go.com/International/4-charged-after-fully-functional-solid-gold-toilet/story?id=104685050

47. Marshall, Alex, 'Two Men Found Guilty in Theft of $6 Million Gold Toilet', *The New York Times*, 18 March 2025, www.nytimes.com/2025/03/18/arts/design/gold-toilet-theft-verdict-maurizio-cattelan.html.

48. Guggenheim Museum, 'Maurizio Cattelan: "America"', https://guggenheim.org/exhibition/maurizio-cattelan-america

49. BBC, 'Museum offers gold toilet to Trump instead of Van Gogh's work - report', 26 January 2018, https://www.bbc.com/news/world-us-canada-42827291

50. Earthworks, 'Environmental Impacts of Gold Mining', https://earthworks.org/issues/environmental-impacts-of-gold-mining/

51. International Labour Organization, 'Child labour in mining and global supply chains', 2019, https://www.ilo.org/sites/default/files/wcmsp5/groups/public/%40asia/%40ro-bangkok/%40ilo-manila/documents/publication/wcms_720743.pdf

52. Sumner, Mark, 'Following a t-shirt from cotton field to landfill shows the true cost of fast cashion', *The Conversation*, 1 December 2020, https://theconversation.com/following-a-t-shirt-from-cotton-field-to-landfill-shows-the-true-cost-of-fast-fashion-127363

53. Statista, 'Land used to produce pne kilogram of food product as of 2018, by type', https://www.statista.com/statistics/1179708/land-use-per-kilogram-of-food-product/

54. Carbon Neutral Group, 'Carbon Emissions – Production of luxury goods', https://www.carbonneutralgroup.co.uk/carbon-emissions-production-of-luxury-goods.html

55. Kahneman, Daniel and Angus Deaton, 2010, 'High income improves evaluation of life but not emotional well-being', *Proceedings of the*

National Academy of Sciences, Vol. 107, https:// doi.org/10.1073/pnas.1011492107

56. Torras Vives, Gemma et al., 'Of dragons, data and clouds: Bhutan's journey into carbon markets, technology, and a resilient future.', *World Bank Blogs*, 19 October 2023, https://blogs.worldbank. org/en/climatechange/dragons-data-and-clouds-bhutans-journey-carbon-markets-technology-and-resilient

57. United Nations, 'Happiness: Towards a Holistic Approach to Development', The General Assembly, 2013, https://unstats.un.org/unsD/broaderprogress/pdf/Happiness%20towards%20a%20holistic%20approach%20to%20development%20(A-67-697).pdf

58. Bizikova, Livia et al., 'Moving Beyond GDP Through Comprehensive Wealth: Findings for Ethiopia, Indonesia, and Trinidad and Tobago', *International Institute for Sustainable Development*, 17 May 2024, https://www.iisd.org/publications/ report/comprehensive-wealth-moving-beyond-gdp

59. 'Wellbeing Economy Governments Partnership (WEGo)' Wellbeing Economy Alliance (WEAll), https://weall.org/

60. Gelles, David, 'Michael Bloomberg Dials Up a War on Plastics', *The New York Times*, 20 September 2023, https://www.nytimes. com/2023/09/20/climate/michael-blooomberg-climate-petrochemicals.html

61. Ibid.

62. Patagonia, 'Patagonia's Next Chapter: Earth Is Now Our Only Shareholder', 14 September 2022, www.patagoniaworks.com/ press/2022/9/14/patagonias-next-chapter-earth-is-now-our-only-shareholder

63. Guinness, Katherine et al., 'Billionaires are building bunkers and buying islands. But are they prepping for the apocalypse – or pioneering a new feudalism?', *The Conversation*, 1 March 2024, https://theconversation.com/billionaires-are-building-bunkers-and-buying-islands-but-are-they-prepping-for-the-apocalypse-or- pioneering-a-new-feudalism-223987

64. Scheidel, Walter and Steven J. Friesen, 2009, 'The Size of the Economy and the Distribution of Income in the Roman Empire,

Journal of Roman Studies, Vol. 99, doi:10.2139/ssrn.1299313; UBS, 'Global Wealth Report 2024: Crafted Wealth Intelligence', 2024, https://www.ubs.com/global/en/wealth-management/ insights/global-wealth-report.html

65. Zucman, Gabriel, 'A blueprint for a coordinated minimum effective taxation standard for ultra-high-net-worth individuals', *EU Tax Observatory*, https://www.taxobservatory.eu/publication/a-blueprint-for-a-coordinated-minimum-effective-taxation-standard-for-ultra-high-net-worth-individuals/
66. France 24, 'Taxing the Richest: What the G20 Decided', 19 November 2024, https://www.france24.com/en/live-news/20241119-taxing-the-richest-what-the-g20-decided
67. Norges Bank Investment Management, 'The fund's value'. https://www.nbim.no/en/investments/the-funds-value/
68. Fulloon, Sandra, 'The volunteers helping to fix your broken household items – and solve a $13 billion problem', *SBS News*, 18 November 2024, https://www.sbs.com.au/news/small-business-secrets/article/wendy-is-working-to-solve-a-13-billion-waste-problem-in-australia/wws820w73
69. Van Ouwerkerk, Charlotte, 'Dutch inventor's mushroom coffins turn bodies into compost', *PHYS.org*, 17 September 2020, https://phys.org/news/2020-09-dutch-inventor-mushroom-coffins-bodies.html
70. OECD, 'Extended producer responsibility and economic instruments', https://www.oecd.org/en/topics/sub-issues/extended-producer-responsibility-and-economic-instruments.html
71. Building and Construction Authority Singapore Government Agency, 'Prefabricated Prefinished Volumetric Construction (PPVC)', https://www1.bca.gov.sg/buildsg/productivity/design-for-manufacturing-and-assembly-dfma/prefabricated-prefinished-volumetric-construction-ppvc
72. Canon, Gabrielle, 'California couple whose gender-reveal party sparked a wildfire charged with 30 crimes', *The Guardian*, 22 July 2021, https://www.theguardian.com/us-news/2021/jul/21/couple-gender-reveal-party-wildfire-charged

73. Yosra, Missaoui, 2024, 'How Social Media Shapes Our Happiness: Exploring the Mediating Effects of Social Comparison and Materialism', *International Journal of Marketing Studies*, Vol. 16, doi:10.5539/ijms.v16n1p75
74. Lin, Liu Yi et al., 2016, 'Association between social media use and depression among U.S. young adults', *Depression and Anxiety*, Vol. 33, https://doi.org/10.1002/da.22466

Chapter 13

1. Westerfeld, Scott (@ScottWesterfeld), 'Plot idea: 97% of the world's scientists contrive an environmental crisis, but are exposed by a plucky band of billionaires & oil companies', X, 21 March 2014, x.com/ScottWesterfeld/status/446805144781348865
2. The Evening Independent, 'Jottings by the Rambler', 20 April 1917, https://books.google.com.au/books?id=6LwNAAAAIBAJ&pg=PA4&article_id=3979,6112125
3. Tye, Larry, *The Father of Spin: Edward L. Bernays and the Birth of Public Relations*, Crown Publishers, New York, 1998
4. Tag Heuer, 'A Wristed Development: What Stopped the Pocket Watch?', https://magazine.tagheuer.com/en/2021/11/04/a-wristed-development-what-stopped-the-pocket-watch-part-2/
5. University of Ottawa, 'Torches of Freedom Campaign', *Digital History – Histoire Numérique*, https://omeka.uottawa.ca/jmccutcheon/exhibits/show/american-women-in-tobacco-adve/torches-of-freedom-campaign
6. Gunderman, Richard, 'The manipulation of the American mind: Edward Bernays and the birth of public relations', *The Conversation*, 9 July 2015, http://theconversation.com/the-manipulation-of-the-american-mind-edward-bernays-and-the-birth-of-public-relations-44393
7. Bernays, Edward, *Propaganda*. Ig Publishing, Brooklyn, 2004
8. Brown and Williamson Records; Master Settlement Agreement, 1969, http://legacy.library.ucsf.edu/tid/tgy93f00
9. Nevis, Allan, et al., *Energy and Man: A Symposium*, Appleton-Century-Crofts, Inc. New York, 1960

10. Ibid.
11. Franta, Benjamin, 'What Big Oil knew about climate change, in its own words', *The Conversation*, 28 October 2021, https://theconversation.com/what-big-oil-knew-about-climate-change-in-its-own-words-170642
12. Supran, Geoffrey et al., 2023, 'Assessing ExxonMobil's global warming projections', *Science*, Vol. 379, doi:10.1126/science.abk0063
13. Banerjee, Neela et al., 'Exxon's Own Research Confirmed Fossil Fuels' Role in Global Warming Decades Ago', *Inside Climate News*, 16 September 2015, https://insideclimatenews.org/news/16092015/exxons-own-research-confirmed-fossil-fuels-role-in-global-warming/
14. Ibid.
15. Supran, Geoffrey et al., 2023, 'Assessing ExxonMobil's global warming projections', Science, Vol. 379, doi:10.1126/science.abk0063
16. Climate Files, '1996 Global Climate Coalition Science and Technology April Minutes and Membership', https://www.climatefiles.com/denial-groups/global-climate-coalition-collection/1996-science-tech-minutes-membership-ipcc-uncertainties/
17. United States Senate Committee on the Budget, 'Denial, Disinformation, and Doublespeak: Big Oil's Evolving Efforts to Avoid Accountability for Climate Change', 1 May 2024, https://www.budget.senate.gov/hearings/denial-disinformation-and-doublespeak_big-oils-evolving-efforts-to-avoid-accountability-for-climate-change
18. DocumentCloud, '1988 Exxon Memo on the Greenhouse Effect', https://www.documentcloud.org/documents/3024180-1998-Exxon-Memo-on-the-Greenhouse-Effect/
19. Climate Files, '1989 Presentation to Exxon Board of Directors on Greenhouse Gas Effects', https://www.climatefiles.com/exxonmobil/1989-presentation-exxon-board-directors-greenhouse-gas-effects/
20. Climate Files, '1991 Information Council for the Environment Climate Denial Ad Campaign', https://www.climatefiles.com/denial-groups/ice-ad-campaign/

21. Climate Files, '1998 American Petroleum Institute Global Climate Science Communications Team Action Plan', https://www.climate-files.com/trade-group/american-petroleum-institute/1998-global-climate-science-communications-team-action-plan/
22. Ibid.
23. Banerjee, Neela et al., 'Exxon's Own Research Confirmed Fossil Fuels' Role in Global Warming Decades Ago', *Inside Climate News*, 16 September 2015, https://insideclimatenews.org/news/16092015/exxons-own-research-confirmed-fossil-fuels-role-in-global-warming/
24. Supran, Geoffrey and Naomi Oreskes, 'What Exxon Mobil Didn't Say About Climate Change', *The New York Times*, 22 August 2017, https://www.nytimes.com/2017/08/22/opinion/exxon-climate-change-.html
25. Supran, Geoffrey and Naomi Oreskes, 'The forgotten oil ads that told us climate change was nothing', *The Guardian*, 18 November 2021, https://www.theguardian.com/environment/2021/nov/18/the-forgotten-oil-ads-that-told-us-climate-change-was-nothing
26. Ibid.
27. Tewari, D. D. 'The chipko: the dialectics of economics and environment.' Dialectical anthropology 20.2 (1995): 133–168.
28. Nuwer, Rachel, 'Global Warming Ad Quickly Dropped', *The New York Times*, 5 May 2012, https://www.nytimes.com/2012/05/06/us/heartland-institute-pulls-its-global-warming-ad.html
29. Institute of Public Affairs, 'Strangled by Tree Huggers' 11 November 2024, https://ipa.org.au/ipa-review-articles/strangled-by-tree-huggers
30. Walker, Tony R. and Lexi Fequet, 2023 'Current trends of unsustainable plastic production and micro(nano)plastic pollution', *TrAC Trends in Analytical Chemistry*, Vol. 160, https://doi.org/10.1016/j.trac.2023.116984
31. International Energy Agency, 'The Future of Petrochemicals', 2018, https://www.iea.org/reports/the-future-of-petrochemicals
32. Ibid.
33. The Ocean Cleanup, 'The Great Pacific Garbage Patch', https://theoceancleanup.com/great-pacific-garbage-patch/

34. Allen, Davis et al., 'The Fraud of Plastic Recycling', *Center for Climate Integrity*, 2024, https://climateintegrity.org/uploads/media/Fraud-of-Plastic-Recycling-2024.pdf
35. United Nations Environment Programme, 'Plastic Pollution', https://unep.org/interactive/beat-plastic-pollution/
36. Ritchie, Hannah et al., 'Plastic Pollution' *Our World in Data*, https://ourworldindata.org/plastic-pollution
37. Sullivan, Laura, 'How Big Oil Misled the Public Into Believing Plastic Would Be Recycled', *NPR*, 11 September 2020, https://www.npr.org/2020/09/11/897692090/how-big-oil-misled-the-public-into-believing-plastic-would-be-recycled
38. Allen, Davis et al., 'The Fraud of Plastic Recycling', *Center for Climate Integrity*, 2024, https://climateintegrity.org/uploads/media/Fraud-of-Plastic-Recycling-2024.pdf
39. Tabuchi, Hiroko, 'In Video, Exxon Lobbyist Describes Efforts to Undercut Climate Action', *The New York Times*, 30 June 2021, https://www.nytimes.com/2021/06/30/climate/exxon-greenpeace-lobbyist-video.html
40. Tabuchi, Hiroko, 'Inside the Plastic Industry's Battle to Win Over Hearts and Minds', *The New York Times*, 27 November 2024, https://www.nytimes.com/2024/11/27/climate/plastic-industry-internal-documents.html
41. Complaint, 20 June, 2024, Climate Change Litigation Databases, https://climatecasechart.com/wp-content/uploads/case-documents/2024/20240220_docket-2024CH01024-_complaint.pdf
42. 'Office of the Attorney General, 'Attorney General Bonta Sues ExxonMobil for Deceiving the Public on Recyclability of Plastic Products', 23 September 2024, https://oag.ca.gov/news/press-releases/attorney-general-bonta-sues-exxonmobil-deceiving-public-recyclability-plastic
43. Sabin Center for Climate Change Law, People v. PepsiCo, Inc., 2024, https://climatecasechart.com/case/people-v-pepsico-inc/
44. Sabin Center for Climate Change Law, City of New York v. Exxon Mobil Corp, 2021, https://climatecasechart.com/case/

city-of-new-york-v-exxon-mobil-corp/; Center for Climate Integrity, 'Kansas County Sues Big Oil, Plastics Producers for Recycling Deception', 2 December 2024, https://climateintegrity.org/news/view/kansas-county-sues-big-oil-plastics-producers-for-recycling-deception; Sabin Center for Climate Change Law, Bucks County v. BP p.l.c., 2024, https://climatecasechart.com/case/bucks-county-v-bp-plc/

45. Climate Change Litigation Databases, https://climatecasechart.com/
46. Global Fossil Fuel Divestment Commitments Database, https://divestmentdatabase.org/
47. Christie, Anna, 'Battle for the Board: Climate Rebellion at Exxon Marks a New Era of Shareholder Activism', *Oxford Business Law Blog*, 12 July 2021, https://blogs.law.ox.ac.uk/business-law-blog/blog/2021/07/battle-board-climate-rebellion-exxon-marks-new-era-shareholder
48. Rebekkah Markey-Towler, 'AGL & Mike Cannon-Brookes: A case study of investor leadership on climate change?', *University of Melbourne*, 21 June 2022, https://www.unimelb.edu.au/ data/
49. assets/pdf_file/0007/4152652/Working-Paper_21-June-2022.pdf 49 Henderson, Andrew (@andrwfhenderson), 'I pledge not to spill 4.9 million barrels of oil into the Gulf of Mexico', X, 25 October, 2019, https://x.com/andrwfhenderson/status/1187386101960454146.
50. Shell Sustainability, 'People', https://www.shell.com/sustainability/people.html
51. Exxon Mobil, 'ExxonMobil Chairman and CEO, Darren Woods, talks about reframing the climate challenge during the APEC CEO Summit', https://corporate.exxonmobil.com/news/viewpoints/reframing-the-climate-challenge
52. ANTAR, 'Rio Tinto's destruction of Juukan Gorge', 22 May 2025, https://antar.org.au/issues/cultural-heritage/the-destruction-of-juukan-gorge/; Clark, Cassandra J. et al., 2022, 'Unconventional Oil and Gas Development Exposure and Risk of Childhood Acute Lymphoblastic Leukemia: A Case–Control Study in Pennsylvania, 2009–2017', *Environmental Health Perspectives*, Vol. 130, https://doi.org/10.1289/EHP11092;Union of Concerned Scientists, 'How the Fossil Fuel Industry Harassed Climate Scientist Michael Mann',

12 October 2017, https://www.ucsusa.org/resources/how-fossil-fuel-industry-harassed-climate-scientist-michael-mann; Chang, Andres et al., 'Dollars vs. Democracy: Inside the Fossil Fuel Industry's Playbook to Suppress Protest and Dissent in the United States' *Greenpeace*, 2023, https://www.greenpeace.org/static/planet4-usa-stateless/2024/11/c9e33463-dollars-vs-democracy-report-2023.pdf; InfluenceMap, 'Big Oil's Big History of Blocking Climate Action', 26 October 2021, https://influencemap.org/pressrelease/Big-Oil-s-Big-History-of-Blocking-Climate-Action-da8364692983184477b5f4347e77d761

53. International Renewable Energy Agency, 'Off-Grid Renewable Energy Solutions to Expand Electricity Access: An Opportunity Not to Be Missed', 2019, https://www.irena.org/-/media/Files/IRENA/Agency/Publication/2019/Jan/IRENA_Off-grid_RE_Access_2019.pdf

54. InfluenceMap, 'Big Oil's Real Agenda on Climate Change', 2019, https://climatecasechart.com/wp-content/uploads/non-us-case-documents/2019/20190919_Case-No.-CHR-NI-2016-0001_na-1.pdf

55. ExxonMobil, 'Transforming Transportation: Lower Emission Fuels', https://corporate.exxonmobil.com/what-we-do/transforming-transportation#Loweremissionfuels

56. International Energy Agency, 'Investment in clean energy this year is set to be twice the amount going to fossil fuels', 6 June 2024, https://www.iea.org/news/investment-in-clean-energy-this-year-is-set-to-be-twice-the-amount-going-to-fossil-fuels

57. Engineers Australia, 'Making a clean transition', 2024, https://www.engineersaustralia.org.au/sites/default/files/2024-09/transferable-engineering-skills-clean-energy-transition.pdf

58. Tobacco Industry Research Committee, 'A Frank Statement to Cigarette Smokers', *The New York Times*, 4 Jan. 1954. Reproduced in the American Newspaper Repository, David M. Rubenstein Rare Book & Manuscript Library, Duke University, https://exhibits.library.duke.edu/items/show/11866.

Chapter 14

1. Reagan, Ronald, 25 January 1983, *Address Before a Joint Session of the Congress on the State of the Union*, House Chamber of the Capitol, www.reaganlibrary.gov/archives/speech/address-joint-session-congress-state-union
2. Kilvert, Nick, 'Carp herpes was meant to get rid of Australia's worst aquatic pest. So what happened to it?' *ABC News*, 1 April 2023, https://www.abc.net.au/news/science/2023-04-01/carp-herpes-barnaby-joyce-pest-biological-control-ecology/102149754
3. Domeier, Michael, 2011, 'Revisiting Spawning Aggregations: Definitions and Challenges' *Fish & Fisheries Series*, Vol. 35, doi:10.1007/978-94-007-1980-4_1
4. Kilvert, Nick, 'Carp herpes was meant to get rid of Australia's worst aquatic pest. So what happened to it?' *ABC News*, 1 April 2023, https://www.abc.net.au/news/science/2023-04-01/carp-herpes-barnaby-joyce-pest-biological-control-ecology/102149754
5. Stuart, Ivor et al., 'Exploding carp numbers are 'like a house of horrors' for our rivers. Is it time to unleash carp herpes?' *The Conversation*, 23 January 2023, https://theconversation.com/exploding-carp-numbers-are-like-a-house-of-horrors-for-our-rivers-is-it-time-to-unleash-carp-herpes-198067
6. World Intellectual Property Organization, 'Political Constitution of the Republic of Costa Rica', https://www.wipo.int/wipolex/en/text/219959
7. United Nations Environment Programme, 'Costa Rica named 'UN Champion of the Earth' for pioneering role in fighting climate change', https://www.unep.org/news-and-stories/press-release/costa-rica-named-un-champion-earth-pioneering-role-fighting-climate
8. Fernandez, Ileana, 'The Legacy of Women's Suffrage in Costa Rica 74 Years On', *The Tico Times*, 31 July 2024, https://ticotimes.net/2024/07/31/the-legacy-of-womens-suffrage-in-costa-rica-74-years-on
9. Food and Agriculture Organization of the United Nations, 'Costa Rica', https://www.fao.org/4/y4632e/y4632e0a.htm

10. Ibid.
11. Green Policy Platform, 'Payment for Ecosystem Services in Costa Rica', 2014 https://www.greenpolicyplatform.org/sites/default/files/downloads/best-practices/GGBP%20Case%20Study%20Series_Costa%20Rica_Payment%20for%20Ecosystem%20Services.pdf
12. Vargas Corrales, Andrea et al., 2022 'Exposure to common-use pesticides, manganese, lead, and thyroid function among pregnant women from the Infants' Environmental Health (ISA) study, Costa Rica', *Science of the Total Environment*, Vol. 810, https://doi.org/10.1016/j.scitotenv.2021.151288
13. World Bank Group, 'Climate Risk Country Profile: Costa Rica', 2021, https://climateknowledgeportal.worldbank.org/sites/default/files/country-profiles/15989-WB_Costa%20Rica%20Country%20Profile-WEB.pdf
14. World Intellectual Property Organization, 'Political Constitution of the Republic of Costa Rica', https://www.wipo.int/wipolex/en/text/219959
15. Rodricks, S., mainly based on Bennet & Henninger 2010; TEEBCase: Enabling the legal framework for PES, Costa Rica (2010), available at: TEEBweb.org
16. Pagiola, Stefano. 2008. 'Designing payments for environmental services in theory and practice: An overview of the issues', *Ecological Economics*, Vol. 65, https://doi.org/10.1016/j.ecolecon.2008.03.011
17. International Energy Agency, 'Costa Rica: Renewables', https://www.iea.org/countries/costa-rica/renewables
18. Vincenzi, Diego, 'How Costa Rica made its seas work for fishermen and for sharks', *World Economic Forum*, 29 April 2024, https://www.weforum.org/stories/2024/04/how-costa-rica-fought-to-make-its-seas-a-place-for-fishermen-and-for-sharks/
19. Abdallah, S. et al., 'The 2024 Happy Planet Index', *Hot or Cool Institute*, 2024, https://happyplanetindex.org/HPI_2024_report.pdf
20. Happy Planet Index, 'Costa Rica', https://happyplanetindex.org/countries/CRI/
21. Durocher, Bettina and Carlos Camacho Nassar, 'The Indigenous

World 2023: Costa Rica', *International Work Group for Indigenous Affairs*, 27 March 2023, https://iwgia.org/en/costa-rica/5084-iw-2023-costa-rica.html
22. Pearce, Fred, "Lauded as Green Model, Costa Rica Faces Unrest in Its Forests', *Yale Environment 360*, 21 March 2023, https://e360.yale.edu/features/costa-rica-deforestation-indigenous-lands
23. Konyn, Carol. 'How Costa Rica Reversed Deforestation and Became an Environmental Model', *Earth.org*, 19 October 2021, https://earth.org/how-costa-rica-reversed-deforestation/
24. Arias Sánchez, Oscar, 11 December 1987, *Only Peace Can Write the New History*, Nobel Prize Lecture, https://www.nobelprize.org/prizes/peace/1987/arias/lecture/
25. Arias Sánchez, Óscar, 'Declaring Peace with Nature', *First Forum*, https://firstforum.org/wp-content/uploads/2021/05/Report_00062.pdf
26. Australian Government: Department of Climate Change, Energy, the Environment and Water, 'Australian energy production - fuel type', https://www.energy.gov.au/energy-data/australian-energy-statistics/data-charts/australian-energy-production-fuel-type
27. United Nations Environment Programme, 'Emissions Gap Report 2024', https://www.unep.org/resources/emissions-gap-report-2024
28. Ibid.
29. United Nations Environment Programme, 'State of Finance for Nature 2023', https://www.unep.org/resources/state-finance-nature-2023
30. Ibid.
31. World Economic Forum, 'The Global Risks Report 2024', https://www.weforum.org/publications/global-risks-report-2024/
32. Tian, Nan et al., 'Trends in World Military Expenditure, 2023' *Stockholm International Peace Research Institute*, 2024, https://www.sipri.org/sites/default/files/2024-04/2404_fs_milex_2023.pdf
33. Sen, Basav, 'How the U.S. Transportation System Fuels Inequality', *Inequality.org*, 27 January 2022, https://inequality.org/article/public-transit-inequality/
34. Brandon, Elissaveta M., 'How automakers insidiously shaped our cities

for cars', *Fast Company*, 1 September 2022, https://www.fastcompany.com/90781961/how-automakers-insidiously-shaped-our-cities-for-cars

35. Begley, Patrick, 'In Canberra, lobbyists outnumber politicians three to one. Now there are growing calls for stronger regulation', *ABC News*, 13 November 2023, www.abc.net.au/news/2023-11-13/lobbyists-outnumber-politicians-code-of-conduct-regulation/103090798

36. Botrel, Clara Almeida et al., 2024, 'Understanding the lobbying actions taken by the Australian renewable energy industry', *Journal of Cleaner Production*, Vol. 434, https://doi.org/10.1016/j.jclepro.2023.139674

37. Waters, Larissa, 20 March 2024, *Removing the Fossil Fuel Industry's Influence on Politics and Parliament*, the Senate at Australia Institute's Climate Integrity Summit, https://australiainstitute.org.au/post/removing-the-fossil-fuel-industrys-influence-on-politics-and-parliament-senator-larissa-waters/

38. Ibid.

39. European Environment Agency, 'Fossil fuel subsidies in Europe', 30 January 2025, https://www.eea.europa.eu/en/analysis/indicators/fossil-fuel-subsidies

40. Abdullah, Halimah, 'Reagan and Thatcher: 'Political soulmates'', *CNN*, 9 April 2013, https://edition.cnn.com/2013/04/08/politics/thatcher-reagan/index.html

41. Farman, Joseph C. et al., 1985, 'Large losses of total ozone in Antarctica reveal seasonal ClOx/NOx interaction', *Nature*, Vol. 315, https://nora.nerc.ac.uk/id/eprint/523353/

42. 'Susan Solomon: Pioneering Atmospheric Scientist', NOAA 200th: Top Tens: History Makers, National Oceanic and Atmospheric Administration, (this webpage has been removed under the Trump Administration)

43. American Chemical Society, 'Chlorofluorocarbons and Ozone Depletion', 18 April 2017, https://www.acs.org/education/whatischemistry/landmarks/cfcs-ozone.html

44. Farman, Joseph C. et al., 1985, 'Large losses of total ozone in

Antarctica reveal seasonal ClOx/NOx interaction', *Nature*, Vol. 315, https://nora.nerc.ac.uk/id/eprint/523353/
45. Torkinton, Simon and Madeleine North, 'The ozone layer is on the path to recovery: Here's how the world made it happen', *World Economic Forum*, 28 November 2024, https://www.weforum.org/stories/2024/11/ozone-layer-hole-update-nasa/
46. United States Environmental Protection Agency, 'Stratospheric Ozone Protection', December 2017, https://www.epa.gov/sites/default/files/2017-12/documents/mp30_report_final_508v3.pdf
47. Mecklin, John, 'Remembering George Shultz: an interview with a key figure in ending the Cold War', *Bulletin of the Atomic Scientists*, 7 February 2021, https://thebulletin.org/2021/02/remembering-george-shultz-an-interview-with-a-key-figure-in-ending-the-cold-war/
48. Dizikes, Peter, 'George Shultz: "Climate is changing," and we need more action', *MIT Energy Initiative*, 2 October 2014, https://energy.mit.edu/news/george-shultz-climate-is-changing-and-we-need-more-action/
49. United Nations Environment Programme, 'About Montreal Protocol', https://www.unep.org/ozonaction/who-we-are/about-montreal-protocol
50. Council on Foreign Relations, 'How Are International Agreements Helping Fight Global Warming?', 17 September 2024, https://education.cfr.org/learn/reading/international-agreements-climate; Climate Action Tracker, https://climateactiontracker.org/countries/
51. Convention on Biological Diversity, 'Global Biodiversity Outlook 5', https://www.cbd.int/gbo5

Chapter 15

1. Equity Generation Lawyers, 'Sharma v Minister for Environment', https://equitygenerationlawyers.com/case/sharma-v-minister-for-environment/
2. New South Wales Department of Planning, Industry and Environment, 'Vickery Extension Project: State Significant Development Assessment SSD 7480', 2020, https://www.ipcn.nsw.gov.au/sites/default/files/pac/projects/2020/03/vickery-extension-project/

SOURCES

referral-from-the-department-of-planning-industry-and-environment/dpie-final-assessment-report.pdf

3. Equity Generation Lawyers, 'Sharma v Minister for Environment', https://equitygenerationlawyers.com/case/sharma-v-minister-for-environment/
4. Federal Court of Australia, 'Minister for the Environment v Sharma [2022] FCAFC 35', https://www.judgments.fedcourt.gov.au/judgments/Judgments/fca/full/2022/2022fcafc0035
5. Duty of Care, https://adutyofcare.davidpocock.com.au/
6. Sabin Center for Climate Change Law, Urgenda Foundation v. State of the Netherlands, 2015, https://climatecasechart.com/non-us-case/urgenda-foundation-v-kingdom-of-the-netherlands/
7. Sahoutara, Naeem, 'Seven-year-old girl takes on federal, Sindh governments', The Express Tribune, 29 June 2016, https://tribune.com.pk/story/1133023/seven-year-old-girl-takes-federal-sindh-governments
8. Sabin Center for Climate Change Law, Future Generations v. Ministry of Environment & Others, 2018, https://climatecasechart.com/non-us-case/future-generation-v-ministry-environment-others/
9. Sabin Center for Climate Change Law, Milieudefensie et al. v. Royal Dutch Shell plc., 2019, https://climatecasechart.com/non-us-case/milieudefensie-et-al-v-royal-dutch-shell-plc/
10. Bell-James, Justine, '"This case has made legal history': young Australians just won a human rights case against an enormous coal mine', The Conversation, 25 November 2022, https://theconversation.com/this-case-has-made-legal-history-young-australians-just-won-a-human-rights-case-against-an-enormous-coal-mine-195350
11. Global Witness, 'Almost 2,000 land and environmental defenders killed between 2012 and 2022 for protecting the planet', 13 September 2023, https://www.globalwitness.org/en/press-releases/almost-2000-land-and-environmental-defenders-killed-between-2012-and-2022-protecting-planet/
12. Parker, Ali, 'Virtually all rape victims are denied justice: Here is the roadmap to failure', Saunders Law, 3 February 2023, https://www.saunders.co.uk/news/

virtually-all-rape-victims-are-denied-justice-here-is-the-roadmap-to-failure/; Hutchens, Gareth, 'Australia leads the world in arresting climate and environment protesters', ABC News, 15 December 2024, https://www.abc.net.au/news/2024-12-15/australia-leads-world-in-arresting-climate-environment-activists/104721294; Early, Catherine, 'Activist arrests in UK 'three times global average'', The Ecologist, 12 December 2024, https://theecologist.org/2024/dec/12/activist-arrests-uk-three-times-global-average; Herald Sun, "It's Disturbing': Sex Charges Among the Least Proven', https://www.heraldsun.com.au/leader/ sexual-offences-charges-among-the-least-proven-in-court/news-story/06408e0434f8c32dab2447e45a1125c7; RAINN, 'The Criminal Justice System: Statistics' https://rainn.org/statistics/criminal-justice-system; United States Environmental Protection Agency, 'Enforcement and Compliance Assurance Annual Results for Fiscal Year 2022', https://www.epa.gov/system/files/documents/2025-03/eoy2022.pdf

13. Climate Rights International, 'Western Democracies: Stop Crackdowns on Climate Protesters', 9 September 2024, https://cri. org/western-democracies-stop-crackdowns-climate-protesters/
14. Global Witness, 'Harsh jail time for climate activists threatens democracy and climate action', 23 July 2024, https://www. globalwitness.org/en/press-releases/harsh-jail-time-climate-activists-threatens-democracy-and-climate-action/
15. Monbiot, George, 'A record sentence for a Zoom call, arrests for those holding signs outside. This is a blight on British democracy', *The Guardian*, 19 July 2024, https://www.theguardian.com/ commentisfree/article/2024/jul/19/protest-democracy-labour-tories-laws
16. Gayle, Damien, 'Contempt, gagging and UN intervention: inside the UK's wildest climate trial', *The Guardian*, 13 July 2024,
17. https://www.theguardian.com/environment/article/2024/jul/12/contempt-gagging-un-intervention-uk-wildest-climate-trial-just-stop-oil
18. The Guardian, 'Activists arrested on suspicion of contempt after London court protest', 3 July 2024, https://www.theguardian.com/

uk-news/article/2024/jul/02/activists-arrested-on-suspicion-of-contempt-after-london-court-protest

19. Human Rights Watch, 'Australia: Climate Protesters' Rights Violated', 22 June 2022, https://www.hrw.org/news/2022/06/22/australia-climate-protesters-rights-violated
20. McNeill, Sophie, 'Australia's Crackdown on Climate Activists', *Human Rights Watch*, 29 May 2023, https://www.hrw.org/news/2023/05/30/australias-crackdown-climate-activists
21. Australian Religious Response to Climate Change, 'I Got Arrested at 97 for My Grandkids', https://www.arrcc.org.au/i_got_arrested_at_97_for_my_grandkids
22. ABC News, 'Protest penalty changes pass upper house of SA parliament after marathon debate', 31 May 2023, https://www.abc.net.au/news/2023-05-31/sa-upper-house-passes-protest-law-changes/102413948
23. Parkin, Andrew, 2023 'South Australia January to June 2023', *Australian Journal of Politics and History*, Vol. 69, https://doi.org/10.1111/ajph.12955
24. SBS News, 'Protesters to face harsher punishment after 'outrageous' disruption', 18 May 2023, https://www.sbs.com.au/news/article/protesters-to-face-harsher-punishment-after-outrageous-disruption/fhycgo8c4
25. Monbiot, George, 'A record sentence for a Zoom call, arrests for those holding signs outside. This is a blight on British democracy', *The Guardian*, 19 July 2024, https://www.theguardian.com/commentisfree/article/2024/jul/19/protest-democracy-labour-tories-laws
26. The Wilderness Society, 'About us', https://www.wilderness.org.au/about National Museum of Australia, 'Franklin Dam and the Greens', https://www.nma.gov.au/defining-moments/resources/franklin-dam-greens
27. The Wilderness Society, 'Our story', https://www.wilderness.org.au/about/story
28. Foxen, Julia, 'Landmark climate change hearings represent largest

ever case before UN world court', *United Nations*, 2 December 2024, https://news.un.org/en/story/2024/12/1157671

Chapter 16

1. Wong, Kate, 'We Are in the Golden Age of Bird-Watching', *Scientific American*, 25 April 2024, www.scientificamerican.com/article/we-are-in-the-golden-age-of-bird-watching/
2. SOS Save Our Songbirds, 'National survey: 1 in 3 U.S. adults reports watching birds', https://www.sossaveoursongbirds.org/bird-news/national-survey-1-in-3-us-adults-reports-watching-birds
3. Birdforum, 9 'Two people break 10,000 species, and on the same day? Can it be?', February 2024, https://www.birdforum.net/threads/two-people-break-10-000-species-and-on-the-same-day-can-it-be.452504/
4. Trezza, Joe, 'With an Orange-Tufted Spiderhunter, Birder Breaks Record for Sightings', *The New York Times*, 11 March 2024, https://www.nytimes.com/2024/03/11/science/birds-birdwatching-record-kaestner.html
5. Tumblr, https://www.tumblr.com/roach-works/742140030798168064/jason-mann-sounds-like-the-kind-of-name-a-rogue-ai
6. Tumblr, https://www.tumblr.com/great-and-small/742166071312842752/official-statement-to-the-birdwatching-community
7. 'One of the Best Birdwatchers in the World Breaks World Record.' Manakin Nature Tours, https://www.manakinnaturetours.com/one-of-the-best-birdwatchers-in-the-world-breaks-world-record/. Accessed 14 Jan. 2025; Manakin Nature Tours has since deleted this article and issued a statement in response to the scandal.
8. Tumblr, https://snafusheltoneyes.tumblr.com/page/4
9. Birdforum, 'Two people break 10,000 species, and on the same day? Can it be?', 9 February 2024, https://www.birdforum.net/threads/two-people-break-10-000-species-and-on-the-same-day-can-it-be.452504/
10. Ibid.

SOURCES

11. Kaestner, Peter, 'The Final Stretch: Planning My Way to 10,000 Birds', American Birding Association, 9 January 2024, https://www.aba.org/the-final-stretch/
12. Trezza, Joe, 'With an Orange-Tufted Spiderhunter, Birder Breaks Record for Sightings', *The New York Times*, 11 March 2024, https://www.nytimes.com/2024/03/11/science/birds-birdwatching-record-kaestner.html
13. Mann, Jason, 'Two people break 10,000 species, and on the same day? Can it be?' *Birdforum*, 10 February 2024, https://www.birdforum.net/threads/two-people-break-10-000-species-and-on-the-same-day-can-it-be.452504/post-4593071
14. Ibid.
15. Manakin Nature Tours, 'Official Statement to the Birdwatching Community', 11 February 2024, https://www.manakinnaturetours.com/official-statement-to-the-birdwatching-community/
16. Angus, Georgia, 'Birdwatching changes the way you look at the world – it truly is the gateway drug to environmental awareness', *The Guardian*, 7 January 2024, https://www.theguardian.com/australia-news/2024/jan/07/birdwatching-changes-the-way-you-look-at-the-world-it-truly-is-the-gateway-drug-to-environmental-awareness
17. Menges, Eric, 2024, 'Bird Watching as a Conservation Driver', *Natural Areas Journal*, Vol. 44, https://doi.org/10.3375/2162-4399-44.3.127
18. eBird, '2024 Year in Review: eBird, Merlin, Macaulay Library, and Birds of the World', 28 December 2024, https://ebird.org/news/2024-year-in-review
19. eBird, 'A World of Positive Impact: Results from Global Big Day 2024', 16 May 2024, https://ebird.org/news/global-big-day-2024-results
20. Milgrom, Melissa, 'The Big Year According to Birders', *Audubon*, 7 November 2011, https://www.audubon.org/news/the-big-year-according-birders
21. Grand View Research, 'Birdwatching Tourism Market Size, Share & Trends Analysis Report by Traveler Type (Hard Core Birders, Enthusiastic Birders), by Age Group, by Group, by Region, and

Segment Forecasts, 2024–2030', https://www.grandviewresearch.com/industry-analysis/birdwatching-tourism-market-report

22. U.S. Fish & Wildlife Service, 'Birding in the United States: A Demographic and Economic Analysis', 2024, https://www.fws.gov/sites/default/files/documents/2024-11/2022-birding-in-the-us-demographic-and-economic-analysis.pdf

23. North American Bird Conservation Initiative, 'Shared Outcomes for Birds and People: Relevancy Toolkit 2.0', 2021, https://nabci-us.org/wp-content/uploads/2021/01/Bird_Conservation_and_Human_Values_01-20-21.pdf

24. Colombia Ministry of Commerce, Industry, and Tourism, 'Colombia, the Land of Beauty, Broke a New Tourism Record by Welcoming Nearly 6.7 Million Non-Resident Visitors in 2024', https://www.mincit.gov.co/prensa/noticias/turismo/colombia-rompio-nuevo-record-en-turismo-en-2024

25. Wills, Santiago, 'Bird-watching offers potential for conservation & economy in Colombia's Guaviare', *Mongabay*, 15 December 2023, https://news.mongabay.com/2023/12/bird-watching-offers-potential-for-conservation-economy-in-colombias-guaviare/

26. Moreno-Salazar, Noemí, et al., 'Colombia, Country of Birds: Strategy 2030 – A Flight Toward Conservation', 2023, https://media.audubon.org/2023-05/Audubon_COLOMBIA%20PAIS%20DE%20LAS%20AVES_Digital%20Pliego.pdf

27. Ferreyra, Javiera, 'Chile Prepares to include birds in its GDP', *Audubon*, 7 March 2023, https://www.audubon.org/news/chile-prepares-include-birds-its-gdp